AF720244

Challenges and Opportunities in Sustainable Agricultural Systems

Challenges and Opportunities in Sustainable Agricultural Systems

Edited by Farrell Waltz

SYRAWOOD
PUBLISHING HOUSE

New York

Published by Syrawood Publishing House,
750 Third Avenue, 9th Floor,
New York, NY 10017, USA
www.syrawoodpublishinghouse.com

Challenges and Opportunities in Sustainable Agricultural Systems
Edited by Farrell Waltz

© 2023 Syrawood Publishing House

International Standard Book Number: 978-1-64740-414-7 (Hardback)

This book contains information obtained from authentic and highly regarded sources. Copyright for all individual chapters remain with the respective authors as indicated. All chapters are published with permission under the Creative Commons Attribution License or equivalent. A wide variety of references are listed. Permission and sources are indicated; for detailed attributions, please refer to the permissions page and list of contributors. Reasonable efforts have been made to publish reliable data and information, but the authors, editors and publisher cannot assume any responsibility for the validity of all materials or the consequences of their use.

Trademark Notice: Registered trademark of products or corporate names are used only for explanation and identification without intent to infringe.

Cataloging-in-publication Data

Challenges and opportunities in sustainable agricultural systems / edited by Farrell Waltz.
p. cm.
Includes bibliographical references and index.
ISBN 978-1-64740-414-7
1. Sustainable agriculture. 2. Agriculture. 3. Cropping systems. I. Waltz, Farrell.
S494.5.S86 A37 2023
630--dc23

TABLE OF CONTENTS

PREFACE

The main aim of this book is to educate learners and enhance their research focus by presenting diverse topics covering this vast field. This is an advanced book which compiles significant studies by distinguished experts in the area of analysis. This book addresses successive solutions to the challenges arising in the area of application, along with it; the book provides scope for future developments.

Sustainable agriculture is an integrated system of plant production practices that aim to satisfy the present food requirements of the society without compromising the needs of the future generations. It also seeks to make efficient use of non-renewable resources to minimize water use, promote soil health, and minimize pollution levels on the farm. Some of the major sustainable agricultural practices (SAPs) and systems that are practiced all over the world include agroforestry, conservation farming, contour farming, covering-mulching, crop-rotation, floating-farming, organic-farming, rainwater-harvesting and vermicomposting. Agroforestry refers to a farming practice wherein forests and shrubs are integrated within animal and crop farming system in order to obtain social, economic and environmental benefits. Conservation farming is an approach to sustainable agriculture which involves minimal disturbance of soil and diversification of plant species. This book includes some of the vital pieces of works being conducted across the world, on various topics related to sustainable agriculture. It unravels the various challenges and opportunities related to sustainable agricultural systems. This book will prove to be immensely beneficial to students and researchers studying sustainable agricultural systems.

It was a great honour to edit this book, though there were challenges, as it involved a lot of communication and networking between me and the editorial team. However, the end result was this all-inclusive book covering diverse themes in the field.

Finally, it is important to acknowledge the efforts of the contributors for their excellent chapters, through which a wide variety of issues have been addressed. I would also like to thank my colleagues for their valuable feedback during the making of this book.

Editor

1

Sustainable Agricultural Systems: A Bibliometrics Analysis of Ecological Modernization Approach

Lucia Rocchi *, Antonio Boggia and Luisa Paolotti

Department of Agricultural, Food and Environmental Science, University of Perugia, Borgo XX Giugno, 74, 06121 Perugia, Italy; antonio.boggia@unipg.it (A.B.); luisa.paolotti@gmail.com (L.P.)

* Correspondence: lucia.rocchi@unipg.it

Abstract: Agricultural systems need to be more sustainable. Nowadays, the global food production has a remarkable impact in terms of greenhouse gases production, terrestrial acidification, eutrophication and land consumption. Moreover, one of the effects of unsustainable agriculture is depletion of ecosystem services (ES) on which agricultural systems themselves are dependent. Alternative agricultural systems are possible: the aim of this review was to analyze one of these alternative systems, i.e., strong ecological modernization, to understand relevant topics and current state of the art connected with it. As a result that strong ecological modernization can be considered an ecosystem services-based farming system, we focused on the development of topics related to ecosystem services but also to diversification. The review has been conducted applying a bibliometrics approach to recognize the main papers, authors, organization and countries, as well as trend topics and main themes investigated. The results showed that basic research content involves agrobiodiversity, agroecology and diversified farming systems. At the same time, the review revealed a lack about the social and economic dimensions of sustainability that need to be addressed for promoting a true transition to a strong ecological modernization.

Keywords: ecological modernization; bibliometrics; ecosystem services; agrobiodiversity; agroecology

---

1. Introduction

Modern agriculture is characterized by an intensification of large-scale activities to achieve higher output levels. This approach has been particularly evident during the last 50 years, due to a growing earth population which needed to be fed [1]. The results of such an intensification are several and serious negative environmental externalities [2]. Nowadays, the global food production has a remarkable impact in terms of greenhouse gases production, terrestrial acidification, eutrophication and land consumption, in particular if considering animal sectors [3–8]. Therefore, one of the primary challenges of modern agriculture is to reduce external inputs and environmental impacts, while still being productive enough. Moreover, one of the effects of intensification of agriculture is depletion of ecosystem services (ES) [9], which can be a huge problem for agricultural production. There is clear recognition that agricultural systems are dependent on ES for producing goods and at the same time a source themselves of important ES, including a wide range of joint non-marketed ones [10]. The flows of ecosystem services depend directly on how agricultural ecosystems are managed and upon the diversity, composition and functioning of remaining natural ecosystems in the landscape [11]. Finding a balance between production and consumption of ES is the way to reach a long-term sustainability for agriculture.

Ecological modernization of agriculture has been promoted as a possible solution since the 1990s [12]. There are two forms of ecological modernization of agriculture: weak and strong [13,14]. Weak ecological modernization aims to improve input use efficiency, mainly using new technologies, with the scope to minimize the environmental impacts of farming systems. Weak modernization does not criticize the specialization or intensification of agricultural. According to Duru et al. [15] it is the dominant pathway between the two. Strong ecological modernization supports the increase of agrobiodiversity and adds the importance of diversification to some principles of the weak approach (e.g., waste recycling or input-use optimization/reduction) [14,15]. Diversification allows developing ecosystem services, favoring above all non-market ones at field, farm and landscape levels [14,15]. Strong ecological modernization currently exists only as a niche and there is not a clear definition for it, neither a clear distinction from the weak one [15].

The aim of this paper was to analyze the subject of strong ecological modernization, in order to understand relevant topics and current state of the art connected with it. In particular, we would like to investigate how research has developed until now the different themes related to strong ecological modernization, which are the most investigated topics and which the most active research fields. In addition, we focused on the aspect of diversification, within strong ecological modernization, for investigating how important diversification is for boosting the ecosystem services in agriculture.

In order to reach these aims, we conducted a comprehensive literature review on the trends and applications of strong ecological modernization. In particular, we focused on the development of topics related to diversification and ecosystem services, such as intercropping, multicropping and agroforestry. The review has been conducted applying a bibliometric approach to recognize the main papers, authors, organization and countries connected with it. Then, we applied cluster analysis on a keyword network, based on the abstracts of the papers to better understand frontiers and trending topics.

The structure of this article is as follows. Section 2 introduces the data collection and research methods. Section 3 shows the research results analyzing three aspects: the research overview, the main research's basic content, then the trending topics. Section 4 summarizes the overall situation and proposes suggestions for future research.

2. Materials and Methods

2.1. Data Collection

Data collection has been based on the Web of Science (WoS) core collection platform, considering the period 1985–2020, aware that 2020 is an incomplete year. WoS can be considered the most authoritative scientific and technical literature indexing database as well as an ideal data source for bibliometric investigations [16]. As strong ecological modernization aims mainly to develop diversified farming systems, to manage and support ecosystem services in general and agrobiodiversity in particular [14], the following research string has been applied: TS = ("ecosystem servic*") AND TS = (cropdiversity) OR TS = (agroecology) OR TS = (diversified farming system) OR TS = (ecoagriculture) OR TS = (multicropping) OR TS = (agrobiodiversity). The keywords chosen have been selected to catch the involvement of ecosystem services in agriculture, as well as all the approach which enhance agrobiodiversity. A total of 3573 publications were found; then we excluded non-English works and selected only journal articles as they represent original scientific development, for obtaining 2407 papers which are the core of the analysis.

2.2. Analytical Methods

Bibliometric analysis has been carried out using CiteSpace ver. 5.7 [17,18]. CiteSpace is a free Java-based application for visualizing and analyzing citations and contents in scientific literature, which combines cluster analysis, social network analysis and many other methods. During the last years, several authors applied CiteSpace to very different fields such as rural landscape [19], climate change and tourism [20] or freshwater pollution by microplastics [21].

The 2407 selected articles were imported into CiteSpace, with each time slice set to one year. As the highest number of publications in a single year within the specified time span was 397 articles in 2019, we specified Top N = 397 to enable full visual analysis of the data, otherwise the software selects the Top N of 50 for generating the network. Research frontiers and trending topics with strong timeliness were searched across the whole analysis period (1985-2020), and the remaining factors were set by default in the software.

The main feature of CiteSpace is the network analysis, presented in the form of nodes (N) and lines (E). The size of nodes reflects the frequency of relevant data references or occurrences, while the lines indicate the relationship between the nodes, and the thickness of the lines reflects the strength of links between data [19]. The analysis can be performed in several ways, allowing also to choose a single node type or multiple concurrent node types. The default node type is Cited References, which allows the creation of the Co-cited references and the networks are made of co-cited references. Other interesting networks are the Collaborative (node type Author), Co-citation (node type Reference), Co-occurrence (node type Category).

The analysis of the core has been performed following the protocol developed by Wu et al. [19]. Firstly, we performed a network analysis, which allows to explore the knowledge domain (i.e., numbers of publications, countries distribution, cooperation network). For some domains (i.e., countries) betweenness centrality analysis has been performed. The term 'betweenness' derived by the fact that it measures the connection of an arbitrary pair of nodes in the network in which the node is part of the connecting path [18]. A high value of betweenness centrality means that the node connects two or more large groups of nodes with the node itself in-between. For instance, if a country has a high value of betweenness centrality, probability it collaborates with several other countries. CiteSpace highlights nodes with high betweenness centrality by means of purple trims, while the strength of betweenness centrality is the thickness: the thicker, the stronger.

Then, we performed a Co-occurrence analysis and time zone chart of keywords and a co-citation clustering analysis of the research core. The clustering effect was measured using modularity and silhouette. Modularity is a quality index and can be defined as a normalized tradeoff between edges covered by clusters and squared cluster degree sum [22]. To measure the degree of modularity, the Q value can be used. If $Q \geq 0.3$ then the modularization of the network is significant. The higher the Q value, the better the quality of clustering. Silhouette (S) refers to a method of interpretation and validation of consistency within clusters of data. It provides a succinct measure of the homogeneity of networks. If $S \geq 0.5$, then the result of clustering is reasonable; as S moves closer to 1, the homogeneity of the network increases. However, note that in case of a small cluster size, a high homogeneity does not have a high significance but maybe it reflects the citing behavior or preferences of one single work [18].

Finally, we performed burst detection on citation, authors and noun-term. The burst detection in CiteSpace is based on Kleinberg's algorithm and allows to capture a huge change of frequency of an entity over a specific time [23]. In CiteSpace, citation and occurrence burst are both supported. Performing the citation burst analysis is useful to find evidence of an association among a specific publication and a surge of citations. Applied to the author- type node burst-detection shows those authors who have rapidly increased the number of publications. Burst-detection of noun-term identifies the emergent terms regardless of how often their host articles are cited, for identifying the trends.

3. Results

3.1. Overview of Research

3.1.1. Overview of Papers Published

Over the considered period (1985–2020), the number of published research shows a growing trend, although in some years there are minor reductions (Figure 1a). Since 2013, there was a rapid rise of the published papers, proving a recent increasing interest for the topic. A particularly high growth rate is between 2012 and 2013 (+39%), while the largest number of articles (397) was published in 2019. Note that almost the half of the papers have been published in the last four years. The citations started to grow markedly since 1999 in a very rapid matter (Figure 1b).

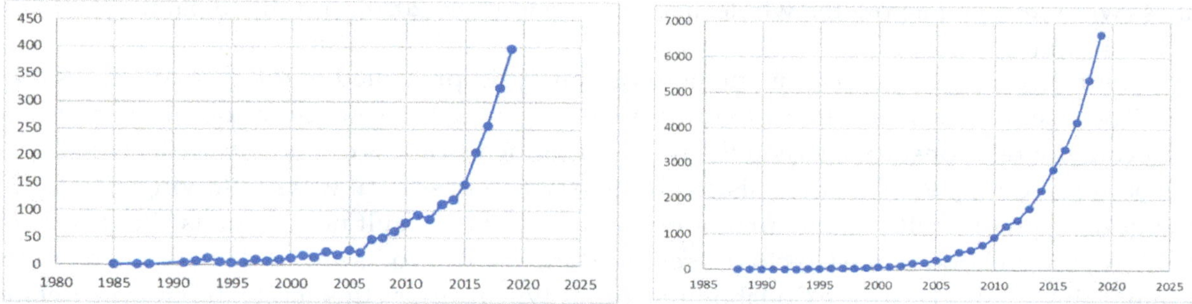

(**a**) (**b**)

Figure 1. Overview of papers published and citations: (**a**) published paper; (**b**) Citation. In both the graphs the year 2020 is not reported, as it is an incomplete year, which generates an unrealistic trend.

3.1.2. Authors: Core Articles and Citations

Using authors as node type, CiteSpace produced a network about author cooperation, resulting in 418 nodes and 783 lines connecting them. Among authors, 17 published at least ten articles. Altieri and Mendez, who published the largest number of articles, had 18 articles included in the WoS database, followed by Pascual and Zimmerer with 17. The academic background of the most productive authors is mainly environmental sciences, ecology and agriculture. There are some relatively dominant academic associations (Figure 2).

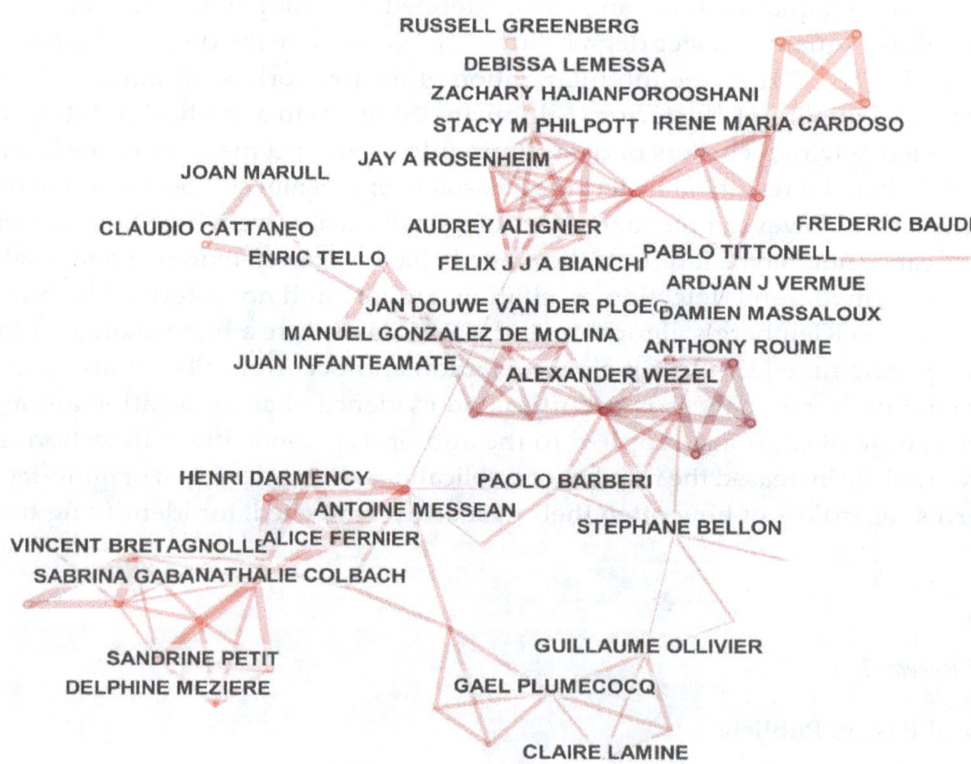

Figure 2. Authors network.

We performed a burst detection applied to the node type of author, to see those authors who have rapidly increased the number of publications (Figure 3). The majority of the authors (13 out of 16) in Figure 3 have their burst starting from 2013, which is also the year that identified a clear growth in the total number of publications. Mendez, who has the strongest burst, is also one of the two most productive authors.

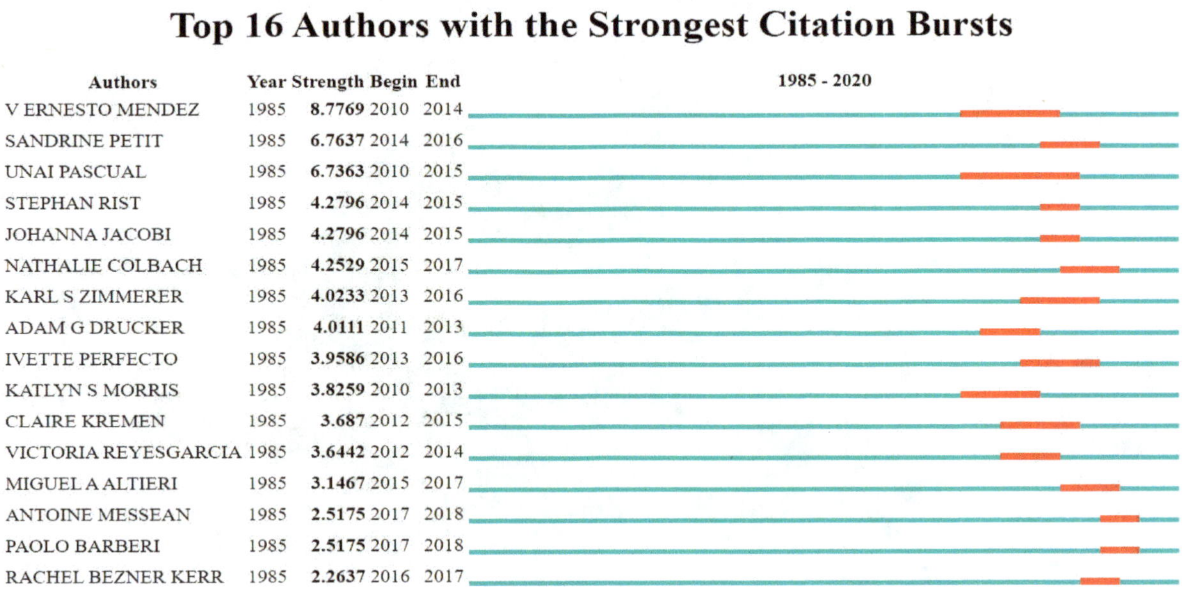

Figure 3. Authors with the strongest citation bursts, sorted by burst strength.

3.1.3. Distribution of Research Disciplines

Using the node type Category, it is possible to build a network of the subject categories. The network produced is formed by 113 nodes and 625 lines (Figure 4). Totally, 87 categories have been identified. The category Agriculture exceeded 1000 papers; Environmental sciences and ecology exceeded 800 papers. The betweenness-centrality value obtained more than 0.1 in 39 disciplines (e.g., Agriculture, Environmental sciences and ecology, Plant science and Business and Economics). Moreover, some categories, like Biotechnology and Applied microbiology, Energy and Fuel, Business and Economics and Plant Sciences, have a relatively high betweenness-centrality value but few publications: probably, these disciplines play a pivotal role in the construction of the interdisciplinary collaboration and research system.

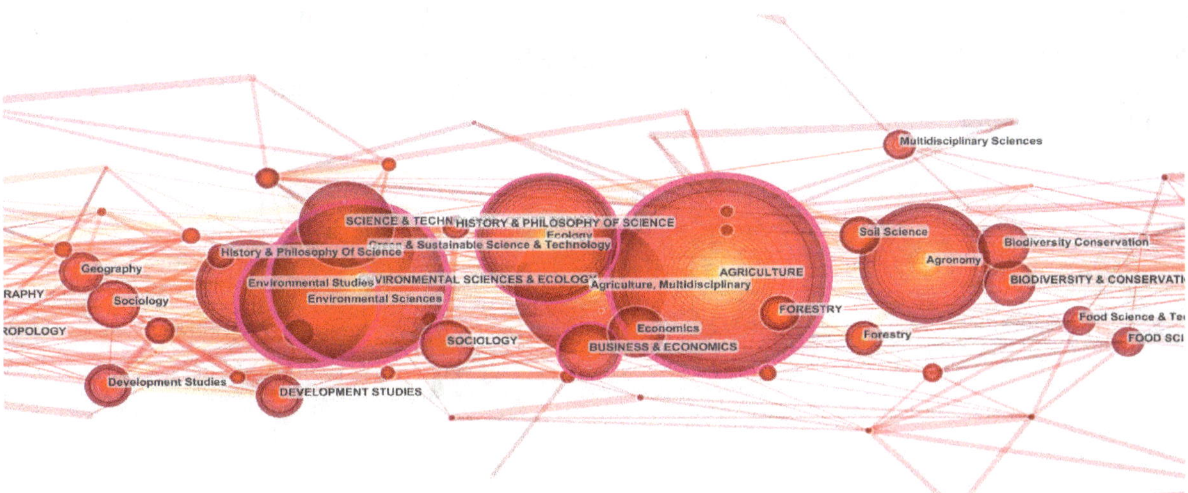

Figure 4. Network of the subject categories; purple trims highlight betweeness centrality.

3.1.4. Countries and Organization

The cooperation network of countries showed 86 nodes and 854 lines connecting the nodes (Figure 5). The most productive country is USA (703 papers), the second country, France, has half of the papers of USA (342). England, Italy and Brazil are also very productive.

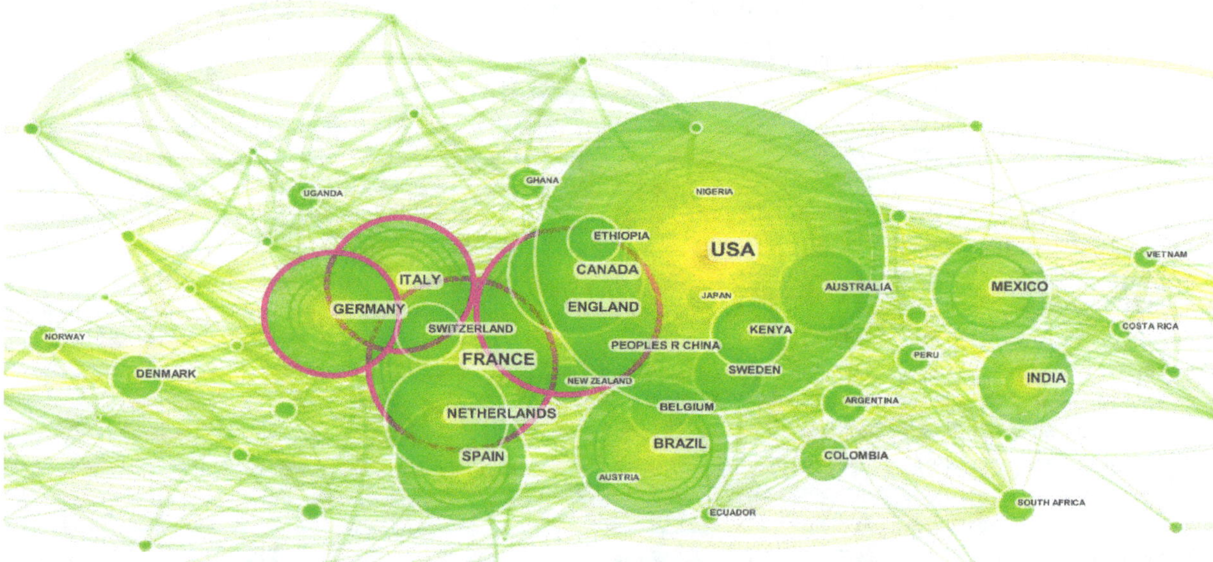

Figure 5. Countries cooperation network.

Germany, France, Italy and England have the highest betweenness centrality, as it is highlighted by the thick purple lines around the node. In addition, Ethiopia and Argentina are among the 10 nodes with the highest betweenness centrality, although they have a relative low number of papers in comparison to the countries of the same group. Note that 5 of the best 10 countries for betweenness centrality are part of the European Union. The common institutional and research framework provided by the European Union may produce a strong research collaboration across its countries, which emerges through betweenness centrality. This aspect seems to be supported also by the organizations network (N = 404; E = 1322), in which the European institutions and universities have the higher level of betweenness (Figure 6). INRA (Institut national de la recherche agronomique— National Institute of agronomic research) plays a crucial role both in terms of number of papers (almost 40% of the France publications) and of betweenness centrality. Therefore, INRA plays a pivot role among different other research institutes, probably European. On the contrary, the USA contribution in terms of papers is scattered across several institutions. The three most productive universities are University of California, Berkeley, Cornell University and University of California, Davis. The sum of their works is around 20% of all USA papers.

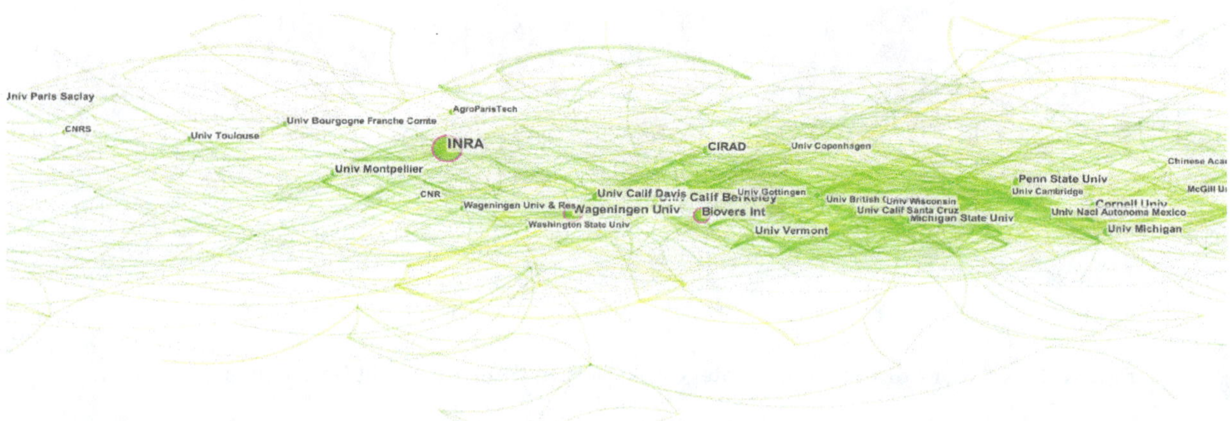

Figure 6. Main organizations.

3.2. *Main Themes*

3.2.1. Keywords Analysis

For better understanding the main themes faced in the ecological modernization approach, we performed a keywords co-occurrence analysis and a burst detection applied to the keywords. After the elimination of the retrieval terms, the five most used keywords are 'agroecology' (934), 'agrobiodiversity' (384), 'biodiversity' (369), 'agriculture' (362) and 'management' (344). Considering also when they first appeared in the literature corpus (Figure 7), 'agrobiodiversity' and 'biodiversity' are more recent then other keywords, including 'diversity' and 'management'. Due to visualization reasons, Figure 7 is based on the keywords' frequency of the top ten papers for each year.

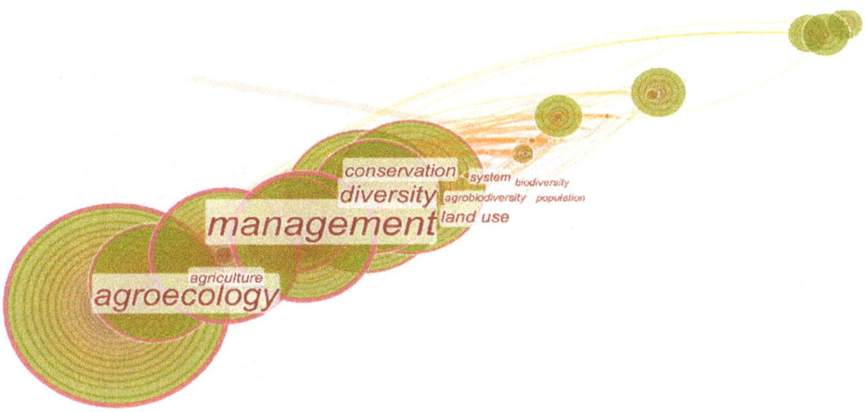

Figure 7. Keyword time zone map.

All the keywords have a low or very low level of between centrality; however, the keyword holding the highest between centrality level is 'agricultural landscape'. This result can be interpreted in two ways. The first is considering the definition of agricultural landscape itself. Since the agriculture landscape is the result of interactions between farming activities and the natural assets in a certain area, it plays a central role among several topics linked to ecological modernization. Instead, using the Tscharntke et al. [24] interpretation, we can affirm that the landscape perspective is the only possible way to relate agrobiodiversity and ecosystem services management to the management of the land use and to agriculture.

Detection of the burst applied to keywords shows the fast growing of topics over time: burst keywords refers to keywords heavily cited by articles over a period. Figure 8 reports such bursts ranked by the beginning of the burst. 'agroecology', which is also the keyword with the strongest burst, appeared in 1992, followed by 'sustainable agriculture' the next year. Although the second is weaker in terms of burst, it is more persistent and it last for almost two decades. On the contrary, 'organic farming' presents the shortest burst. The second and third keywords according to the strength of the burst are 'agrobiodiversity' and 'diversity', proving that diversification is a main topic in strong ecological modernization. None of the keywords identified persists till nowadays.

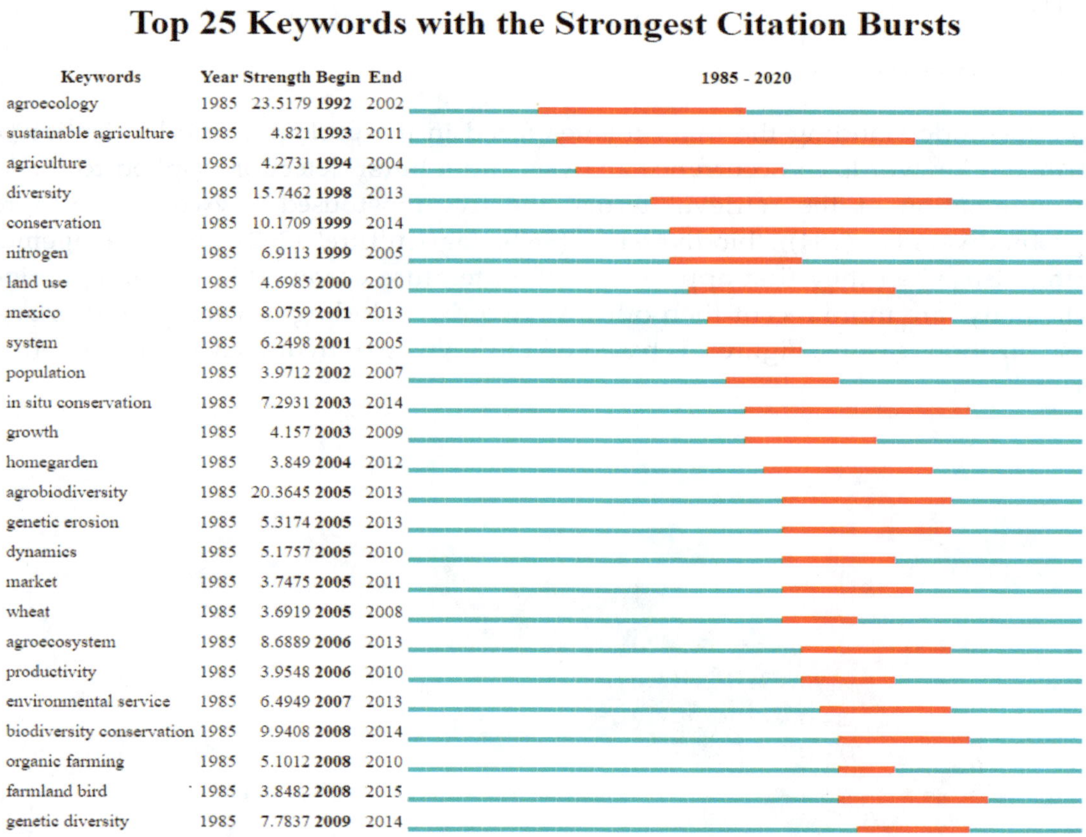

Figure 8. Keyword citation burst.

3.2.2. Clusters Analysis

In order to deepen the analysis of the themes, a co-citation cluster analysis has been performed using as nodes the cited references. The network produced is built up by 1390 nodes and 6137 lines. The cluster analysis has been processed using log-likelihood ratios (LLR) algorithm among the ones available in CiteSpace as it usually gives the best result in terms of uniqueness and coverage [18], and labels given according to the abstract. The quality of the clustering is given by Q = 0.7212 and an average silhouette equal to S = 0.3423. These values mean that the network is reasonably divided into loosely couple clusters but the homogeneity of them on average is not very high. Figure 9 reports the results of the clustering. Clusters are numbered starting from cluster #0, which is the one with the largest number of references. Each cluster can be considered as a theme, a topic or a line of research. Then we selected five clusters, according to the dimension and silhouette values, for going in deep in the analysis (Table 1). The most cited papers have been read and, using also the trends highlighted by the keyword's analysis, we identified five lines of research, which can be considered as predominant.

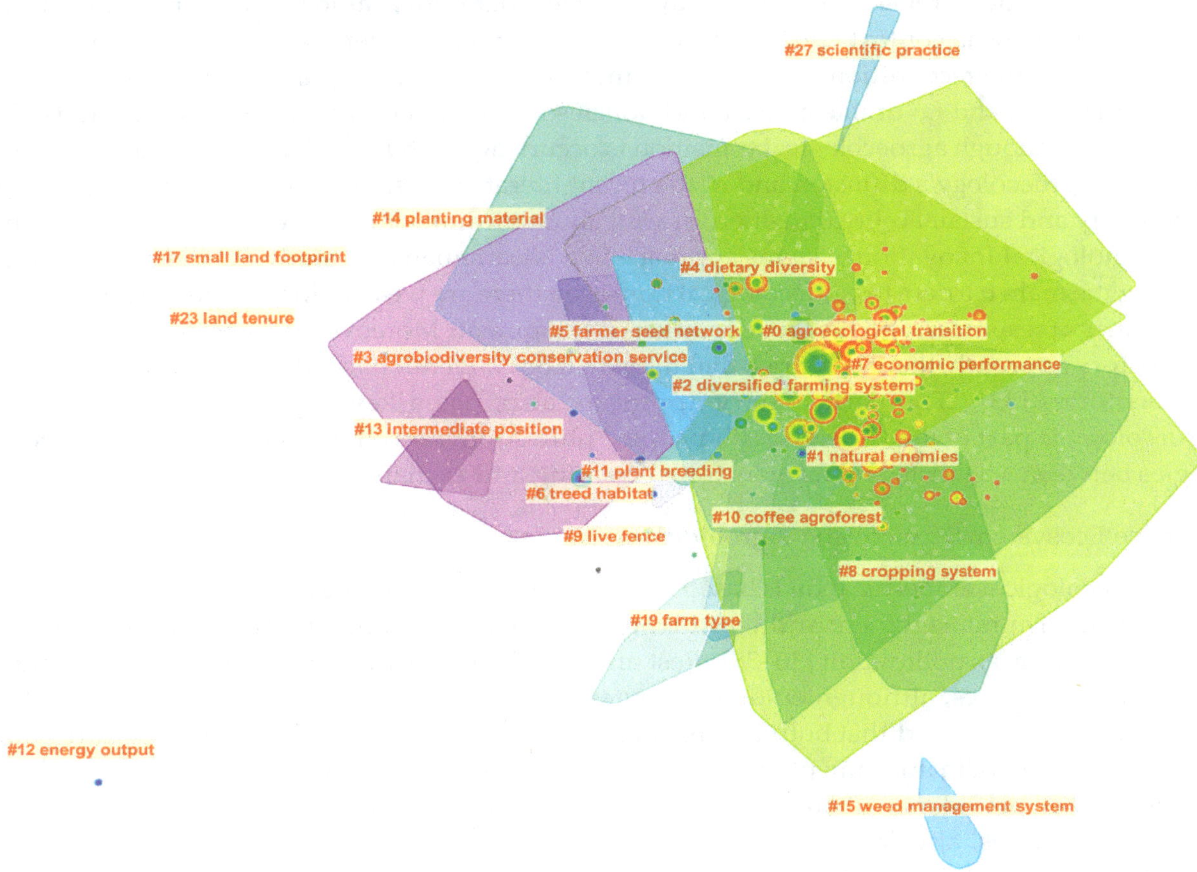

Figure 9. Clustering of co-citation references.

Table 1. High-frequency co-words of co-citation clustering (top 5).

Cluster ID	Size	Silhouette	Mean (year)	Core Noun Terms
#0	200	0.683	2013	agroecology transition; food sovereignty; agroecological practice; agroecological paradigm; Latin America; sustainability transition
#1	157	0.729	2013	natural vegetation; biological control; organic farming; agricultural intensification; habitat amount
#2	117	0.745	2008	carbon sequestration; soil quality; conventional farming system; diversified farming practice; water-holding capacity
#3	108	0.823	2006	cultural landscape; conservation area; on-farm conservation; conservation service; policy option
#4	104	0.752	2013	crop diversity; food group; semi-arid area; land use; livestock diversity

#0 Cluster Agroecology

Initially started as grassroots movements, nowadays agroecology is recognized as a promised alternative to intensified agricultural approach [25]. Moreover, it can be seen as a way for the transition toward a more sustainable food system, in which innovation crosses the farm borders, settings as combinations of farming practices, agriculture type and value chain [26]. Altieri and Toledo described the agroecology transition as a threefold 'agroecological revolution', applied in an epistemological, technical and social direction [27]. The core of such revolution leans on the regeneration and conservation of agrobiodiversity as well as on input reduction. In addition, the role of the peasants and local movements is crucial for success [27,28]. The agronomic interpretation of agroecology only as a practice may lead to a misinterpretation of the social dimensions of agroecology, which allows to criticize those food systems approaches that undervalue the farmers' contribution in determining their social context [29,30]. Moreover, some crucial drivers which allow the agroecology transition include the recognition of a crisis that motivates the search for alternatives, the presence of social organization, and favorable market and policies in the direction of the food sovereignty [31].

However, although agroecological transition is occurring, with Latin America leading [27], there is a problem of agroecology's diffusion and in the recent literature there is a deep analysis of the barriers to agroecology and sustainability transition, as well about the limits to the diversification approach [26]. Using Technological Innovation Systems approach in a Nicaraguan case study, Schiller et al. [32] found that among the main barriers to agroecological transition there are weak guidance, insufficient capacities and quantities of resources, and the lack of market development. Moreover, if the transition does not use a participatory approach, there is the risk that these will be conventionalized and co-opted, especially in larger territories [33]. The participatory approach, also, has a key importance in the transition, as agroecology is part of the "science with people" philosophy, and there is the need of collaborative strategies between scientists and farmers for co-creating and exchanging knowledge [31,34,35].

#1 Enhance Ecosystem Services and Agrobiodiversity

Strong ecological modernization is based on agrobiodiversity and the ecosystem services provided. Both the local farm management and the surrounding landscape potentially affect the provision of ecosystem services, as pollination [36–38]. Pest and weed control and pollination can be augmented by diversified practices, although evidence suggests that these practices may often be insufficient. Lichtenberg et al. [39] found that both organic farming and in-field plant diversification carry out the strongest effects on pollinators and predators, suggesting how these management schemes can facilitate ecosystem service providers without augmenting pest populations. However, to conserve high-quality habitats and services, also large spatial area plans are needed, integrating multiple conservation approaches at landscape scale [37]. Although some works are presented (as [40]), few papers are still present about the full range of benefits and costs of a productive diversified agroecosystem in which several species, vegetables and crops, interact.

#2 Diversified Farm Systems

Diversified Farming Systems (DFS) can be defined as all the practices at farm and landscape levels that intentionally include functional biodiversity at multiple spatial and/or temporal scales, with the aim to maintain ecosystem services [41]. In this definition, livestock integration, non-crop planting, polycultures are all included. In DFS, there is a reciprocity between agriculture models promoted by the provision of ecosystem services and biodiversity promotion through a diversification approach [41]. Kremen and Miles [42] proved that diversified farming systems sustained agrobiodiversity and other ecosystem services, as soil quality, carbon sequestration and water-holding capacity in comparison to the conventional farms. Moreover, DFS can provide greater food security, as it contributes to the self-consumption, improved income of farmers and reduce vulnerability against climate change and variability in comparison to conventional system, even in case of livestock productions [41,43,44]. This is thanks to the fact that DFS gives more possibility to the farmers of adopting different strategies because of the strict connection with both traditional and contemporary knowledge, cultures, practices [41,45]. However, the social dimension of diversified systems has been less studied than the economic and ecological ones [46,47].

In comparison to other systems, in particular to conventional, the improvement of diversified farm system received less funding, although these produced fewer environmental and social harms [42]. DFS is not only significantly less promoted than other systems [47] but its promotion is not always effective because of a lack of connection with the environmental and agroecological concerns embodied within the DFS concept itself [42]. DFS needs an institutional environment that specifically promotes diversified agricultural practices with a broad approach [45,46]. van Zonneveld et al. [48] identified a weak point in the lack of support and decision making on how to diversify farm system to fit in the agroecological paradigm, particularly in case of smallholders, as well as the delay in policy action [45].

#3 Agrobiodiversity Conservations

The erosion of agrobiodiversity can be defined as the replacement of the diverse existing pool of local plant and animal genetic resources with a smaller range of them, despite their crucial role in contributing to sustainability of agriculture systems [49]. Around the world, peasants cultivate crop genetic diversity, enabling crops adaptation to changing environmental condition [50,51]. In situ conservation of biodiversity by smallholder is proved to be a successful strategy [50,52,53]. Such process is undermined by the market expansion and homogenization of the field [50].

To support and enlarge in situ conservation, payment schemes, as payments for agrobiodiversity conservation services, can be useful, although some challenge for the success, as the institutional framework, are present and market niche should be developed [49,54]. Moreover, a huge attention to the conservation goals should be paid because trade-off between them may happen [55]. Some social aspects, such as inclusion, participation and planning, are crucial for agrobiodiversity management [56], as well as migration activities [50,56]. In addition, the governance framework must have an active and positive role in conservation, in order to encourage farmers about exchange and conservation of genetic resources [57].

#4 Dietary Diversity

Crop diversification and livestock integration mean also dietary diversity in several situations for smallholder farm households in developing countries, although empirical evidence of such a link is still rare [58]. Some authors highlight the contribution of wild edible species from arid and semi- arid areas, in particular, for a healthy diet [59,60]; although, these results are still under investigation [61].

Motivations to crop diversity can largely vary, however, there is a centrality of consumption-related traits in semi-subsistence farmers [62]. Market transactions tend to reduce the role of farm diversity for household nutrition, while having positive effects on dietary diversity [58,62].

3.3. Trending Topics in the Past and Current Literature

In order to map the trending topic in the corpus, an analysis of the high-frequency words and their bursts has been performed. Figure 10 reports 20 groups of terms with the highest frequency, ranked according to the year in which the burst started.

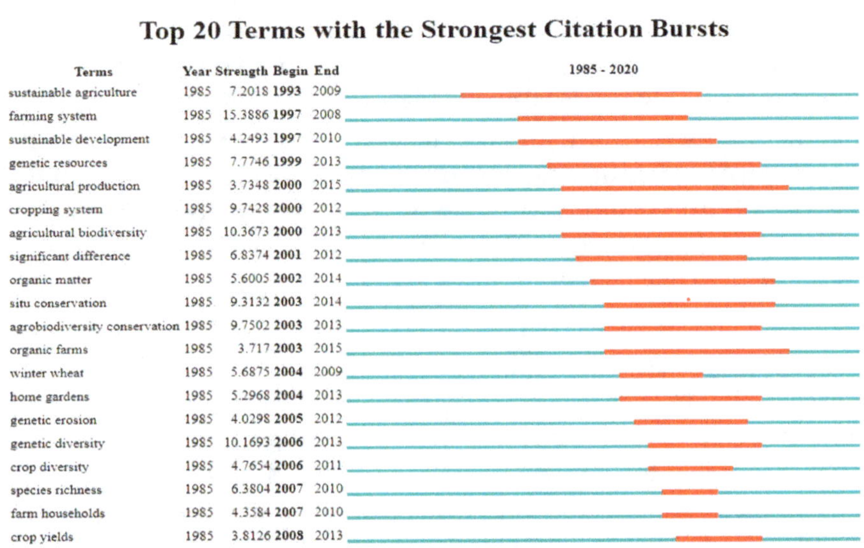

Figure 10. High-frequency terms and their burst time.

Most of the bursts detected appear after 2000. Before that date, the trending topics have been related in particular to 'sustainable agriculture', which stands for 16 years, and 'farming system' in general, which shows the strongest burst. Clearly, the majority of the terms are somehow linked to agrobiodiversity and its conservation: 7 out of 20 terms. Landscape related terms do not appear among the trend topics, although in the analysis we identified a clear pivotal role for it. Moreover, diversification of crop systems seems to be marginal in comparison to other topics.

4. Discussion

The research questions of this study were: (1) What is the state-of-the-art connected with strong ecological modernization? (2) What are the relevant topics and research fields connected with strong ecological modernization? (3) Which of them are the most investigated so far?

As for the first question, the analysis showed that, even if strong ecological modernization is considered a niche sector, this may be true with regard to field practice, which is not yet widespread. With reference to research, we found that things are changing, with ever greater acceleration. Studies, experiments, research projects dedicated to the theme of strong ecological modernization have shown a remarkable, fast and constant growth over time. Research interest in this topic is high, well spread across countries and sectors and improving their quantitative presence. Both agriculture and environmental sciences devoted attention to those themes and the presence of long and overlapping bursts can be read as prove of a constant, not occasional or discontinuous interest of the scientific communities.

To find the relevant topics and research fields connected with strong ecological modernization, we used cluster analysis. As a result, we identified some topics and areas of research, developed in the considered period. All the five clusters identified are linked to the diversification issue. Each cluster identified a different connotation of diversification, although all connected. In particular, for agroecology approach diversity means not only enhance biodiversity through the promotion of polycultures and/or agroforestry patterns, but also diversity of approach, strategy and techniques [63]. Multiple land use strategy in agroecology is a key point, as it is the only way to ensure the resilience of the agroecosystem [63,64]. As a matter of fact, the integration between natural systems and agroecosystems is a key element for the resilience of the territories, for the realization of effective models of strong ecological modernization, to continuously feed the capacity of agricultural systems to provide their wide range of ecosystem services, from the provisioning, to the regulating, to the supporting, up to the cultural ones. At same time, biodiversity conservation does not mean only conservation of local varieties and species but also biological pest control and soil quality, thanks to the biotic activities [63]. In addition, the diversification present in the Diversified Farming Systems includes the management of agrobiodiversity at multiple spatial and/or temporal scales, also explicitly using agroecological knowledge and approach [41]. However, in DFS, the focus is more on the role of the farmer in defining practices from the field level to the landscape one [41].

Agroecology is the theme with the greatest research activity, followed by the enhancement of ecosystem services and agrobiodiversity. Agrobiodiversity is the diversification link across the different studies. Burst analysis of both keywords and terms highlight the overabundant presence of terms related to agrobiodiversity (i.e., genetic conservation/erosion/diversification; in situ conservation, agrobiodiversity, biodiversity conservation, etc.). This is not surprising as its capacity to generate critical ecosystem services to agriculture is proven [65].

The two themes of agroecology and enhancement of ecosystem services and agrobiodiversity are strongly linked to each other. In fact, agroecology is the key for the conservation of agrobiodiversity and, as a consequence, for the maintenance and strengthening of ecosystem services. The reference to an agroecology well placed in the context of sustainability, therefore not linked only to agronomic aspects, is evident [66]. The economic and above all social implications of agroecology, and the needs for political and regulatory guidelines are clearly discussed. The agroecological paradigm is well designed, and it can be synthesized as follows:

- First the system and then the techniques;
- strong reduction of external inputs;
- prevention (e.g., soil health, crop adversity);
- diversification and agroecosystem services;
- transdisciplinary and participatory innovation.

As a result, from the studies it emerges clearly that the agroecological approach, based on conscious use of agrobiodiversity, is the best strategy for a long-term sustainability of agri-food systems [67].

5. Conclusions

Sustainable agriculture is a key argument in the current literature. However, the perfect solution to achieve it has been not yet found. Among the different approaches, strong ecological modernization seems to be one of the most promising. In comparison to other approaches, as the weak ecological modernization, its framework seems quite scattered, including approaches and themes very different.

In this framework, agroecology seems playing a great role, probably linked also to its capacity to include not only ecological and agronomic aspects but also social and economic issues. The lack of a proper governance level or of decision support for decision makers emerges in a very clear way from literature; in particular, in relation to funding and supporting framework.

The use of CiteSpace helped us in identifying in a very objective way a picture of strong ecological modernization, without artificial screening. Bibliometric analysis revealed some interesting information, as the presence of a clear network of countries and organization working on those topics in EU. Cluster analysis highlighted the presence of some clear pathways for research. Keyword analysis and noun-term detection stressed the effort of the literature core for sustainable agriculture. However, the analysis showed that although the ecological and environmental dimensions are well developed in the literature, to the connection with to the topics of ecosystem services and agrobiodiversity, social and economic research are still needed. As it emerged that strong ecological modernization is also linked to themes such as food sovereignty and diet diversity, a deeper reflection on the social and economics pillars is required for the progress of the approach itself. Moreover, this supplementary and further analysis is required not only in developing countries but also in developed countries where other types of farming approaches receive more support although they ensure a lower level of resilience to the whole food production system.

Author Contributions: L.R. contributed to conceptualization, methodology, software, validation, formal analysis, literature search, data curation, and writing—original draft preparation. A.B. contributed to conceptualization, supervision, and writing—review and editing. L.P. contributed literature search, writing—original draft preparation and writing—review and editing. All authors have read and agreed to the published version of the manuscript.

References

1. Gerland, P.; Raftery, A.E.; Ševčíková, H.; Li, N.; Gu, D.; Spoorenberg, T.; Alkema, L.; Fosdick, B.K.; Chunn, J.; Lalic, N.; et al. World population stabilization unlikely this century. *Science* **2014**, *346*, 234–237. [CrossRef] [PubMed]
2. Tilman, D.; Cassman, K.G.; Matson, P.A.; Naylor, R.; Polasky, S. Agricultural sustainability and intensive production practices. *Nature* **2002**, *418*, 671–677. [CrossRef] [PubMed]
3. Godfray, H.C.J.; Beddington, J.R.; Crute, I.R.; Haddad, L.; Lawrence, D.; Muir, J.F.; Pretty, J.; Robinson, S.; Thomas, S.M.; Toulmin, C. Food security: The challenge of feeding 9 billion people. *Science* **2010**, *327*, 812–818. [CrossRef] [PubMed]
4. Bajželj, B.; Richards, K.S.; Allwood, J.M.; Smith, P.; Dennis, J.S.; Curmi, E.; Gilligan, C.A. Importance of food-demand management for climate mitigation. *Nat. Clim. Chang.* **2014**, *4*, 924–929. [CrossRef]

5. Mottet, A.; de Haan, C.; Falcucci, A.; Tempio, G.; Opio, C.; Gerber, P. Livestock: On our plates or eating at our table? A new analysis of the feed/food debate. *Glob. Food Secur.* **2017**, *14*, 1–8. [CrossRef]
6. Poore, J.; Nemecek, T. Reducing food's environmental impacts through producers and consumers. *Science* **2018**, *360*, 987–992. [CrossRef]
7. Springmann, M.; Clark, M.; Mason-D'Croz, D.; Wiebe, K.; Bodirsky, B.L.; Lassaletta, L.; de Vries, W.; Vermeulen, S.J.; Herrero, M.; Carlson, K.M.; et al. Options for keeping the food system within environmental limits. *Nature* **2018**, *562*, 519–525. [CrossRef]
8. Willett, W.; Rockström, J.; Loken, B.; Springmann, M.; Lang, T.; Vermeulen, S.; Garnett, T.; Tilman, D.; DeClerck, F.; Wood, A.; et al. Food in the Anthropocene: The EAT–Lancet Commission on healthy diets from sustainable food systems. *Lancet* **2019**, *393*, 447–492. [CrossRef]
9. Bommarco, R.; Kleijn, D.; Potts, S.G. Ecological intensification: Harnessing ecosystem services for food security. *Trends Ecol. Evol.* **2013**, *28*, 230–238. [CrossRef]
10. Palm, C.; Blanco-Canqui, H.; DeClerck, F.; Gatere, L.; Grace, P. Conservation agriculture and ecosystem services: An overview. *Agric. Ecosyst. Environ.* **2014**, *187*, 87–105. [CrossRef]
11. Zhang, H.; Gao, Y.; Hua, Y.; Zhang, Y.; Liu, K. Assessing and mapping recreationists' perceived social values for ecosystem services in the Qinling Mountains, China. *Ecosyst. Serv.* **2017**, *3*, 7. [CrossRef]
12. Mastrangelo, M.E.; Aguiar, S. Are Ecological Modernization Narratives Useful for Understanding and Steering Social-Ecological Change in the Argentine Chaco? *Sustainability* **2019**, *11*, 3593. [CrossRef]
13. Horlings, L.G.; Marsden, T.K. Towards the real green revolution? Exploring the conceptual dimensions of a new ecological modernisation of agriculture that could "feed the world". *Glob. Environ. Chang.* **2011**, *21*, 441–452. [CrossRef]
14. Duru, M.; Therond, O. Livestock system sustainability and resilience in intensive production zones: Which form of ecological modernization? *Reg. Environ. Chang.* **2015**, *15*, 1651–1665. [CrossRef]
15. Duru, M.; Therond, O.; Fares, M. Designing agroecological transitions; A review. *Agron. Sustain. Dev.* **2015**, *35*, 1237–1257. [CrossRef]
16. van Leeuwen, T. The application of bibliometric analyses in the evaluation of social science research. Who benefits from it, and why it is still feasible. *Scientometrics* **2006**, *66*, 133–154. [CrossRef]
17. CiteSpace: Visualizing Patterns and Trends in Scientific Literature. Available online: http://cluster.cis.drexel.edu/~{}cchen/citespace/ (accessed on 24 September 2020).
18. Chen, C. CiteSpace II: Detecting and visualizing emerging trends and transient patterns in scientific literature. *J. Am. Soc. Inf. Sci. Technol.* **2006**, *57*, 359–377. [CrossRef]
19. Wu, Y.; Wang, H.; Wang, Z.; Zhang, B.; Meyer, B.C. Knowledge mapping analysis of rural landscape using CiteSpace. *Sustainability* **2020**, *12*, 66. [CrossRef]
20. Fang, Y.; Yin, J.; Wu, B. Climate change and tourism: A scientometric analysis using CiteSpace. *J. Sustain. Tour.* **2018**, *26*, 108–126. [CrossRef]
21. Yao, L.; Hui, L.; Yang, Z.; Chen, X.; Xiao, A. Freshwater microplastics pollution: Detecting and visualizing emerging trends based on Citespace II. *Chemosphere* **2020**, *245*, 125627. [CrossRef]
22. Brandes, U.; Delling, D.; Gaertler, M.; Görke, R.; Hoefer, M.; Nikoloski, Z.; Wagner, D. On modularity clustering. *IEEE Trans. Knowl. Data Eng.* **2008**, *20*, 172–188. [CrossRef]
23. Kleinberg, J. Bursty and Hierarchical Structure in Streams. *Data Min. Knowl. Discov.* **2003**, *7*, 373–397. [CrossRef]
24. Tscharntke, T.; Klein, A.M.; Kruess, A.; Steffan-Dewenter, I.; Thies, C. Landscape perspectives on agricultural intensification and biodiversity—Ecosystem service management. *Ecol. Lett.* **2005**, *8*, 857–874. [CrossRef]
25. Altieri, M.A.; Nicholls, C.I. Agroecology: A brief account of its origins and currents of thought in Latin America. *Agroecol. Sustain. Food Syst.* **2017**, *41*, 231–237. [CrossRef]
26. Morel, K.; Revoyron, E.; San Cristobal, M.; Baret, P.V. Innovating within or outside dominant food systems? Different challenges for contrasting crop diversification strategies in Europe. *PLoS ONE* **2020**, *15*, e0229910. [CrossRef]
27. Altieri, M.A.; Toledo, V.M. The agroecological revolution in Latin America: Rescuing nature, ensuring food sovereignty and empowering peasants. *J. Peasant Stud.* **2011**, *38*, 587–612. [CrossRef]
28. Rosset, P.M.; Sosa, B.M.; Jaime, A.M.R.; Lozano, D.R.Á. The Campesino-to-Campesino agroecology movement of ANAP in Cuba: Social process methodology in the construction of sustainable peasant agriculture and food sovereignty. *J. Peasant Stud.* **2011**, *38*, 161–191. [CrossRef]

29. Tessier, L.; Bijttebier, J.; Marchand, F.; Baret, P.V. Pathways of action followed by Flemish beef farmers–an integrative view on agroecology as a practice. *Agroecol. Sustain. Food Syst.* **2020**. [CrossRef]
30. Gliessman, S. Agroecology: Growing the roots of resistance. *Agroecol. Sustain. Food Syst.* **2013**, *37*, 19–31. [CrossRef]
31. Mier y Terán Giménez Cacho, M.; Giraldo, O.F.; Aldasoro, M.; Morales, H.; Ferguson, B.G.; Rosset, P.; Khadse, A.; Campos, C. Bringing agroecology to scale: Key drivers and emblematic cases. *Agroecol. Sustain. Food Syst.* **2018**, *42*, 637–665. [CrossRef]
32. Schiller, K.J.F.; Klerkx, L.; Poortvliet, P.M.; Godek, W. Exploring barriers to the agroecological transition in Nicaragua: A Technological Innovation Systems Approach. *Agroecol. Sustain. Food Syst.* **2020**, *44*, 88–132. [CrossRef]
33. López-García, D.; García-García, V.; Sampedro-Ortega, Y.; Pomar-León, A.; Tendero-Acin, G.; Sastre-Morató, A.; Correro-Humanes, A. Exploring the contradictions of scaling: Action plans for agroecological transition in metropolitan environments. *Agroecol. Sustain. Food Syst.* **2020**, *44*, 467–489. [CrossRef]
34. Levidow, L.; Sansolo, D.; Schiavinatto, M. Agroecological practices as territorial development: An analytical schema from Brazilian case studies. *J. Peasant Stud.* **2019**. [CrossRef]
35. Cuéllar-Padilla, M.; Calle-Collado, Á. Can we find solutions with people? Participatory action research with small organic producers in Andalusia. *J. Rural Stud.* **2011**, *27*, 372–383. [CrossRef]
36. Kennedy, C.M.; Lonsdorf, E.; Neel, M.C.; Williams, N.M.; Ricketts, T.H.; Winfree, R.; Bommarco, R.; Brittain, C.; Burley, A.L.; Cariveau, D.; et al. A global quantitative synthesis of local and landscape effects on wild bee pollinators in agroecosystems. *Ecol. Lett.* **2013**, *16*, 584–599. [CrossRef]
37. Gonthier, D.J.; Ennis, K.K.; Farinas, S.; Hsieh, H.-Y.; Iverson, A.L.; Batáry, P.; Rudolphi, J.; Tscharntke, T.; Cardinale, B.J.; Perfecto, I. Biodiversity conservation in agriculture requires a multi-scale approach. *Proc. R. Soc. B Biol. Sci.* **2014**, *281*, 20141358. [CrossRef]
38. Wood, S.A.; Karp, D.S.; DeClerck, F.; Kremen, C.; Naeem, S.; Palm, C.A. Functional traits in agriculture: Agrobiodiversity and ecosystem services. *Trends Ecol. Evol.* **2015**, *30*, 531–539. [CrossRef]
39. Lichtenberg, E.M.; Kennedy, C.M.; Kremen, C.; Batáry, P.; Berendse, F.; Bommarco, R.; Bosque-Pérez, N.A.; Carvalheiro, L.G.; Snyder, W.E.; Williams, N.M.; et al. A global synthesis of the effects of diversified farming systems on arthropod diversity within fields and across agricultural landscapes. *Glob. Chang. Biol.* **2017**, *23*, 4946–4957. [CrossRef]
40. Saunders, M.E.; Peisley, R.K.; Rader, R.; Luck, G.W. Pollinators, pests, and predators: Recognizing ecological trade-offs in agroecosystems. *Ambio* **2016**, *45*, 4–14. [CrossRef]
41. Kremen, C.; Iles, A.; Bacon, C. Diversified Farming Systems: An Agroecological, Systems-based Alternative to Modern Industrial Agriculture. *Ecol. Soc.* **2012**, *17*, 44. [CrossRef]
42. Kremen, C.; Miles, A. Ecosystem Services in Biologically Diversified versus Conventional Farming Systems: Benefits, Externalities, and Trade-Offs. *Ecol. Soc.* **2012**, *17*, 40. [CrossRef]
43. Bellon, M.R.; Kotu, B.H.; Azzarri, C.; Caracciolo, F. To diversify or not to diversify, that is the question. Pursuing agricultural development for smallholder farmers in marginal areas of Ghana. *World Dev.* **2020**, *125*, 104682. [CrossRef] [PubMed]
44. Martin, G.; Magne, M.A. Agricultural diversity to increase adaptive capacity and reduce vulnerability of livestock systems against weather variability—A farm-scale simulation study. *Agric. Ecosyst. Environ.* **2015**, *199*, 301–311. [CrossRef]
45. Iles, A.; Marsh, R. Nurturing Diversified Farming Systems in Industrialized Countries: How Public Policy Can Contribute. *Ecol. Soc.* **2012**, *17*, 42. [CrossRef]
46. Bacon, C.M.; Getz, C.; Kraus, S.; Montenegro, M.; Holland, K. The Social Dimensions of Sustainability and Change in Diversified Farming Systems. *Ecol. Soc.* **2012**, *17*, 41. [CrossRef]
47. Díaz, S.; Quétier, F.; Cáceres, D.M.; Trainor, S.F.; Pérez-Harguindeguy, N.; Bret-Harte, M.S.; Finegan, B.; Peña-Claros, M.; Poorter, L. Linking functional diversity and social actor strategies in a framework for interdisciplinary analysis of nature's benefits to society. *Proc. Natl. Acad. Sci. USA* **2011**, *108*, 895–902. [CrossRef]
48. Van Zonneveld, M.; Turmel, M.S.; Hellin, J. Decision-Making to Diversify Farm Systems for Climate Change Adaptation. *Front. Sustain. Food Syst.* **2020**, *4*, 32. [CrossRef]

49. Narloch, U.; Drucker, A.G.; Pascual, U. Payments for agrobiodiversity conservation services for sustained on-farm utilization of plant and animal genetic resources. *Ecol. Econ.* **2011**, *70*, 1837–1845. [CrossRef]
50. Isakson, S.R. No hay ganancia en la milpa: The agrarian question, food sovereignty, and the on-farm conservation of agrobiodiversity in the Guatemalan highlands. *J. Peasant Stud.* **2009**, *36*, 725–759. [CrossRef]
51. Keleman, A.; Hellin, J.; Bellon, M.R. Maize diversity, rural development policy, and farmers' practices: Lessons from Chiapas, Mexico. *Geogr. J.* **2009**, *175*, 52–70. [CrossRef]
52. Zimmerer, K.S. The compatibility of agricultural intensification in a global hotspot of smallholder agrobiodiversity (Bolivia). *Proc. Natl. Acad. Sci. USA* **2013**, *110*, 2769–2774. [CrossRef] [PubMed]
53. Kerr, R.B. Lost and Found Crops: Agrobiodiversity, Indigenous Knowledge, and a Feminist Political Ecology of Sorghum and Finger Millet in Northern Malawi. *Ann. Assoc. Am. Geogr.* **2014**, *104*, 577–593. [CrossRef]
54. Krishna, V.V.; Drucker, A.G.; Pascual, U.; Raghu, P.T.; King, E.D.I.O. Estimating compensation payments for on-farm conservation of agricultural biodiversity in developing countries. *Ecol. Econ.* **2013**, *87*, 110–123. [CrossRef]
55. Narloch, U.; Pascual, U.; Drucker, A.G. Cost-effectiveness targeting under multiple conservation goals and equity considerations in the Andes. *Environ. Conserv.* **2011**, *38*, 417–425. [CrossRef]
56. Zimmerer, K.S. Conserving agrobiodiversity amid global change, migration, and nontraditional livelihood networks: The dynamic uses of cultural landscape knowledge. *Ecol. Soc.* **2014**, *19*, 1. [CrossRef]
57. Stromberg, P.M.; Pascual, U.; Bellon, M.R. Seed systems and farmers' seed choices: The case of maize in the Peruvian Amazon. *Hum. Ecol.* **2010**, *38*, 539–553. [CrossRef]
58. Krishna, V. Farm Production Diversity and Dietary Diversity in Developing Countries. *Open Access J.* **2015**, *112*, 10657–10662. [CrossRef]
59. Ng'Endo, M.; Keding, G.B.; Bhagwat, S.; Kehlenbeck, K. Variability of on-farm food plant diversity and its contribution to food security: A case study of smallholder farming households in western Kenya. *Agroecol. Sustain. Food Syst.* **2015**, *39*, 1071–1103. [CrossRef]
60. Segnon, A.C.; Achigan-Dako, E.G. Comparative analysis of diversity and utilization of edible plants in arid and semi-arid areas in Benin. *J. Ethnobiol. Ethnomed.* **2014**, *10*, 80. [CrossRef]
61. Powell, B.; Thilsted, S.H.; Ickowitz, A.; Termote, C.; Sunderland, T.; Herforth, A. Improving diets with wild and cultivated biodiversity from across the landscape. *Food Secur.* **2015**, *7*, 535–554. [CrossRef]
62. Nordhagen, S.; Pascual, U.; Drucker, A.G. Feeding the Household, Growing the Business, or Just Showing Off? Farmers' Motivations for Crop Diversity Choices in Papua New Guinea. *Ecol. Econ.* **2017**, *137*, 99–109. [CrossRef]
63. Altieri, M.A. Agroecology: The science of natural resource management for poor farmers in marginal environments. *Agric. Ecosyst. Environ.* **2002**, *93*, 1–24. [CrossRef]
64. Nicholls, C.I.; Altieri, M.A. Pathways for the amplification of agroecology. *Agroecol. Sustain. Food Syst.* **2018**, *42*, 1170–1193. [CrossRef]
65. Zhang, W.; Ricketts, T.H.; Kremen, C.; Carney, K.; Swinton, S.M. Ecosystem services and dis-services to agriculture. *Ecol. Econ.* **2007**, *64*, 253–260. [CrossRef]
66. Migliorini, P.; Gkisakis, V.; Gonzalvez, V.; Raigón, M.D.; Bàrberi, P. Agroecology in mediterranean Europe: Genesis, state and perspectives. *Sustainability* **2018**, *10*, 2724. [CrossRef]
67. Picchi, M.S.; Bocci, G.; Petacchi, R.; Entling, M.H. Effects of local and landscape factors on spiders and olive fruit flies. *Agric. Ecosyst. Environ.* **2016**, *222*, 138–147. [CrossRef]

2

Arbuscular Mycorrhizal Fungi Modulate the Crop Performance and Metabolic Profile of Saffron in Soilless Cultivation

Matteo Caser [1,†,*], Sonia Demasi [1,†], Íris Marisa Maxaieie Victorino [2,3,4], Dario Donno [1], Antonella Faccio [2], Erica Lumini [2], Valeria Bianciotto [2] and Valentina Scariot [1,2]

1 Department of Agricultural, Forest and Food Sciences, University of Torino. Largo Paolo Braccini 2, 10095 Grugliasco (TO), Italy; sonia.demasi@unito.it (S.D.); dario.donno@unito.it (D.D.); valentina.scariot@unito.it (V.S.)

2 Institute for Sustainable Plant Protection, National Research Council, Viale Mattioli 25, 10125 Torino, Italy; rismarisamazaieie.victorino@unito.it (Í.M.M.V.); antonella.faccio@ipsp.cnr.it (A.F.); erica.lumini@ipsp.cnr.it (E.L.); valeria.bianciotto@ipsp.cnr.it (V.B.)

3 Department of Life Sciences and Systems Biology, University of Turin. Viale Mattioli 25, 10125 Torino, Italy

4 Biological Science Department, Science Faculty, Eduardo Mondlane University (UEM), Av. Julius Nyerere-Campus Universitário, Maputo 3453, Mozambique

* Correspondence: matteo.caser@unito.it

† Both authors contributed equally to this work.

Abstract: Saffron (*Crocus sativus* L.) is cultivated worldwide. Its stigmas represent the highest-priced spice and contain bioactive compounds beneficial for human health. Saffron cultivation commonly occurs in open field, and spice yield can vary greatly, from 0.15 to 1.5 g m^{-2}, based on several agronomic and climatic factors. In this study, we evaluated saffron cultivation in soilless systems, where plants can benefit from a wealth of nutrients without competition with pathogens or stresses related to nutrient-soil interaction. In addition, as plant nutrient and water uptake can be enhanced by the symbiosis with arbuscular mycorrhizal fungi (AMF), we also tested two inocula: a single species (*Rhizophagus intraradices*) or a mixture of *R. intraradices* and *Funneliformis mosseae*. After one cultivation cycle, we evaluated the spice yield, quality (ISO category), antioxidant activity, and bioactive compound contents of saffron produced in soilless systems and the effect of the applied AMF inocula. Spice yield in soilless systems (0.55 g m^{-2}) was on average with that produced in open field, while presented a superior content of several health-promoting compounds, such as polyphenols, anthocyanins, vitamin C, and elevated antioxidant activity. The AMF symbiosis with saffron roots was verified by light and transmission electron microscopy. Inoculated corms showed larger replacement corms (+50% ca.). Corms inoculated with *R. intraradices* performed better than those inoculated with the mix in terms of spice quality (+90% ca.) and antioxidant activity (+88% ca.). Conversely, the mixture of *R. intraradices* and *F. mosseae* increased the polyphenol content (+343% ca.). Thus, soilless systems appeared as an effective alternative cultivation strategy for the production of high quality saffron. Further benefits can be obtained by the application of targeted AMF-based biostimulants.

Keywords: biostimulants; *Crocus sativus*; *Funneliformis mosseae*; glasshouse; protected cultivation; *Rhizophagus intraradices*; substrate

1. Introduction

Crocus sativus L. (saffron) is a flowering plant belonging to the Iridaceae family [1], grown for its red scarlet stigmas that represent the world's highest-priced spice. The market price for high quality saffron can reach 15,000–20,000 € kg^{-1} [2]. This species is widely cultivated in several countries, such as Iran, Italy, Spain, Morocco, France, Greece, China, India and Mexico [3], with an annual spice production that exceeds 220,000 kg [4].The importance and notoriety of saffron, used since ancient times as a dye, ingredient for the preparation of spirits, and condiment for food, is due to the substances contained in the spice, primarily crocins, picrocrocin and safranal [5,6]. These compounds confer the saffron's unique colour, taste, and aroma, and can also have positive biological effects. Saffron active constituents, such as carotenoids (i.e., crocins), polyphenols, and vitamins showed significant antioxidant activity [7–12]. Furthermore, saffron extracts exhibit anti-carcinogenic, anti-depressive, anti-hyperglycemic, hypoglycemic, and memory-enhancing effects [3,13]. *Crocus sativus* is a highly hand labour-intensive crop, mainly during flower harvesting and stigma separation. It is traditionally cultivated in small and flat plots, wherein mechanisation is not economically sustainable due to the harvest type and short flowering period [5,8]. Five hundred hand labour hours are needed to obtain 1 kg of dried saffron [4,5]. Saffron cultivation can be carried out on an annual or multi-year cycle [14,15]. Annual cultivation guarantees the effective control of plant diseases with a more accurate corm selection. On the contrary, in a multi-year cycle (e.g., 3–4 years in Spain, 4–5 years in Italy, and 6–8 years in India and Greece) [14], corm multiplication and the size of replacement corms in the ground can decrease drastically over the third year [15]. Environment and cultivation management affect flower induction in *C. sativus* [5,16–18]. In Mediterranean environments, flower induction occurs from early spring to mid-summer, while flower emergence occurs from early- to late-autumn. Differences in the time required for flower initiation have mostly been attributed to the corm size [19]. To produce flowers, the *C. sativus* corm diameter needs to be greater than 1 cm [20]. As the corm increases, flowering increases [16,21] and occurs in advance [22]. Commercially, a 2.5–3.5 cm diameter corm appears to be the most common size used to have full flowering already during the first cultivation cycle [23]. To increase saffron yield and quality, and to reduce production costs, flowering modulation through cultivation in soilless systems has been proposed [6,19,24]. In this cultivation system, plants are grown without the use of soil as a rooting medium and are supplied of inorganic nutrients via the irrigation water [25], and thus can benefit of a wealth of nutrients without competition with pathogens or stresses related to nutrient-soil interaction [26]. However, at present, only limited and controversial reports of saffron soilless cultivation under protected conditions are present in the literature. Molina et al. [18] reported that, in a glasshouse, temperatures may be responsible for production differences in terms of flower induction and flowering duration. Maggio et al. [19] showed that, in southern Italy, cultivation in a cold glasshouse on vermiculite and perlite-based substrates positively affected the yield and number of replacement corms. Similarly, Helal Beigi et al. [27] found that cocopeat and perlite substrates enhanced corm dry weight. While Souret and Weathers [28] and Mollafilabi et al. [24] concluded that soilless cultivation in experiments carried out in France and Iran, respectively significantly decreased the spice yield, in comparison to open field cultivation.

Plant performance in soilless systems may be improved through use of biostimulants, i.e., any natural substance or microorganism applied to plants with the aim to enhance nutrition efficiency, abiotic stress tolerance and/or crop quality traits, regardless of its nutrients content [29], with a consequent decrease of chemicals and increase of sustainability of the production system [30]. Soil microorganisms such as arbuscular mycorrhizal fungi (AMF) are collecting growing interest as biostimulants. They can form mutualistic symbiosis with about 80% of land plant species, including several crops [31]. Across the interface between the plant and the fungus, carbohydrates and mineral nutrients (i.e., N, P, Zn and B) are exchanged [32]. Thus, AMF can alleviate the limitation in plant growth caused by an inadequate nutrient supply and can improve tolerance to biotic and abiotic stress [33]. Additionally, there is evidence to indicate that AMF symbiosis may have a positive impact on crop quality [34]. Increased yield of essential oils, terpenes and polyphenols, and enhanced antioxidant activity were induced by AMF symbiosis in several medicinal and aromatic plants (MAPs) [12,35–38]. This higher concentration of bioactive molecules makes AMF-hosted plants generally more attractive for the pharmaceutical and food industries [39].

The positive effects of AMF on corm growth, spice yield, and the nutraceutical compound content of *C. sativus* have already been reported in open field trials [12,40–42]. However, so far little is known of the proper saffron AMF inocula application and effects in soilless conditions, where plants are cultivated in pots filled with sterilised substrates that are free of AM fungal propagules or highly reduced in AMF diversity [43]. In the meta-analysis performed by Berruti et al. [31], it has been observed that the fungal colonization gain in inoculated plants was significantly more frequent in the greenhouses than in the open-field conditions, even if the effectiveness of AMF inoculation on shoot biomass and yield was equally successful.

Thus, in the literature, saffron cultivation on soilless systems has been proposed for spice production, but no comparison with open field has been reported. While, the effects of AMF-based biostimulants have been investigated only in open fields. To evaluate if saffron cultivation in soilless systems and AMF application may improve crop performance, spice yield and quality, and modulate bioactive compounds content, we cultivated saffron on soilless systems, applying two AMF inocula, and we compared results with those obtained in a previous open field-based trial [12].

2. Materials and Methods

2.1. Plant Material and Soilless Cultivation

Crocus sativus corms with horizontal diameters of 2.5–3.5 cm, provided by the Azienda agricola "Les épices Vda" di Alessandro Putzolu (Chatillon, AO, Italy), were planted during the last 10 days of August 2017 in the experimental heated glasshouse of the Department of Agricultural Forest and Food Sciences (DISAFA) of the University of Torino (Italy, 45°06′23.21″N Lat, 7°57′82.83″E Long; 300 m a.s.l.). Corms were cultivated in pots (4 L, 14 cm diameter and 17 cm height; two corms per pot; density of 91 corms m^{-2}) filled with sterile quartz sand (2 L per pot; bulk density of 1.2 kg m^{-3}) on a layer of sterilised expanded clay (1 L per pot; bulk density of 300 kg m^{-3}) for a total weight of about 1.5 kg. During the flowering period, the average temperatures were 22 °C during the day and 14 °C during the night.

Two inocula (MycAgro Lab, Breteniére, FR) were used in this experiment: one composed of a single fungus *Rhizophagus intraradices* (Ri) and one composed of *R. intraradices* and *Funneliformis mosseae* (Ri+Fm). Both inocula consisted of AMF spores and inorganic substrate (calcined clay, vermiculite and zeolite). Inocula treatments were compared to a control without any formulation (AMF-). Ten grams of each inoculum were inserted into each vase. The treatment was placed under each corm in order to guarantee contact between the inoculum and the roots, therefore, favouring mutualistic symbiosis. Corms were not treated for fungal pathogens and cultivation lasted one cycle (August 2017–April 2018).

A complete randomised block design was used, with a total of 48 pots in two experimental plot units (24 pots per unit) and three treatments (8 pots per treatment). Irrigation water (pH 7.4, EC 505 μS cm) was added weekly (250 mL per pot) with a drip system. The corms were fertilised by fertigation (N:K 13:46; VIGORFLOR, AL.FE. srl, MN, Italy) every 2 weeks starting from the emergence of the spate, in quantities of 1.5 g L^{-1} of water.

2.2. Determination of Flower Production, Stigma Yield and Corm Growth

At flowering (October and November 2017), the number of flowers produced daily per corm and the yield of spice (i.e., stigmas dried at 40 °C for 8 h in an oven) were measured. The spice yield was calculated by weighting the mg of saffron produced per pot (area equal to 196 cm^2) and comparing the values to g of spice per square meter (m^2). At the end of the vegetative period (April 2018), corms were lifted, rid of topsoil, cleaned and de-tunicated, then the number, size and weight of replacement corms were determined.

2.3. Preparation of the Saffron Extract

The saffron aqueous extracts were prepared according to Caser et al. [12]. Briefly, 50 mg of powdered saffron were suspended into 5 mL of deionised water. After stirring (1000 rpm) for 1 hour at room temperature (circa 21 °C) in the dark, the solution was filtered with polytetrafluoroethylene (PTFE, VWR International, Milano, Italy) filters with a 25 mm diameter and 0.45 μm pore size. The saffron extract was then diluted 1:10 with deionised water to obtain the working solution. Each sample was prepared in triplicate.

2.4. Determination of Saffron Quality by ISO 3632

Saffron aqueous extracts were analysed with a spectrophotometer (Ultrospec 2100 Pro, Ultrospec 2100 pro, Amersham Biosciences, Uppsala, Sweden) to determine the content of picrocrocin, safranal, and crocin to have the information on the bitterness, the flavouring strength, and the colouring strength [44]. Data were related to the dry matter percentage and expressed as the absorbance of a 1% aqueous solution of dried saffron at 257, 330 and 440 nm respectively, using a 1 cm pathway quartz cell [A1% 1 cm (λ max)] and calculated according to the following formula [45]:

$$\text{A1\%1cm } (\lambda \text{ max}) = \text{D} \times 10000/\text{m} \times (100-\text{wMV}) \tag{1}$$

where D is the specific absorbance; m is the mass of the evaluated solution in grams; and wMV is the moisture expressed as a percentage mass fraction of the sample.

Moisture content (wMV) was determined using the following formula:

$$\text{wMV} = (\text{m0}-\text{m1}) \times (100/\text{m0})\% \tag{2}$$

where m0 is the mass, in grams, of the saffron portion before drying; and m1 is the mass, in grams, of the dry residue after incubation, performed in an oven for 16 h at 103 ± 2 °C.

All analytical steps were conducted in the dark to prevent analyte degradation.

2.5. Determination of Bioactive Compounds by HPLC

Bioactive compounds were determined by means of four high performance liquid chromatography-diode array detection (HPLC–DAD) methods (Table 1; [46]) using an Agilent 1200 High-Performance Liquid Chromatograph coupled to an Agilent UV-Vis diode array detector (Agilent Technologies, Santa Clara, CA, USA). Phytochemical separation was achieved with a Kinetex C18 column (4.6 × 150 mm, 5 μm, Phenomenex, Torrance, CA, USA) using several mobile phases for compound identification and recording UV spectra at different wavelengths, based on HPLC methods, as previously tested and validated [47], with some modifications. UV spectra were recorded at 330 nm (α), 280 nm (β), 310 and 441 nm (γ), and 261 and 348 nm (δ).

All single compounds were identified by a comparison and combination of their retention times and UV spectra with those of authentic standards under the same chromatographic conditions.

Table 1. Characteristics of the HPLC methods applied to analyse the bioactive compounds present in the studied saffron samples.

HPLC Method	Class	Standard	Stationary Phase	Mobile Phase	Flow (mL min^{-1})	Time (min)
α	Cinnamic acids	Caffeic acid Chlorogenic acid Coumaric acid Cerulic acid	KINETEX–C18 column (4.6 × 150 mm, 5 μm)	A: 10mM KH_2PO_4/H_3PO_4, pH = 2.8 B: CH_3CN	1.5	20 + 2 (CT)
	Flavanols	Hyperoside Isoquercitrin Quercetin Wuercitrin Rutin				
β	Benzoic acids	Ellagic acid Gallic acid		A: $H_2O/CH_3OH/HCOOH$ (5:95:0.1 v/v/v), pH = 2.5 B: $CH_3OH/HCOOH$ (100:0.1 v/v)	0.6	23 + 2 (CT)
	Catechins	Catechin Epicatechin				
γ	Carotenoids	Crocin I Crocin II Safranal		A: H_2O B: CH_3CN	0.6	35 + 10 (CT)
δ	Vitamin C	Ascorbic acid Dehydroascorbic acid		A: 5 mM $C_{16}H_{33}N(CH_3)_3Br$/50 mM KH_2PO_4, pH=2.5 B: CH_3OH	0.9	10 + 5 (CT)

CT = conditioning time; Method α—gradient analysis: 5% B to 21% B in 17 min + 21% B in 3 min + 2 min of conditioning time—wavelength: 330 nm; Method β—gradient analysis: 3% B to 85% B in 22 min + 85% B in 1 min + 2 min of conditioning time—wavelength: 280 nm; Method γ—gradient analysis: 5% B to 95% B in 30 min + 95% B to 5% B in 5 min + 10 min of conditioning time—wavelengths: 310 nm + 441 nm; Method δ—isocratic analysis: 10 min + 5 min of conditioning time—wavelengths: 261 nm + 348 nm.

2.6. Phytochemical Characterisation

The phytochemical characterisation of each sample was performed as previously described by Caser et al. [48,49]. Briefly, the total anthocyanin content (TAC) was determined using the pH-differential method. Saffron solution was added to pH 1 and pH 4.5 buffer solutions. The absorbance of samples was determined at 515 and 700 nm after 15 min of equilibration. The results were expressed as milligrams of cyanidin 3-O-glucoside (C3G) per 100 grams of dry weight (mg_{C3G} $100g^{-1}$ DW). The total phenol content (TPC) was measured using the Folin–Ciocalteau phenolic method at 765 nm. The results were expressed as mg of gallic acid equivalents (GAE) per 100 g of dry weight (DW; mg_{GAE} $100g^{-1}$ DW). The antioxidant activity (AOA) was determined at 595 nm using the ferric reducing antioxidant power (FRAP) method and at 734 nm using the 2,2′-azinobis (3-ethylbenzothiazoline-6-sulphonic acid; ABTS) method. Results were expressed as millimoles of ferrous iron (Fe^{2+}) equivalents per kilogram of dry weight (mmol Fe^{2+} kg^{-1} DW) and as μmol of Trolox equivalents per gram of dry weight (μmol TE g^{-1} DW), respectively. All analyses were performed in three replicates and the absorbances were read using a spectrophotometer (Ultrospec 2100 Pro, Ultrospec 2100 pro, Amersham Biosciences, Uppsala, Sweden).

2.7. AMF Evaluation

On the base of saffron highly mycorrhization level (70 to 90% mycorrhizal intensity) previously reported [12], we randomly selected saffron roots in April 2018. Then, the root segments were processed for observation in light and under transmission electron microscopy. Root segments were excised under a stereomicroscope and quickly fixed in 2.5% glutaraldehyde in 0.1 M cacodilate buffer (pH 7.2) for 2 hours at room temperature and overnight at 4 °C. The samples were then post-fixed in 1% OsO4 in the same buffer and dehydrated in an ascending series of ethanol to 100%, incubated in two changes of absolute acetone and infiltrated in Epon-Araldite resin [50]. The resin was polimerised for 24 h at 60 °C. Semi-thin (1 μm) sections were then stained with 1% toluidine blue and ultra-thin (70 nm) sections were counter-stained with uranyl acetate and lead citrate [51], and used for electron microscopy analyses under a Philips CM10 transmission electron microscope.

2.8. Chemicals and Reagents

Sodium carbonate, Folin–Ciocalteu phenol reagent, sodium acetate, citric acid, hydrochloric acid, iron (III) chloride hexahydrate, 2,4,6-tripyridyl-S-triazine (TPTZ) and 1,2-phenylenediamine dihydrochloride (OPDA) were purchased from Sigma Aldrich (St. Louis, MO, USA), whereas acetic acid was purchased from Fluka Biochemika (Buchs, Switzerland). Ethylenediaminetetraacetic acid (EDTA) disodium salt was purchased from AMRESCO (Solon, OH, USA), whereas sodium fluoride was purchased from Riedel-de Haen (Seelze, Germany). Ethanol, acetone, sodium citrate and lead nitrate were purchased from Fluka Biochemika. Analytic HPLC grade solvents, methanol and formic acid were purchased from Sigma Aldrich and Fluka Biochemika, respectively; potassium dihydrogen phosphate, ammonium dihydrogen phosphate and phosphoric acid were also purchased from Sigma Aldrich. Milli-Q ultrapure water was produced by Sartorius Stedium Biotech mod. Arium (Sartorius, Goettingen, Germany). Cetyltrimethylammonium bromide (cetrimide) was purchased from Extrasynthése (Genay, France), whereas 1,2-phenylenediamine dihydrochloride (OPDA) was purchased from Sigma Aldrich. All polyphenolic and terpenic standards were purchased from Sigma Aldrich. The organic acids were purchased from Fluka Biochemika, whereas ascorbic acid and dehydroascorbic acid were purchased from Extrasynthése. All chemicals specific for electron and optical microscopy were purchased from Electron Microscopy Sciences (Newark, PA, USA), i.e., glutharaldeyde, cacodylate buffer, osmium tetroxide, epon/araldite resin, toluidine "O" and uranyl acetate.

2.9. Statistical Analysis

An arcsin transformation was performed on all percentage incidence data before statistical analysis in order to improve the homogeneity of the variance (Levene test). All the analysed data were checked for the normality of variance. For all the analysed parameters, mean differences were computed using a one-way ANOVA with a Tukey *post hoc* test ($p \leq 0.05$). Mean comparisons between data obtained in soilless and those from the first growing season of a previous work conducted in open field [12] cultivations were performed using an independent samples t-test. All analyses were performed using SPSS 24.0 Inc. software (SPSS Inc., Chicago, IL, USA).

3. Results and Discussion

3.1. Crop Performance, Quality and Secondary Metabolite Content of Saffron in Soilless Cultivation

Soilless cultivation in a glasshouse has been recently proposed as an alternative method to open field cultivation for saffron. Maggio et al. [19] and Gresta et al. [6] reported that, by controlling growth conditions, flowering could be modulated, extended and considerably increased, compared with open field cultivation. In the present study, under protected conditions, flowering had the same duration (ca. 22 days) compared to cultivation of the same corms planted on the same days in a northwestern Italian open field [12], but the saffron flowering moved forward about 20 days (from 5 October 2017 to 23–30 October 2017), in agreement with Gresta et al. [6]. Since, for the flower emergence, corms required to be transferred from 23–27 °C to 17 °C [18], the most likely reason for this results is related to the fact that, in a glasshouse, the lowering of seasonal temperatures takes place more slowly than in an open field. In addition to the temperature lowering, Gresta et al. [52] indicated the soil water content as another environmental component that can trigger flowering. However, as in these two studies object of comparison, the cultivation occurred in different substrates (quartz sand vs soil), it appears not possible to make speculations.

Saffron yield can vary from 0.15 to 1.5 g m^{-2}, based on planting density, plantation age (from one to six year crop cycles), and climatic conditions during the crop season [1]. In this study, an average of 0.55 g m^{-2} was obtained, indicating a profitable production already during the first year.

This yield was similar to what obtained cultivating the same corms at a density of 39 corms m^{-2} in a northwestern Italian open field [12] and superior to that obtained in south Italy under similar glasshouse conditions by Gresta et al. [6] (corm density equal to 40 corms m^{-2}; 0.46 g m^{-2}) with corms coming from Sardinia (Italy). With similar corm density to our work, Cavusoglu and Erkel [53] and Maggio et al. [19] obtained much higher yields (0.88 g m^{-2} and 2.34 g m^{-2}, respectively) in glasshouses located in Turkey and south Italy. In Iranian open fields, at a corm density similar to our study, Mollafilabi et al. [24] and Koocheki and Seyyedi [54] obtained an average spice yield of 0.48 g m^{-2}. As affirmed by Gresta et al. [52], to trigger saffron flowering, a not yet fully understood combination of temperature and soil water content is needed.

In addition to the spice yield, another economically important attribute of saffron is the number of replacement corms. The obtained values (2.63 replacement corms $corm^{-1}$) are lower of those obtained by Maggio et al. [19] in soilless cultivation in a cold glasshouse in south Italy, by using peat and perlite (1:1) substrates, where corms produced from 3.0 to 4.5 replacement corms per corm. In addition to a different substrate, these authors also incubated corms in the dark for 83 days before planting. Thus, the combination of these two factors could have guarantee a superior result. Comparing to open field experiments that used corms with similar size to our study, results were in agreement with those from our trial in northwest Italy [12], and the trials performed by Turhan et al. [55] in Turkey (2.32 replacement corms $corm^{-1}$), while superior to those obtained by Koocheki and Seyyedi [54] in Iranian fields (1.32 replacement corms $corm^{-1}$).

Guidelines for the analyses of the main compounds that contribute to the sensory profile of saffron have been established by ISO 3632 regulations [44]. These regulations define procedures to determine these compounds by spectrophotometric analyses and have established the limits by which saffron quality is classified into three different categories (first, second and third). Specifically, the saffron produced under soilless conditions belongs to the highest quality, i.e., first category, for all the studied parameters.

The evaluation of antioxidant activity is generally considered as an important method to evaluate the nutraceutical properties of food, as indicated in other previous studies [30]. Apart from crocins, Karimi et al. [56] and Asdaq and Inamdar [57] highlighted that phenols and flavonols are responsible for the antioxidant potential of saffron. Overall, the saffron produced in soilless systems showed a very high TPC (4445.4 mg_{GAE} $100g^{-1}$ DW), more than the saffron cultivated in other sites in the Alps (range between 1340 and 2355 mg_{GAE} $100g^{-1}$ DW) [12], Lebanon (160 mg_{GAE} $100g^{-1}$ DW) [58], and India (828 mg_{GAE} $100g^{-1}$ DW) [8]. In terms of antioxidant activity, FRAP values were superior to those of Iranian and Italian samples (circa 570 and 1250 mmol Fe^{2+} kg^{-1}) [12,56] and ABTS values were comparable to those found in Italian and Greek saffron by Caser et al. [12] and Ordoudi et al. [59].

3.2. AMF Colonisation

In our study, the presence of AMF and their colonisation of saffron roots were confirmed by observations using light microscopy (Figure 1) and transmission electron microscopy (TEM; Figure 2) on semi-thin and thin sections, respectively. Observations on semi-thin sections, stained in blue, show that the saffron roots are mycorrhised when inoculated with both inocula (Figure 1A–C), confirming the mycorrhizal intensity described in Caser et al. [12]. At the level of the cortical root parenchyma, the typical mycorrhizal arbuscular fungal structures have been highlighted (insets Figure 1A,C). Figure 1 shows the presence of intercellular and intracellular hyphae (Figure 1C) and arbuscules (Figure 1A,B). No fungal structures were found in the roots of the control treatments (Figure 1D).

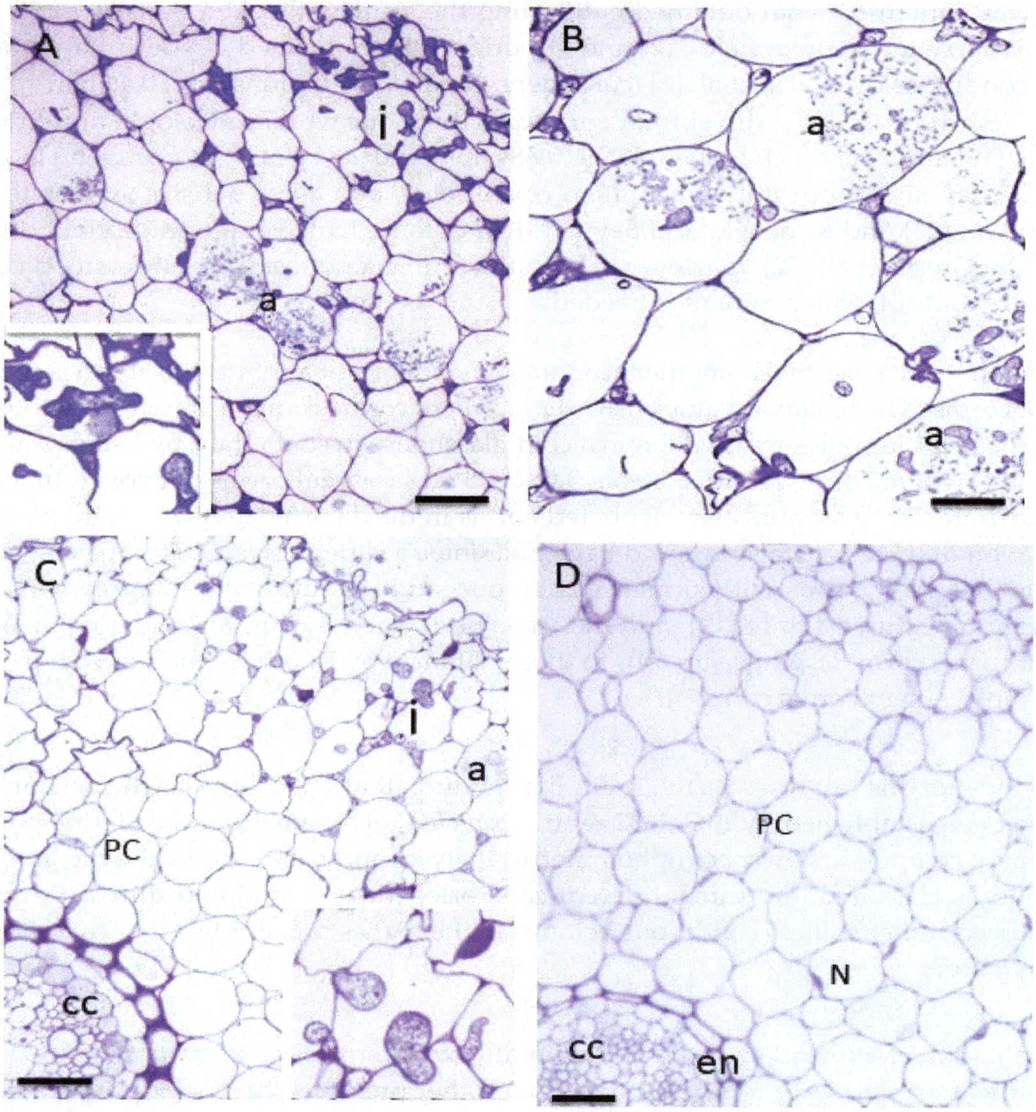

Figure 1. Light microscope images of semi-thin sections of *Crocus sativus* roots inoculated with *Rhizophagus intraradices* and *Funneliformis mosseae* (Ri+Fm, **A**), *R. intraradices* alone (Ri, **B** and **C**) or the control (arbuscular mycorrhizal fungi (AMF)-, **D**), stained with toluidine blue. At the level of the cortical cells, note the presence of intercellular and intracellular hyphae (i) and arbuscules (a). Magnification in insets A and C shows details of the intracellular hyphae. Cortical parenchyma (PC) cells with nucleus (N) are indicated. No fungal structure is present between and inside the root cells in AMF-roots (D). Note the central cylinder (cc) and the endodermide (en). Bars are 20 μm in A, C and D, and 10 μm in B.

Here, the host plasma membrane invaginates and proliferates around all the developing intracellular fungal structures, and cell wall material is laid down between this membrane and the fungal cell surface. The exchange of molecules between the fungal and plant cytoplasm takes place both through their plasma membranes and their cell walls; a functional compartment, known as the symbiotic interface, is thus defined. At the electron microscope level, as seen in Figure 2A,C (arrows), this new apoplastic space, based on membrane proliferation, is evident around the intracellular and arbusculated hyphae of the AMF penetrated inside the saffron root cortical cells. On the basis of TEM observations, we can conclude that the mycorrhizae, formed between saffron roots and the two species of AM fungi in the inocula used in pot experiments, are alive and functionally active.

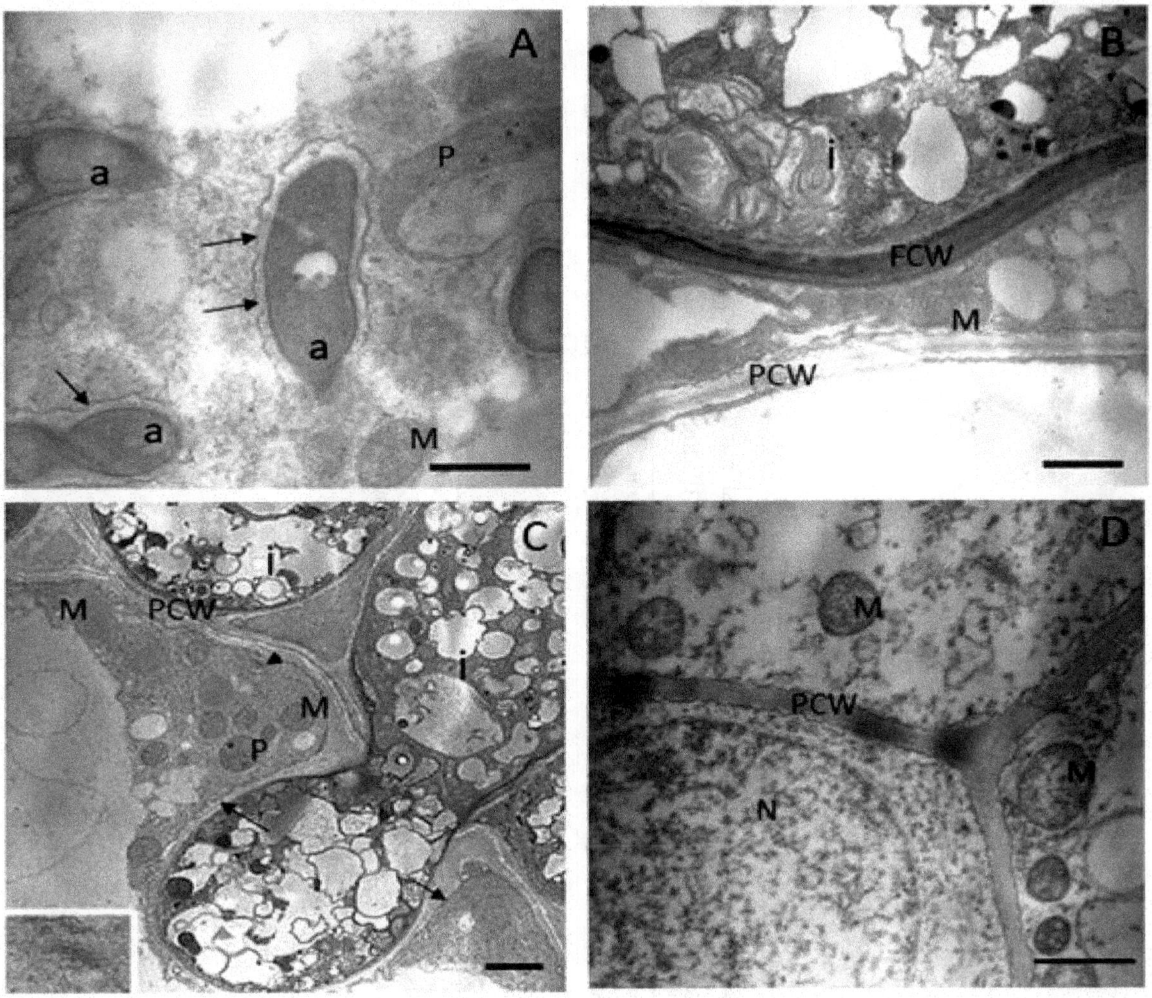

Figure 2. Transmission electron microscopy images of thin sections of saffron roots colonised by *Rhizophagus intraradices* and *Funneliformis mosseae* (Ri+Fm, **A**), *R. intraradices* alone (Ri, **B** and **C**) or the control (AMF-, **D**). In details, a: fungal arbuscule; N: nucleus; M: mithocondria; P: plastids; i: fungal hyphae; PCW: plant cell wall; FCW: fungal cell wall; arrow: plant plasmamembrane; arrowhead and inset: Golgi apparatus. The bar is 1 μm in A, B, C and D.

3.3. Impact of AMF on Saffron in Soilless Cultivation

3.3.1. Crop Performance and Quality Classification

In the present study, slight differences in flowering time and production were detected between treated corms (Figure 3 and Table 2). Both applied inocula (Ri and Ri+Fm) anticipated saffron flowering time of one week, compared to untreated corms (AMF-; 23 October vs. 30 October), whereas the flowering peaks and end of flowering occurred in about the same number of days (6–9 November and 11–13 November, respectively).

No significant differences were observed between the treatments in terms of the number of flowers corm^{-1} and the obtained mg of spice flower^{-1} (Table 2). Very few reports about the effective role of AMF in saffron flowering and yield are available in the literature, and only under open field conditions. Aimo et al. [40] and Caser et al. [12] indicated a positive role of AMF on the saffron productive performance, with an increase in flower production (+68% and +138%, respectively, compared to the untreated corms) using AMF species belonging to the genus *Glomus*.

Table 2. Effects of AMF inoculum composed of *Rhizophagus intraradices* alone (Ri), *R. intraradices* and *Funneliformis mosseae* (Ri+Fm) or the control (AMF-) on yield performances (flower corm^{-1} and saffron flower^{-1}), growth (number of replacement corms corm^{-1}; replacement corm size and weight variation between the end and beginning of the trial) and mean absorbance values for picrocrocin, safranal and crocin of saffron samples obtained during glasshouse cultivation.

	Yield		Replacement corm			Quality category (ISO3632 [44])		
Treatment	**Flower corm^{-1} (n)**	**Saffron flower^{-1} (mg)**	**Size (%)**	**Corm^{-1} (n)**	**Weight (%)**	**Picrocrocin ($A^{1\%}_{1cm}$ (λ257)**	**Safranal ($A^{1\%}_{1cm}$ (λ330)**	**Crocin ($A^{1\%}_{1cm}$ (λ440)**
Ri	0.84 ± 0.62	6.8 ± 1.3	45.8 ± 4.6a	2.71 ± 1.53	7.8 ± 5.6	143.8 ± 4.6(I)$^{\beta}$a	61.0 ± 5.3(I)a	422.6 ± 4.1(I)a
Ri+Fm	0.66 ± 0.60	6.0 ± 1.4	54.6 ± 6.2a	2.25 ± 0.95	8.6 ± 3.8	124.3 ± 3.9(I)c	30.7 ± 3.4(I)c	164.2 ± 3.8(III)c
AMF-	0.97 ± 0.53	6.6 ± 0.4	33.1 ± 6.8b	2.63 ± 1.06	12.6 ± 5.1	135.9 ± 3.4(I)b	54.3 ± 6.7(I)b	324.7 ± 5.9(I)b
p	ns	ns	***	ns	ns	***	***	***

Mean values with the same letter are not statistically different at $p \leq 0.05$, according to a Tukey *post hoc* test. The statistical relevance of 'Between-Subjects Effects' tests (***$p < 0.001$, ns = not significant). $^{\beta}$ The quality category (ISO3632) is indicated in brackets. The limits for the first (I) quality category are: picrocrocin>70; safranal 20–50; crocins>200. ISO3632 limits for the second (II) quality category are: picrocrocin >55; safranal 20–50; crocins >170. ISO3632 limits for the third (III) category are: picrocrocin>40; safranal 20–50; crocins>120.

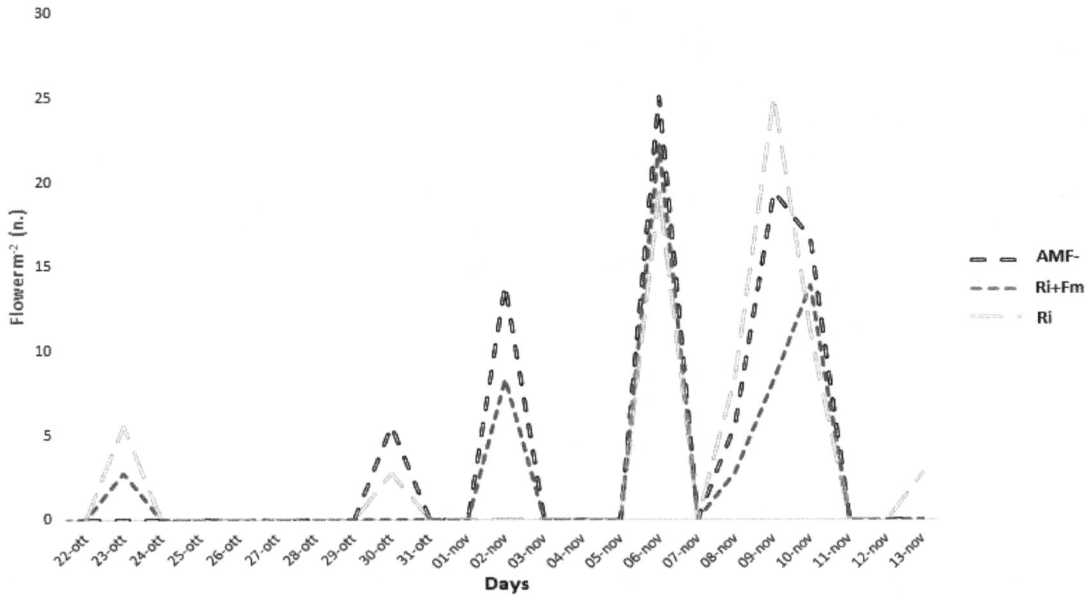

Figure 3. Effects of AMF inoculum composed of *Rhizophagus intraradices* alone (Ri), *R. intraradices* and *Funneliformis mosseae* (Ri+Fm) or the control (AMF-) on the flowering calendar of *Crocus sativus* corms and the daily number of picked flowers m^{-2} during soilless cultivation.

Both of the AMF inocula increased the size of replacement corms in comparison to untreated corms (Table 2), suggesting a positive effect on flower production for the following cultivation cycle, in agreement with Aimo et al. [40] and Mohebi-Anabat et al. [39]. Corm size is indeed a major factor in bulbous plants to determine the flowering capacity and production of new replacement corms [5,42].

Saffron quality greatly depends on the growing conditions [12,60]. In the present study, among the AMF inocula, *R. intraradices* alone significantly increased the content of picrocrocin (bitterness), safranal (flavouring strength) and crocins (colouring strength), in comparison to the other treatments. On the contrary, Ri+Fm significantly reduced the content of these molecules and, thus the quality of the spice, in particular by lowering the crocin content to the third category of ISO 3632. To the best of our knowledge, this is the first report indicating the effect of AMF on the quality (ISO) of saffron obtained by soilless cultivation. The positive role of Ri on the increase of the saffron quality, especially on the content of picrocrocin, was highlighted also in northwestern Italian open field [12]. Thus, the corm inoculation with Ri could further increase the already high quality saffron produced in the Italian Alps [45,61].

3.3.2. Saffron Metabolic Profiling Comparing to Other Foods

In addition to the peculiar organoleptic characteristics, the stigmas of the *C. sativus* flower contain many secondary metabolites with demonstrated pharmacological effects [3,11,62–64]. The identification and quantification of bioactive compounds in saffron and the evaluation of their biological activities are important to gauge their potential efficacy in food and pharmaceutical industries [65]. The range of all chemicals can vary greatly as a result of growing conditions, such as in response to the application of biostimulants [63]. Inoculation with AMF is known to alter the production of secondary metabolites in MAPs, both in roots, shoots, and flowers, even if is not consistent among plant organs [66]. The effects of AMF inocula on the biosynthesis of secondary metabolites in saffron are presented in Table 3. This more in-depth analysis confirmed the results obtained by assessing the spice quality according to ISO3632 guidelines. The single species inoculum Ri significantly increased the content of crocins (crocin I and II), whereas the mix Ri+Fm decreased it; these findings are in agreement with those obtained by Caser et al. [12] under field conditions in a temperate mountain area (north-west Italy), where the saffron obtained by corms inoculated with Ri resulted in superior quality (i.e., quality compared to the ISO standards). Regarding antioxidant activity (AOA), inoculation with Ri resulted in superior values in both used methods (FRAP and ABTS). The AMF inoculum composed of Ri+Fm significantly increased the contents of isoquercitrin and the total phenolic (TPC) compared to Ri, while of ellagic acid in comparison to Ri and AMF-. Differences in results according to the AMF inoculum composition were also observed in other plant species cultivated on different substrates. Among the reviewed studies, it has been found that the single inoculation of *R. intraradices* tend to be more successful for bioactive compounds increase than inoculation experiments with more than one species applied at the same time. In *Echinacea purpurea* Moench. [67] cultivated in a sand and soil (1:1) substrate, *R. intraradices* alone increased more the content of polyphenols than the mixed inoculum, while in *Cynara cardunculus* L. cultivated in sandy soil [68] and *Lactuca sativa* L. cultivated in a mixture of peat, sandy loam soil and calcinated clay (1:1:1) [69] *R. intraradices* enhanced more the antioxidant activity. However, it has not been observed any effect on the accumulation of polyphenols in *Ocimum basilicum* L. cultivated in a sterilised sand and soil (3:1) substrate [70] and in *Salvia officinalis* L. in sand, soil, and expanded clay (1:1:1) [71,72].

Table 3. Bioactive compounds, anthocyanins, total polyphenol content and antioxidant activity (ferric reducing antioxidant power (FRAP) and 2,2′-azinobis(3-ethylbenzothiazoline-6-sulfonic acid) (ABTS); antioxidant activity (AOA)) of the saffron produced via glasshouse cultivation with AMF inocula composed of *Rhizophagus intraradices* alone (Ri), *R. intraradices* and *Funneliformis mosseae* (Ri+Fm) or a control (AMF-).

Class	Compound (mg $100g^{-1}$ DW)	Ri	Ri+Fm	AMF-	*p*
Cinnamic acids	Coumaric acid	23.4 ± 3.5	23.7 ± 2.6	23.7 ± 3.1	ns
Flavonols	Isoquercitrin	1.9 ± 0.3b	2.6 ± 0.2a	2.3 ± 0.3ab	**
	Quercitrin	17.8 ± 4.6	11.6 ± 4.1	19.1 ± 3.6	ns
Benzoic acids	Gallic acid	4.5 ± 1.5	5.1 ± 1.3	4.9 ± 1.4	ns
	Ellagic acid	1.9 ± 0.5b	3.2 ± 0.3a	1.0 ± 0.4b	**
Catechins	Catechin	1.9 ± 0.4	1.6 ± 0.3	1.8 ± 0.3	ns
	Epicatechin	9.8 ± 2.9	5.9 ± 2.1	9.6 ± 2.5	ns
Carotenoids	Safranal	4.0 ± 0.9	4.0 ± 1.2	4.0 ± 0.7	ns
	Crocin I	104.2 ± 8.6a	22.1 ± 6.5c	55.5 ± 8.4b	***
	Crocin II	42.7 ± 9.6a	16.4 ± 3.8b	38.7 ± 12.9ab	**
Vitamin C	Dehydroascorbic acid	28.8 ± 6.5	30.2 ± 4.1	31.8 ± 6.9	ns
	Ascorbic acid	31.1 ± 9.5	36.3 ± 6.7	41.7 ± 4.8	ns
	Total vitamin C	59.9 ± 10.2	66.5 ± 5.9	73.6 ± 8.4	ns
TAC	Anthocyanin (mg_{C3G} $100g^{-1}$ DW)	640.7 ± 84.6b	146.4 ± 29.8c	1654.5 ± 68.4a	*
	Methods				
TPC	Folin–Ciocalteu (mg_{GAE} $100g^{-1}$ DW)	816.5±152.7b	3619.0±400.2a	4445.4±450.2a	***
AOA	FRAP (mmol Fe^{2+} kg^{-1} DW)	3133.9±1524.3a	1383.0±589.7ab	379. 7±128.4b	**
	ABTS (μmol_{TE} g^{-1} DW)	5.4±0.8a	3.6±0.4c	4.5±0.7ab	**

Mean values with the same letter are not statistically different at $p \leq 0.05$, according to a Tukey *post hoc* test. The statistical relevance of 'Between-Subjects Effects' tests (* $p < 0.05$, ** $p < 0.01$, *** $p < 0.001$, ns = not significant).

Karimi et al. [56] and Rahaiee et al. [63] indicated that the antioxidant capacities of saffron might be due to the presence of total phenolics and flavonoids. Based on the obtained results, the content of the bioactive compounds detected in saffron could be compared to other commonly eaten fruits with highly advantageous nutritive properties. Saffron had a higher total phenol content (TPC) and antioxidant activity (AOA) than fresh *Ribes nigrum* L. berries (circa +1000% and +493%, respectively), and fresh (circa +2000% and +1800%, respectively) and dried (circa +900% and +1650%, respectively) *Lycium* spp. fruits [65,73], analysed with the same method. Since saffron showed an antioxidant activity superior than 500 mg_{GAE} $100g^{-1}$ it could be also listed within the health beneficial fruits such as *Rubus glaucus* Berth. and *Prunus serotina* var. Capulì as suggested by Vasco et al. [74]. Its content of vitamin C was similar to what found in *Actinidia deliciosa* (A.Chev.) C.F.Liang & A.R.Ferguson and *Citrus sinensis* (L.) Osb., and even higher than in *Lycium* spp. (+150%) and *Vaccinium* spp. (+580%). Also, the coumaric acid content was superior (+85%) than in *Morus nigra* L. fruits [75] while lower than in *Lycium* spp. fruits, that showed also higher content of gallic acid, ellagic acid, catechin, and epicatechin was generally lower in saffron (on average circa −75%, −70%, −92%, and −95%, respectively) [73,75]. Lastly, the content of anthocyanins, that are suggested to have neuroprotective properties [76], was up to 11654.5 mg_{C3G} $100g^{-1}$ DW, i.e., a value very high in comparison to fresh fruit extracts from *Morus nigra*, *Rubus idaeus* L., and *Fragaria ananassa* D. (80.0, 33.7, and 35.2 mg_{C3G} $100g^{-1}$, respectively) [75].

3.3.3. Soilless Cultivation vs. Open Field

Saffron root colonisation by AMF could be affected by the cultivation conditions related to the substrate composition, root temperature or the presence of antagonistic fungi naturally occurring in the soil [31,40,41,76]. In our recent studies, AM fungal colonisation was noted in *C. sativus* roots inoculated with Ri and Ri+Fm, both in soilless (Figures 1 and 2) and in open field conditions [12]. Figures 4 and 5 report the comparisons of the results obtained by these studies. Compared to open field, in soilless conditions not-inoculated corms (AMF-) showed similar spice yields but with higher quality while, referring to AMF treatments, Ri-inoculated corms produced less spice but with a higher quality, whereas Ri+Fm inoculated corms produced less spice, with a lower quality (i.e., reduction in crocin content).

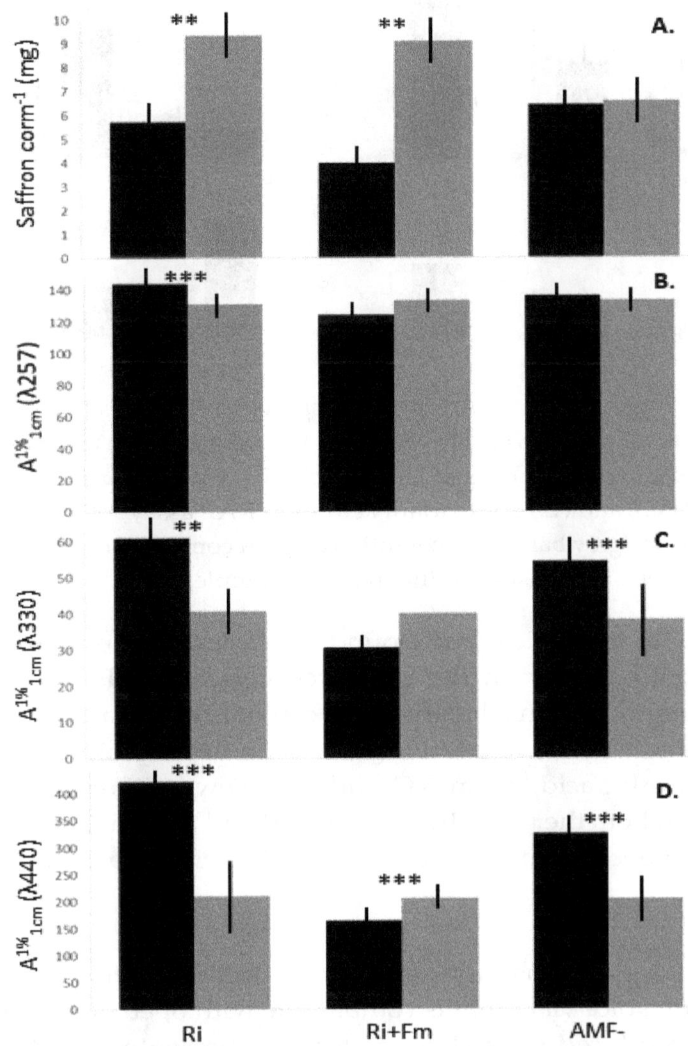

Figure 4. Effects of AMF inoculum consisting of *Rhizophagus intraradices* alone (Ri), *R. intraradices* and *Funneliformis mosseae* (Ri+Fm) or a control (AMF-) on (**A.**) mg of saffron corm^{-1}, (**B.**) picrocrocin, (**C.**) safranal, and (**D.**) crocin of *Crocus sativus* corms cultivated in soilless (black bars) and open field (grey bars, [12]) conditions. Mean comparisons of each treatment in the two cultivation types were performed using an independent samples t-test.

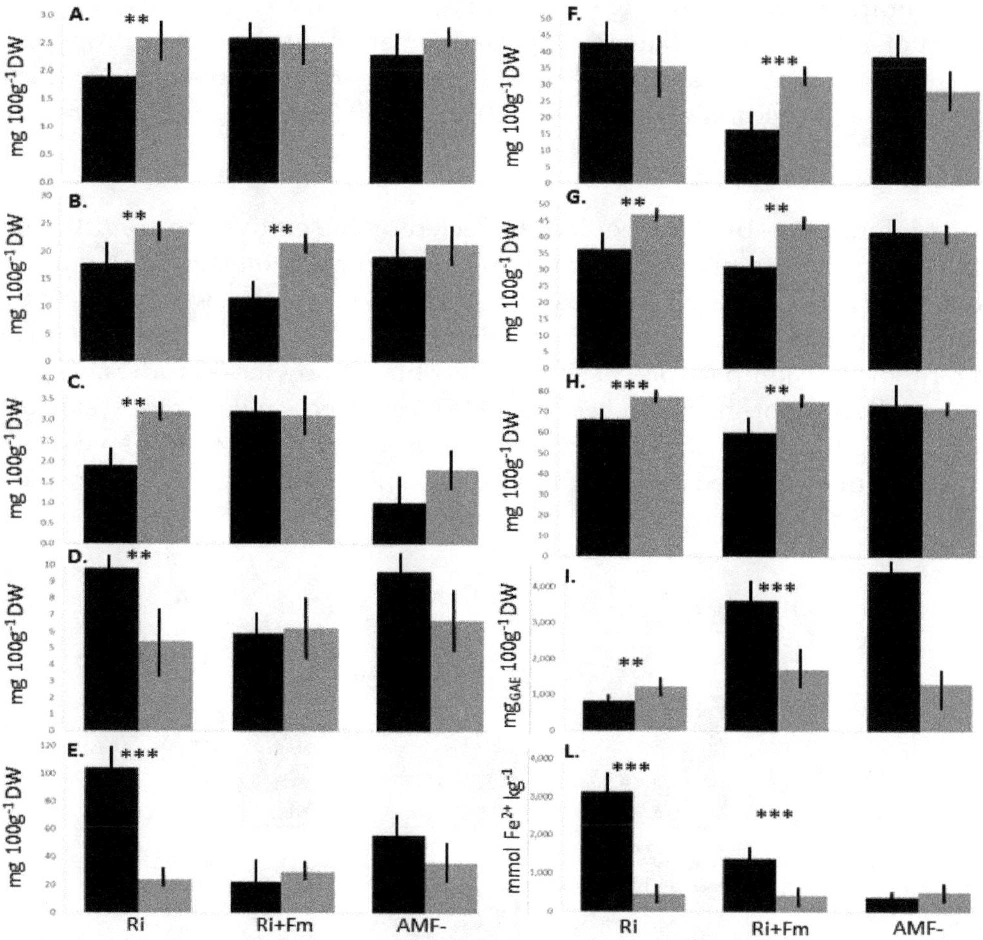

Figure 5. Effects of AMF inoculum consisting of *Rhizophagus intraradices* alone (Ri), *R. intraradices* and *Funneliformis mosseae* (Ri+Fm) or a control (AMF-) on the content of (**A.**) isoquercitrin, (**B.**) quercitrin, (**C.**) ellagic acid, (**D.**) epicatechin, (**E.**) crocin I, (**F.**) crocin II, (**G.**) ascorbic acid, (**H.**) vitamin C, (**I.**) total polyphenol content (TPC), and (L.) antioxidant activity (FRAP assay) of saffron produced in soilless (black bars) and open field (grey bars, [12]) conditions. Mean comparisons of each treatment in the two cultivation types were performed using an independent samples t-test.

With respect to the nutraceutical compounds, the comparisons are presented in Figure 5. No differences were reported between the untreated corms (AMF-), whereas the application of Ri in the soilless condition induced an increase in the contents of epicatechin, crocin I, and antioxidant activity (+80%, +435%, and +675%, respectively), while a decrease in the contents of isoquercitrin, quercitrin, ellagic acid, ascorbic acid, vitamin C, and TPC. Fewer differences were induced by Ri+Fm, which positively stimulated both the total phenolic content and antioxidant activity (+210% and +325%, respectively), but caused a decrease in quercitrin, crocin II, ascorbic acid, and vitamin C.

4. Conclusions

Soilless cultivation in a glasshouse appeared as an effective strategy for the cultivation of saffron with a first-year cultivation spice yield that is comparable with open field production sites. Moreover, the high quality saffron produced via soilless cultivation presented an elevated content of several health-promoting compounds with highly advantageous nutritive properties, such as polyphenols and elevated antioxidant activity. Further studies are needed to define better the methodologies to modulate time and duration of flowering, to improve yield, and to efficiently schedule harvest practices.

Arbuscular mycorrhizal-based products have received great interest in agriculture for their potential to improve crop productivity, nutritional quality, as well as resistance to plant pathogens and numerous environmental stresses. The literature highlights that AMF must be chosen by evaluating different aspects, such as the inoculum type, host plants, and the environmental and growing conditions.

Here, AMF successfully colonised *C. sativus* roots; their effects varied on the basis of inoculum type and cultivation conditions. Among the studied AMF inocula, *R. intraradices* appeared to give more benefits to *C. sativus* than the mix of *R. intraradices* and *F. mosseae*. Specifically, the *R. intraradices* inoculation appeared successful in open field to increase spice yields while in soilless systems to increase the spice quality.

Thus, soilless systems appeared as an effective alternative cultivation strategy for the production of high quality saffron. Further benefits can be obtained by the application of targeted AMF-based biostimulants. A cost-benefit analysis should be performed to assess the economic sustainability.

Author Contributions: M.C., E.L., V.B. and V.S. contributed to the experimental design. M.C., S.D., Í.M.M.V., D.D., A.F. and V.S. acquired and interpreted the data. M.C. drafted the manuscript. V.B. and V.S. conceived the study, coordinated the work and critically revised the manuscript.

Acknowledgments: The authors acknowledge Dario Sacco for statistical assistance and Alessandro Putzolu for technical assistance.

References

1. Gresta, F.; Santonoceto, C.; Avola, G. Crop rotation as an effective strategy for saffron (*Crocus sativus* L.) cultivation. *Sci. Hortic.* **2016**, *211*, 34–39. [CrossRef]
2. Maleki, F.; Kazemi, H.; Siahmarguee, A.; Kamkar, B. Development of a land use suitability model for saffron (*Crocus sativus* L.) cultivation by multi-criteria evaluation and spatial analysis. *Eco Eng.* **2017**, *106*, 140–153. [CrossRef]
3. Giorgi, A.; Scheurer, T. Alpine resources: Assets for a promising future-conclusions from the forum alpinum. *Mount. Res. Dev.* **2015**, *35*, 414–415. [CrossRef]
4. Husaini, A.; Shafiq, A.; Parvaze, S.; Rather, G.; Parray, A.; Shikari, A.; Javid, M. Bioinformatics for saffron (*Crocus sativus* L.) improvement. *Comm. Biomet. Crop. Sci.* **2009**, *4*, 3–8.
5. Gresta, F.; Lombardo, G.; Siracusa, L.; Ruberto, G. Saffron an alternative crop for sustainable agricultural systems. A review. *Agron. Sustain. Dev.* **2008**, *28*, 95–112. [CrossRef]
6. Gresta, F.; Avola, G. Hypothesis on the relation between meteorological data and flowering peaks in saffron (*Crocus sativus L.*) crop. *Acta Hortic.* **2017**, *1184*, 189–194. [CrossRef]
7. Elgazar, A.; Rezq, A.; Bukhari, H. Anti-Hyperglycemic Effect of Saffron Extract in Alloxan-Induced Diabetic Rats. *Europ. J. Biol. Sci.* **2013**, *5*, 14–22.
8. Baba, S.; Mohiudin, T.; Basu, S.; Swarnkar, M.; Malik, A.; Abbas, N.; Singh, A.; Ashraf, N. Comprehensive transcriptome analysis of *Crocus sativus* for discovery and expression of genes involved in apocarotenoid biosynthesis. *BMC Genom.* **2015**, *16*, 698. [CrossRef] [PubMed]
9. Lahmass, I.; Touria, L.; Delporte, C.; Sikdar, S.; Antwerpen, P.; Saalaoui, E.; Mégalizzi, V. The waste of saffron crop, a cheap source of bioactive compounds. *J. Funct. Foods* **2017**, *35*, 341–351. [CrossRef]
10. Cusano, E.; Consonni, R.; Petrakis, E.A.; Astraka, K.; Cagliani, L.R.; Polissiou, M.G. Integrated analytical methodology to investigate bioactive compounds in *Crocus sativus* L. flowers. *Phytochem. Anal. Phytochem. Analysis* **2018**, *29*, 1–11. [CrossRef]
11. Menghini, L.; Leporini, L.; Vecchiotti, G.; Locatelli, M.; Carradori, S.; Ferrante, C.; Zengin, G.; Recinella, L.; Chiavaroli, A.; Leone, S.; et al. *Crocus sativus* L. stigmas and byproducts: Qualitative fingerprint, antioxidant potentials and enzyme inhibitory activities. *Food Res. Int.* **2018**, *109*, 91–98. [CrossRef]
12. Caser, M.; Maxaieie Victorino, Í.M.; Demasi, S.; Berruti, A.; Donno, D.; Lumini, E.; Bianciotto, V.; Scariot, V. Saffron cultivation in marginal Alpine environments: How AMF inoculation modulates yield and bioactive compounds. *Agronomy* **2019**, *9*, 12. [CrossRef]
13. Talaei, A.; Moghadam, H.; Tabassi, S.; Mohajeri, S. Crocin, the main active saffron constituent, as an adjunctive treatment in major depressive disorder: A randomized, double-blind, placebo-controlled, pilot clinical trial. *J. Affect. Disord.* **2015**, *174*, 51–56. [CrossRef]
14. Landi, R. *Lo zafferano: Tradizione e qualità. Camera di commercio: Industria, Artigianato e Agricoltura di Firenze*; CCIAA (Camera di Commercio: Industria, Artigianato e Agricoltura): Firenze, Italy, 2007; Volume 4, p. 65.

15. Lachaud, C. *La Bible du Safranier. Tout savoir sur le Crocus sativus et sur le Safran*, 1st ed.; La Paradisière: Correze, France, 2012; ISBN 978-2-7466-4412-0.
16. De Juan, J.A.; Córcoles, H.L.; Muñoz, R.M.; Picornell, M.R. Yield and yield components of saffron under different cropping systems. *Ind. Crops Prod.* **2009**, *30*, 212–219. [CrossRef]
17. Yarami, N.; Kamgar-Haghighi, A.A.; Sepaskhah, A.R.; Zand-Parsa, S. Determination of the potential evapotranspiration and crop coefficient for saffron using a water-balance lysimeter. *Arch. Agr. Soil Sci.* **2011**, *57*, 727–740. [CrossRef]
18. Molina, R.; Valero, M.; Navarro, Y.; Guardiola, J.; García-Luis, A. Temperature effects on flower formation in saffron (*Crocus sativus* L.). *Sci. Hortic.* **2005**, *103*, 361–379. [CrossRef]
19. Maggio, A.; Raimondi, G.; Martino, A.; De Pascale, S. Soilless cultivation of saffron in mediterranean environment. *Acta Hortic.* **2006**, *718*, 515–522. [CrossRef]
20. Benschop, M. *Crocus*; De Hertogh, A., Le Nard, M., Eds.; The Physiology of Flower Bulbs, Elsevier Science Publishers: Amsterdam, The Netherlands, 1993; pp. 257–272.
21. De Mastro, G.; Ruta, C. Relation between corm size and saffron (*Crocus sativus* L.) flowering. *Acta Hortic.* **1993**, *334*, 512–517. [CrossRef]
22. Douglas, M.H.; Smallfield, B.M.; Wallace, A.R.; McGimpsey, J.A. Saffron (*Crocus sativus* L.): The effect of mother corm size on progeny multiplication, flower and stigma production. *Sci. Hortic.* **2014**, *166*, 50–58. [CrossRef]
23. Kumar, R. Calibration and validation of regression model for non-destructive leaf area estimation of saffron (*Crocus sativus* L.). *Sci. Hortic.* **2009**, *122*, 142–145. [CrossRef]
24. Mollafilabi, A.; Koocheki, A.; Moeinerad, H.; Kooshki, L. Effect of plant density and weight of corm on yield and yield components of saffron (*Crocus sativus* L.) under soil, hydroponic and plastic tunnel cultivation. *Acta Hortic.* **2013**, *997*, 51–58. [CrossRef]
25. Putra, P.A.; Yuliando, H. Soilless culture system to support water use efficiency and product quality: A review. *Agric. Agricult. Sci. Proc.* **2015**, *3*, 283–288. [CrossRef]
26. Augé, R.M.; Toler, H.D.; Saxton, A.M. Arbuscular mycorrhizal symbiosis alters stomatal conductance of host plants more under drought than under amply watered conditions: A meta-analysis. *Mycorrhiza* **2015**, *25*, 13–24. [CrossRef] [PubMed]
27. Helal Beigi, Y.; Khoshgoftar, M.; Shamsi, A.; Zamani, N. Possibility of saffron corm growth in different bed of cultivation in soilless system. In *Proceedings of 1st Congress of Hydroponic in Greenhouse Production*; Isfahan University of Technology: Isfahan, Iran, 2009.
28. Sourell, F.; Weathers, P. The growth of saffron (*Crocus sativus L.*) in aeroponics and hydroponics. *J. Herb. Spices Med. Plants* **2000**, *7*, 113–127.
29. Du Jardin, P. Plant biostimulants: Definition, concept, main categories and regulation. *Sci. Hortic.* **2015**, *196*, 3–14. [CrossRef]
30. Rouphael, Y.; Franken, P.; Schneider, C.; Schwarz, D.; Giovannetti, M.; Agnolucci, M.; De Pascale, S.; Bonini, P.; Colla, G. Arbuscular mycorrhizal fungi acta s biostimulants in horticultural crops. *Sci. Hortic.* **2015**, *196*, 91–108. [CrossRef]
31. Berruti, A.; Lumini, E.; Balestrini, R.; Bianciotto, V. Arbuscular mycorrhizal fungi as natural biofertilizers: Let's benefit from past successes. *Front. Microbiol.* **2016**, *6*, 1559. [CrossRef]
32. Smith, S.E.; Read, D.J. *Mycorrhizal Symbiosis*; Academic Press: London, UK, 1997; Volume 40, p. 154.
33. Todeschini, V.; Lahmidi, N.A.; Mazzucco, E.; Marsano, F.; Gosetti, F.; Robotti, E.; Bona, E.; Massa, N.; Bonneau, L.; Marengo, E.; et al. Impact of beneficial microorganisms on strawberry growth, fruit production, nutritional quality, and volatilome. *Front. Plant. Plant Sci.* **2018**, *9*, 1611. [CrossRef] [PubMed]
34. Pellegrino, E.; Bedini, S. Enhancing ecosystem services in sustainable agriculture: Biofertilization and biofortification of chickpea (*Cicer arietinum* L.) by arbuscular mycorrhizal fungi. *Soil Biol. Biochem.* **2014**, *68*, 429–439. [CrossRef]
35. Zeng, Y.; Guo, L.; Chen, B.; Hao, Z.; Wang, J.; Huang, L.; Yang, G.; Cui, X.; Yang, L.; Wu, Z.; et al. Arbuscular mycorrhizal symbiosis and active ingredients of medicinal plants: Current research status and prospectives. *Mycorrhiza* **2013**, *23*, 253–265. [CrossRef] [PubMed]
36. Giovannetti, M.; Avio, L.; Sbrana, C. *Improvement of nutraceutical value of food by plant symbionts*; Ramawat, K.G., Merillon, J.M., Eds.; Natural Products, Phytochemistry, Botany and Metabolism of Alkaloids, Phenolics and Terpenes; Springer: Berlin/Heidelberg, Germany, 2013; pp. 2641–2662.

37. Swamy, M.K.; Akhtar, M.S.; Sinniah, U.R. *Response of PGPR and AM fungi toward growth and secondary metabolite production in medicinal and aromatic plants*; Hakeem, K., Akhtar, M., Eds.; Plant, Soil and Microbes; Springer: Heidelberg, Germany, 2016.
38. Avio, L.; Turrini, A.; Giovannetti, M.; Sbrana, C. Designing the ideotype mycorrhizal symbionts for the production of healthy food. *Front. Plant Sci.* **2018**, *9*, 1089. [CrossRef] [PubMed]
39. De Sousa Oliveira, M.; da Silva Campos, M.; de Albuquerque, U.P.; da Silva, F.S.B. Arbuscular mycorrhizal fungi (AMF) affects biomolecules content in *Myracrodruon urundeuva* seedlings. *Ind. Crops Prod.* **2013**, *50*, 244–247. [CrossRef]
40. Aimo, S.; Gosetti, F.; D´Agostino, G.; Gamalero, E.; Gianotti, V.; Bottaro, M.; Gennaro, M.; Berta, G. Use of arbuscular mycorrhizal fungi and beneficial soil bacteria to improve yield and quality of saffron (*Crocus sativus* L.). *Acta Hortic.* **2010**, *850*, 159–162. [CrossRef]
41. Lone, R.; Shuab, R.; Koul, K. AMF association and their effect on metabolite mobilization, mineral nutrition and nitrogen assimilating enzymes in saffron (*Crocus sativus*) plant. *J. Plant Nutr.* **2016**, *39*, 1852–1862. [CrossRef]
42. Shajari, M.A.; Moghaddam, P.R.; Ghorbani, R.; Koocheki, A. Increasing saffron (*Crocus sativus* L.) corm size through the mycorrhizal inoculation, humic acid application and irrigation managements. *J. Plant Nutr.* **2018**, *41*, 1047–1064. [CrossRef]
43. Mohebi-Anabat, M.; Riahi, H.; Zanganeh, S.; Sadeghnezhad, E. Effects of arbuscular mycorrhizal inoculation on the growth, photosynthetic pigments and soluble sugar of *Crocus sativus* L. (saffron) in autoclaved soil. *Int. J. Agron. Agric. Res.* **2015**, *6*, 296–304.
44. ISO 3632-1:2011. *Spices—Saffron (Crocus sativus L.)*; Food products SC 7, Spices, culinary herbs and condiments; International Organization for Standardization: Geneva, Switzerland, September 2011.
45. Giorgi, A.; Pentimalli, D.; Giupponi, L.; Panseri, S. Quality traits of saffron (*Crocus sativus* L.) produced in the Italian Alps. *Open Agriculture* **2017**, *2*, 52–57. [CrossRef]
46. Donno, D.; Mellano, M.G.; Prgomet, Z.; Beccaro, G.L. Advances in *Ribes* x *nidigrolaria* Rud. Bauer & A. Bauer fruits as potential source of natural molecules: A preliminary study on physico-chemical traits of an underutilized berry. *Sci. Hortic.* **2018**, *237*, 20–27.
47. Donno, D.; Mellano, M.G.; Prgomet, Z.; Cerutti, A.K.; Beccaro, G.L. Phytochemical characterization and antioxidant activity evaluation of Mediterranean medlar fruit (*Crataegus azarolus* L.): Preliminary study of underutilized genetic resources as a potential source of health-promoting compound for food supplements. *J. Food Nutrit. Res.* **2017**, *56*, 18–31.
48. Caser, M.; D'Angiolillo, F.; Chitarra, W.; Lovisolo, C.; Ruffoni, B.; Pistelli, L.; Pistelli, L.; Scariot, V. Ecophysiological and phytochemical responses of *Salvia sinaloensis* Fern. to drought stress. *Plant Growth Regul.* **2018**, *84*, 383–394. [CrossRef]
49. Caser, M.; Chitarra, W.; D'Angiolillo, F.; Perrone, I.; Demasi, S.; Lovisolo, C.; Pistelli, L.; Pistelli, L.; Scariot, V. Drought stress adaptation modulates plant secondary metabolite production in *Salvia dolomitica* Codd. *Ind. Crop Prod.* **2019**, *129*, 85–96. [CrossRef]
50. Hoch, H.C. *Freeze-substitution of fungi*; Aldrich, H.C., Todd, W.J., Eds.; Ultrastructure techniques of microorganisms; Plenum Press: New York, NY, USA, 1986; pp. 183–211.
51. Reynolds, E.S. The use of lead citrate at high pH as an electron-opaque stain in electron microscopy. *J. Cell. Biol.* **1963**, *17*, 208–212. [CrossRef]
52. Gresta, F.; Avola, G.; Lombardo, G.M.; Siracusa, L.; Ruberto, G. Analysis of flowering, stigmas yield and qualitative traits of saffron (*Crocus sativus* L.) as affected by environmental conditions. *Sci. Hortic.* **2009**, *119*, 320–324.
53. Cavusoglu, A.; Erkel, E. The effect of different planting areas and corm size on yield and harvest period of saffron (*Crocus sativus* L.) in Kocaeli province. *J. Akdeniz Univ. Faculty Agric.* **2005**, *18*, 179–184.
54. Koocheki, A.; Seyyedi, S. How irrigation rounds and mother corm size control saffron yield, quality, daughter corms behavior and phosphorus uptake. *Sci. Hortic.* **2016**, *213*, 132–143. [CrossRef]
55. Turhan, H.; Kahriman, F.; Egesel, C.; Kemal Gul, M. The effects of different growing media on flowering and corm formation of saffron (*Crocus sativus* L.). *Afr. J. Biotech.* **2007**, *6*, 2328–2332.
56. Karimi, E.; Oskoueian, E.; Hendra, R.; Jaafar, H. Evaluation of Crocus sativus L. stigma phenolic and flavonoid compounds and its antioxidant activity. *Molecules* **2010**, *15*, 6244–6256. [CrossRef]

57. Asdaq, S.; Inamdar, M. Potential of garlic and its active constituent, S-allyl cysteine, as antihypertensive and cardioprotective in presence of captopril. *Phytomedicine* **2010**, *17*, 1016–1026. [CrossRef] [PubMed]
58. Makhlouf, H.; Saksouk, M.; Habib, J.; Chahine, R. Determination of antioxidant activity of saffron taken from the flower of *Crocus sativus* grown in Lebanon. *Afr. J. Biotech.* **2011**, *10*, 8093–8100.
59. Ordoudi, S.; Befani, C.; Nenadis, N.; Koliakos, G.; Tsimidou, M. Further Examination of Antiradical Properties of *Crocus sativus* Stigmas Extract Rich in Crocins. *J. Agric. Food Chem.* **2009**, *57*, 3080–3086. [CrossRef]
60. Siracusa, L.; Gresta, F.; Avola, G.; Lombardo, G.M.; Ruberto, G. Influence of corm provenance and environmental condition on yield and apocarotenoid profiles in saffron (*Crocus sativus* L.). *J. Food Comp. Anal.* **2010**, *23*, 394–400. [CrossRef]
61. Manzo, A.; Panseri, S.; Bertoni, D.; Giorgi, A. Economic and qualitative traits of Italian Alps saffron. *J. Mt. Sci.* **2015**, *12*, 1542–1550. [CrossRef]
62. Sánchez-Vioque, R.; Rodríguez-Conde, M.F.; Reina-Ureña, J.V.; Escolano-Tercero, M.A.; Herraiz-Peñalver, D.; Santana-Méridas, O. In vitro antioxidant and metal chelating properties of corm, tepal and leaf from saffron (*Crocus sativus* L.). *Ind. Crop Prod.* **2012**, *39*, 149–153. [CrossRef]
63. Rahaiee, S.; Moini, S.; Hashemi, M.; Shojaosadati, S. Evaluation of antioxidant activities of bioactive compounds and various extracts obtained from saffron (*Crocus sativus* L.): A review. *J. Food Sci. Tech.* **2015**, *52*, 1881–1888. [CrossRef] [PubMed]
64. Shahi, T.; Assadpour, E.; Jafari, S. Main chemical compounds and pharmacological activities of stigmas and tepals of 'red gold'; saffron. *Trends Food Sci. Technol.* **2016**, *58*, 69–78. [CrossRef]
65. Donno, D.; Mellano, M.G.; Raimondo, E.; Cerutti, A.K.; Prgomet, Z.; Beccaro, G.L. Influence of applied drying methods on phytochemical composition in fresh and fried goji fruits by HPLC fingerprint. *Eur. Food Res. Technol.* **2016**, *242*, 1961–1974. [CrossRef]
66. Heidari, M.; Karami, V. Effects of different mycorrhiza species on grain yield, nutrient uptake and oil content of sunflower under water stress. *J. Saudi Soc. Agric. Sci.* **2014**, *13*, 9–13. [CrossRef]
67. Araim, G.; Saleem, A.; Arnason, J.T.; Charest, C. Root colonization by arbuscular mycorrhizal (AM) fungus increases growth and secondary metabolism of purple coneflower, *Echinacea purpurea* (L.) Moench. *J. Agric. Food Chem.* **2009**, *57*, 2255–2258. [CrossRef]
68. Ceccarelli, N.; Curadi, M.; Martelloni, L.; Sbrana, C.; Picciarelli, P.; Giovannetti, M. Mycorrhizal colonization impacts on phenolic content and antioxidant properties of artichoke leaves and flower heads two years after field transplant. *Plant Soil* **2010**, *335*, 311–323. [CrossRef]
69. Avio, L.; Sbrana, C.; Giovannetti, M.; Frassinetti, S. Arbuscular mycorrhizal fungi affect total phenolics content and antioxidant activity in leaves of oak leaf lettuce varieties. *Sci. Hortic.* **2017**, *224*, 265–271. [CrossRef]
70. Toussaint, J.P.; Smith, F.A.; Smith, S.E. Arbuscular mycorrhizal fungi can induce the production of phytochemicals in sweet basil irrespective of phosphorus nutrition. *Mycorrhiza* **2007**, *17*, 291–297. [CrossRef]
71. Nell, M.; Vötsch, M.; Vierheilig, H.; Steinkellner, S.; Zitterl-Eglseer, K.; Franz, C.; Novak, J. Effect of phosphorus uptake on growth and secondary metabolites of garden (*Salvia officinalis* L.). *J. Sci. Food Agric.* **2009**, *89*, 1090–1096. [CrossRef]
72. Geneva, M.P.; Stancheva, I.V.; Boychinova, M.M.; Mincheva, N.H.; Yonova, P.A. Effects of foliar fertilization and arbuscular mycorrhizal colonization on *Salvia officinalis* L. growth, antioxidant capacity, and essential oil composition. *J. Sci. Food Agric.* **2010**, *90*, 696–702.
73. Donno, D.; Beccaro, G.; Mellano, M.; Cerutti, A.; Bounous, G. Goji berry fruit (*Lycium* spp.): Antioxidant compound fingerprint and bioactivity evaluation. *J. Funct. Foods* **2015**, *18*, 1070–1085. [CrossRef]
74. Vasco, C.; Ruales, J.; Kamal-Eldin, A. Total phenolic compounds and antioxidant capacities of majpr fruits from Ecuador. *Food Chem.* **2008**, *111*, 816–823. [CrossRef]
75. Donno, D.; Cerutti, A.K.; Prgomet, I.; Mellano, M.G.; Beccaro, G.L. Foodomics for mulberry fruit (Morus spp.): Analytical fingerprint as antioxidants' and health properties' determination tool. *Food Res. Int.* **2015**, *69*, 179–188. [CrossRef]
76. Ma, H.; Johnson, S.L.; Liu, W.; DaSilva, N.A.; Meschwitz, S.; Dain, J.A.; Seeram, N.P. Evaluation of polyphenol anthocyanin-enriched extracts of blackberry, black raspberry, blueberry, cranberry, res raspberry, and strawberry for free radical scavenging, reactive carbonyl species trapping, anti-glycation, anti-β-amyloid aggregation, and microglial neuroprotective effects. *Int. J. Mol. Sci.* **2018**, *19*, 461.

3

Biostimulants for Plant Growth Promotion and Sustainable Management of Phytoparasitic Nematodes in Vegetable Crops

Trifone D'Addabbo [1,*], Sebastiano Laquale [2], Michele Perniola [3] and Vincenzo Candido [2]

1 Institute for Sustainable Plant Protection, CNR, Via Giovanni Amendola 122/d, 70126 Bari, Italy
2 School of Agricultural, Forest, Food and Environmental Sciences–UNIBAS, Viale dell'Ateneo Lucano 10, 85100 Potenza, Italy; sebastiano.laquale@unibas.it (S.L.); vincenzo.candido@unibas.it (V.C.)
3 Department of European and Mediterranean Cultures, UNIBAS, Via Lanera 20, 75100 Matera, Italy; michele.perniola@unibas.it
* Correspondence: trifone.daddabbo@ipsp.cnr.it

Abstract: The parasitism of root-knot nematodes, *Meloidogyne* spp., can cause heavy yield losses to vegetable crops. Plant biostimulants are often reported for a side-suppressive effect on these pests and many commercial products are increasingly included in sustainable nematode control strategies. Source materials of most biostimulants derived from plant or seaweed raw materials were documented for a reliable suppression of root-knot nematode species, whereas the suppressiveness of microbial biostimulants was found largely variable, as related to the crop and to environmental factors. Chitosan-based biostimulants were also stated for a variable phytonematode suppression, though clearly demonstrated only by a few number of studies. In a preliminary experimental case study, four commercial biostimulants based on quillay extract (QE), sesame oil (SO), seaweeds (SE), or neem seed cake (NC) were comparatively investigated for their effects against the root-knot nematode *M. incognita* on potted tomato. Soil treatments with all the four biostimulants resulted in a significant reduction of nematode eggs and galls on tomato roots, though NC and SO were significantly more suppressive than QE or SE. In addition, almost all biostimulant treatments also resulted in a significant improvement of tomato growth compared to the non-treated control. These preliminary results seem to confirm the literature data and clearly indicate the potential role of biostimulants for a safe nematode management both in organic and integrated crop systems.

Keywords: biostimulants; phytoparasitic nematodes; suppressiveness; sustainable management

1. Introduction

Phytoparasitic nematodes are among the most harmful pests of vegetable crops, responsible for an annual yield loss amounting to 9–15% of the world crop yield [1]. Most of these losses are due to root-knot nematode species, *Meloidogyne* spp., causing poor plant growth and reduced crop yield and quality and reducing plant resistance to other biotic and abiotic stresses [2]. Traditionally, control of these pests relied on soil treatments with synthetic nematicides, but the increasing demand for a higher crop safety to the environment and humans has led to a progressive dismission of these products, giving a strong impulse to the search and the implementation of control strategies based on natural mechanisms, such as the use of plant biostimulants [3].

Plant biostimulants derived from natural materials have been receiving a growing interest by researchers, farmers, and industrial companies, as considered an effective tool for improving crop productivity [4]. The previous unclear and misunderstanding legislation frame led to include among the biostimulants a large variety of products with different activities, such as growth enhancers, plant strengtheners or conditioners, resistance elicitors, as well to registration procedures variable among countries or even within the same country [5]. The uncertain legislative frame resulted in the immission in the market of a large variety of biostimulants stated for a suppressiveness on phytoparasitic nematodes, because of their content of raw materials (plants, seaweeds, microorganisms, and more) widely demonstrated for an activity against phytonematode species [6–8]. However, the recent EU Regulation 2019/1009 [9] has restricted the definition of fertilizing products and biostimulants and, therefore, many of these borderline products are destined to be classified as phytochemicals, dealing with more complex and expensive registration procedures.

Because of the increasing technical and economic relevance of these products, the aim of this study is to provide a review of the main groups of nematode-suppressive plant biostimulants actually available in the market and to indicate their potential for an effective but safe nematode management by a preliminary experimental case study on the root knot nematode *M. incognita* Kofoid et White (Chitw.) on tomato.

2. The State-of-the-Art

2.1. The Market Supply

A survey of the Italian market in 2018 revealed the presence of almost 40 different commercial plant biostimulants/strengtheners declaring a side activity on phytoparasitic nematodes on their labels (Table 1). More than 50% of these commercial products were based on plant raw materials, such as extracts, seed oils or green and seed biomasses, whereas another 25% was represented by seaweed derivatives. There was only one chitosan-based formulate, whereas the remaining others were microbial formulations. Only four products were clearly described as nematotoxic and the activity of other nine formulates was related to nematode repellence, disorientation, or antifeeding effects, whereas the remaining products were generically described as enhancers of plant resistance or of unfavorable soil conditions.

Table 1. Commercial biostimulants reporting an activity against phytoparasitic nematodes available in the Italian market at December 2018.

Commercial Name	Formulation [1]	Raw Materials	Activity [2]
Aegis ™	P	Micorrhizal fungi	1, 4, 5
Alg-a-Mic ™	L	Seaweed extract	1, 4, 5, 7
Algafit ™	L	Seaweed extract	1, 4
Ascogreen ™	L	Seaweed extract	1, 4
Biofence ™	P	Brassica meal	1, 3, 5, 6
Biofence 10 ™	P	Brassica meal	1, 3, 5, 6
Biofence FL ™	L	Brassica extract	1, 2, 4, 6
Bioki ™	p	Neem oil	1, 3, 7
Cogisin ™	L	Plant extracts	1, 2, 4, 7
Ecoessen NP ™	P	Bone meal, neem cake	1, 3, 6
Ekoprop Nemax ™	P	Mycorrhizal fungi	1, 2, 4
Ergo Bio ™	L	Humic and fulvic acids	1, 3, 4, 5, 8
Ergon ™	L	Seaweed extract	1, 4
Fertineem ™	L	Neem oil	1, 4
Force 4 ™	L	Seaweed extract	1, 2, 4, 5
Hunter ™	L	Plant extracts	1, 4
Ilsaneem ™	P	Neem cake	1, 2, 3, 7
Kendal Nem ™	L	Plant extracts	1, 2, 4, 6
Keos Guardian ™	L	Chitosan	1, 5
Micofort ™	P	Micorrhizal fungi	1, 2, 4, 5
Micosat F ™	P	Micorrhizal fungi	1, 4, 5
Micosat Jolly ™	P	Micorrhizal fungi	1, 4, 5
Mychodeep ™	P	Micorrhizal fungi	1, 2, 4

Table 1. *Cont.*

Commercial Name	Formulation [1]	Raw Materials	Activity [2]
Neem Soil ™	**P**	**Neem cake**	**1, 3, 4, 6**
Neem Care FL ™	L	Plant extracts	1, 2, 4
Nema 300 WW ™	L	Plant oils	1, 2, 4
Nemaforce ™	L	Humic and fulvic acids, plant extracts	1, 2, 4, 5, 7
Nematec ™	**L**	**Seaweed extract**	**1, 2, 4**
Nematiller ™	L	Plant extracts	1, 2, 4
Nematon EC ™	L	Sesame oil	1, 4
NeMax ™	**L**	**Sesame oil**	**1, 4**
Nutrich ™	P	Neem and pongamia cake	1, 2, 3, 4
Propoli oleoso ™	L	Propolis oil	1, 4
Rigenera Active	L	Seaweed macerate, plant extracts	1, 2, 4
Sesamin EC ™	L	Sesame oil	2, 4, 5
Tagete ™	L	Tagetes extract	1, 2, 4, 8
Tequil Multi ™	**L**	**Quillay and yucca extracts**	**1, 2, 4, 8**
Tyson ™	L	Propolis oil	1, 4, 8
Xedaneem ™	P	Neem cake	1, 6

[1] L = liquid; D = dry meals, P = pellets, G = granules; [2] 1 = biostimulant; 2 = rooting; 3 = fertilizing; 4 = plant defense enhancement; 5 = increase of soil beneficial microflora; 6 = creation of a nematode-unfavorable environment; 7 = repellence, antifeeding, disorientation; 8 = toxicity. Products applied in the case study are reported in bold.

2.2. The Literature Review

Plant-derived biostimulants previously documented for an activity on phytonematodes were mostly liquid formulations of extracts and oils or, at a less instance, granular or powder seed meal or cake derivatives. A large number of plant biostimulants based on sesame seed oil [10], quillay water extract [11,12], or meals from biomasses or seeds of *Brassicaceae* plants and neem [13–15] were previously demonstrated for a suppressive activity on root-knot nematode populations on field and greenhouse tomato.

Seaweed extracts were found to cause an almost complete mortality of root-knot nematode juveniles and eggs in in vitro studies [16,17], as well as formulations of the extracts from seaweed species *Ascophyllum nodosum* L. and *Ecklonia maxima* Osbeck were reported for an effective control of root-knot nematodes also in soil experiments on tomato [18–20]. In addition to extract derivatives, a strong suppression of *Meloidogyne* spp. infestations on fruit or vegetable crops was described also for soil amendments with biomasses of seaweeds *Uva lactuca* L. and *Spatoglossus schroederi* Agardh (Kützing), may be due to their high content of phenolics and other bioactive compounds [21,22]. In addition to Meloidogyne species, suppressive activity of seaweed products was also detected on nematode parasites economically relevant to tropical or subtropical vegetable crops, such as *Helicotylenchus indicus* Siddiqui, *Belonolaimus longicaudatus* Rau, or *Radopholus similis* Cobb (Thorne) [23–26].

Literature studies are available also on the suppressive activity of chitosan and/or its derivatives, both alone or combined with other suppressive materials (agricultural wastes, plant compounds, biocontrol agents), either on root-knot nematodes [27–30] and other phytoparasitic species i.e., the soybean cyst nematode *Heterodera glycines* Ichinoe and the pinewood parasite *Bursaphelenchus xylophilus* (Steiner et Buhrer) Nickle [31–33].

Most of the microbial biostimulants reported as active on phytoparasitic nematodes were formulations of arbuscular mycorrhizal fungi [34,35]. Suppressiveness to root-knot nematodes of these products, either alone or combined with other microorganisms or plant extracts, was documented both in field and greenhouse [36–39]. Moreover, their activity was demonstrated also on other phytonematode parasites, such as *Nacobbus aberrans* Thorne et Allen or *Helicotylenchus multicinctus* (Cobb) Golden on field banana and greenhouse tomato, respectively [40,41]. In addition to mycorrhizal fungi, formulations of other fungal or bacterial biocontrol agents (*Trichoderma* spp., *Bacillus* spp.) or nitrogen fixers (*Azospirillum* spp., *Azotobacter* spp.) were also reported for controlling *M. incognita* in glasshouse tomato and field sunflower [42–44], or improving crop tolerance to the cyst nematode *Heterodera schachtii* Schmidt and more generically to soil phytoparasitic nematophauna [45,46].

3. An Experimental Case Study

3.1. Materials and Methods

A sandy soil (64.4% sand,18.7% silt, 16.9% clay, 0.8% organic matter, pH 7.5; 18.2% soil average humidity, 23.5% field capacity, 12.9% wilting point), artificially infested with the root-knot nematode *M. incognita* (8 eggs and juveniles mL^{-1} soil) was poured into 2.5 L clay pots. Soil was then treated with three commercial liquid biostimulants derived from quillay (*Quillaja saponaria* Molina) extract (Tequil Multi®, Fertenia) (QE), sesame (*Sesamum indicum* L.) oil (NeMax®, Sumitomo Chemical) (SO) or brown algae (*Laminaria* spp.) extract (AgriPrime Nematec®, BioAtlantis) (SE), and a granular formulation of neem (*Azadirachta indica* Juss) cake (Neem Soil®, Serbios) (NC). QE, SO and SE were applied at transplant and 15 and 30 days later at amounts corresponding to 60, 10, and 2 L ha^{-1}, respectively, whereas NC was incorporated to the soil at a 1000 kg ha^{-1} rate two weeks before transplanting. The same treatments were also provided to pots containing non-infested soil. Soil treated with the nematicide Oxamyl (OX), applied at a 10 L ha^{-1} field rate 3 days before tomato transplant and 15 days later, and non-treated soil, both infested (NT) and non-infested (NI) by *M. incognita*, were used as controls. One-month-old tomato seedlings (cv. Harvester) were transplanted in each pot, providing five replicates for each treatment in comparison.

The pots were arranged in a randomized block design in a plastic greenhouse at 25 °C, where they were maintained for 75 days, receiving a regular irrigation but no additional pesticide or fertilizer treatment. At the end of their permanence in the greenhouse, plants were uprooted and weight of green and root biomass was recorded for each plant. Root gall formation was estimated according to a 0–10 scale [47] and nematode multiplication on tomato roots was determined by extracting eggs and juveniles by the Hussey and Barker's method [48]. Data were statistically analyzed by ANOVA and treatment means were compared by the Fisher's Least Significant Difference Test at $P \leq 0.05$, using PlotIT 3.2 (Scientific Programming Enterprises, Haslett, MI) software.

3.2. Results

The number of nematode eggs and juveniles on tomato roots were always significantly lower in the soil treated with the four biostimulants or OX than in NT soil (Figure 1A). Moreover, the multiplication of *M. incognita* in pots treated with NC or SO was not statistically different from OX and significantly lower than the treatments with QE and SE. Finally, QE resulted to be significantly more suppressive than SE.

Treatments with the four biostimulants and OX also resulted in a significantly lower number of root galls in comparison with NT (Figures 1B and 2). As for nematode eggs and juveniles, the formation of galls in soil treated with NC and SO was statistically lower than QE and SE, though only NC was significantly not different from OX. No statistical difference occurred between the number of galls from QE and SE.

Tomato plant biomass in soil infested by *M. incognita*, either non-treated and treated with the biostimulants or OX, was always significantly lower than NI (Figure 3A). Green biomass from plants in soil treated with QE was significantly larger compared to all the other treatments and NT. Adversely, weight of green biomass from pots treated with the other three formulates was not significantly different from NT and statistically lower than OX.

Weight of the tomato roots from all the treatments but NC was significantly higher than the NT (Figure 3B). Moreover, QE resulted in a root biomass significantly heavier than the other three biostimulants and OX and not different from NI. Finally, SE resulted in a root growth statistically not different from OX but higher compared to NC and SO.

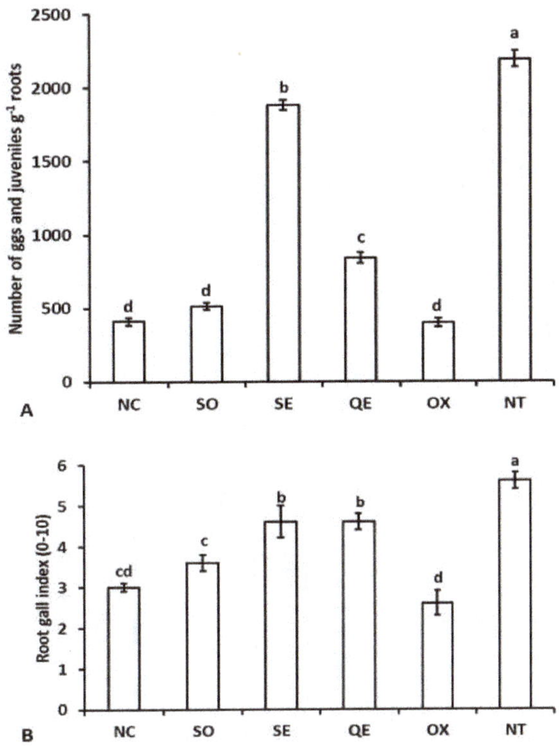

Figure 1. Multiplication of the root-knot nematode *Meloidogyne incognita* (**A**) and gall formation (**B**) on the roots of tomato cv. Harvester in soil non-treated (NT) or treated with commercial biostimulants based on neem cake (NC), sesame oil (SO), seaweed extract (SE), and quillay extract (QE) or with nematicide Oxamyl (OX). Bars tagged with the same letters are not statistically different ($P \leq 0.05$) according to the Least Significant Difference's Test.

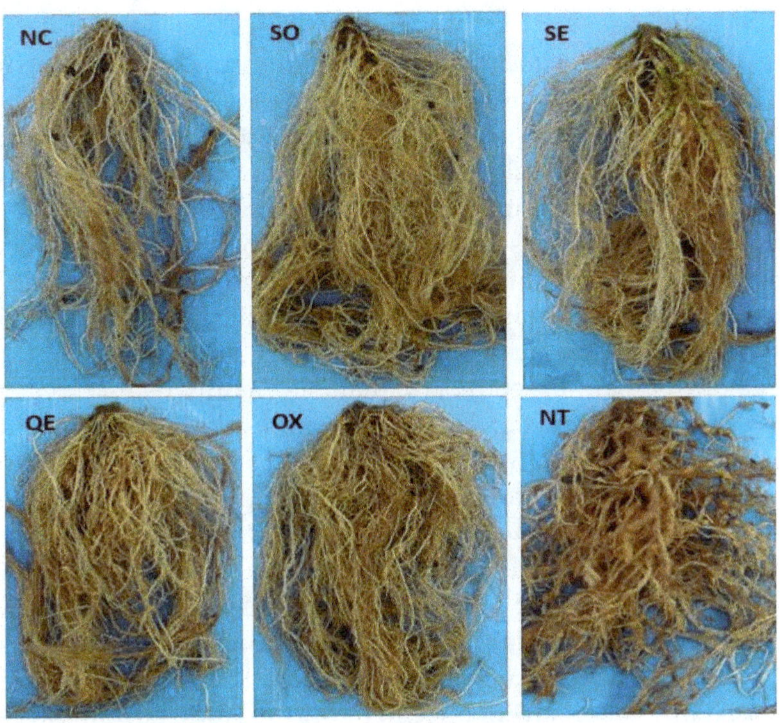

Figure 2. Roots of tomato plants cv. Harvester from soil treated with commercial biostimulants based on neem cake (NC), sesame oil (SE), seaweed extract (SE) and quillay extract (QE) or with nematicide Oxamyl (OX) and from non-treated soil (NT).

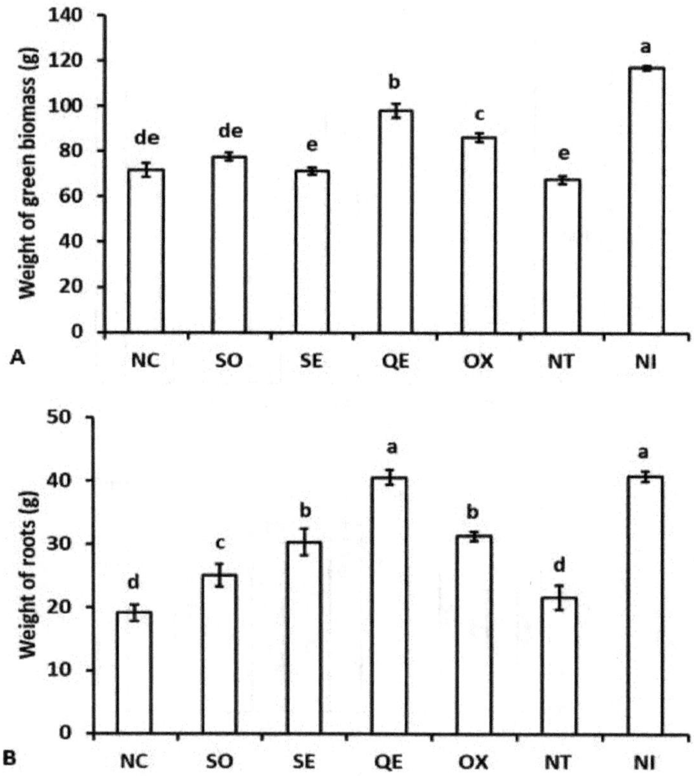

Figure 3. Weight of green biomass (**A**) and roots (**B**) of tomato plants cv. Harvester in soil non-treated (NT) or treated with commercial biostimulants based on neem cake (NC), sesame oil (SO), seaweed extract (SE), and quillay extract (QE) or with nematicide Oxamyl (OX). Bars tagged with the same letters are not statistically different ($P \leq 0.05$) according to the Least Significant Difference's Test.

4. Discussion

The experimental case study indicated that biostimulants can also provide a satisfactory nematode suppression, as confirming previous findings from literature studies. However, these results aim to be only indicative of the potential use of biostimulants in nematode management and need to be validated by future trials in field conditions, as well as different combinations of biostimulants should be also tested to verify a potential synergism among different products.

The mechanisms of biostimulants suppressiveness to nematodes are only partially known or simply hypothesized. Seaweed activity on phytoparasitic nematodes was generally attributed to their content of secondary metabolites, such as steroids, triterpenoids, alkaloids, and phenols, known for a nematicidal activity or as plant resistance elicitors [49,50]. Analogously, the suppressiveness to phytonematode populations of plant-based biostimulants is mainly related to nematotoxic metabolites both preformed in raw plant material (saponins, fatty acids, alkaloids and more) or released during the plant materials degradation in soil [51,52]. Induction of a systemic plant resistance to nematode penetration has been also documented for some active compounds of plant-derived biostimulants, such as neem azadiractin or chestnut (*Castanea sativa* Mill.) tannins [53,54]. Nematode suppression by microbial biostimulants was generally attributed to the induction of crop defense responses to nematode invasion [55,56]. Additional or alternative mechanisms, such as a competition for nutrients and space or the synthesis of nematicidal microbial metabolites have been also suggested [57–59]. The nematicidal effectiveness of chitosan products was generally attributed to the induction of a local or systemic plant resistance [60], though an enhancement of nematode-suppressive rhizospheric bacteria and fungi has been also hypothesized [36,40].

In our study, only QE was confirmed for a biostimulant effect on tomato growth, as limited only to the root biomass for SO and SE or nil for NC. The growth effect of QE can be attributed to the high content of triterpenic saponins, widely acknowledged for significant plant growth regulating properties [61], in *Q. saponaria* extracts.

Chemical composition of plant-based biostimulants can change according to a range of environmental and agronomic factors [62], as well as the nematode suppressiveness of microbial formulations may vary according to microbial strains, crop species/varieties, and environmental conditions [63]. Variable effects on soil phytonematode populations were also documented for chitosan products, as strictly dependent on the molecular weight of raw materials [32,64]. The unstable composition is a serious constraint to the full implementation of biostimulants in nematode management strategies, as leading to a fluctuating activity in field and, consequently, to a difficult certification of nematicidal performances and registration of commercial products [51]. A preliminary standardization of source raw materials and manufacturing processes should ensure constant suppressive performances and a successful market presence to the future commercial plant biostimulants addressed to nematode management. Moreover, preliminary toxicological screenings should be provided for any new biostimulant, as to exclude the presence of compounds with an unknown toxicological profile or the persistence of human pathogens in materials of animal origin [51].

In conclusion, plant biostimulants can also play a relevant role in the future nematode management strategies, as providing an acceptable nematode suppression in addition to their main activity of plant growth and ensuring a full safety to the other biotic soil components. It may be reasonably expected that the Regulation 2019/1009 [9] will lead to the disappearance of products with a direct toxicity to nematodes activity, because of the high costs of their registration as pesticides, as limiting the market to the products working through plant resistance improvement. A stand-alone application of these products can be reasonable only in organic crop systems, where few nematode control tools are available, or in short-cycle crops where the short pre-harvest intervals do not allow the use of most synthetic nematicides. However, a combination with other chemical or nonchemical control tools can justify the application of these products also in conventional crop systems. Benefit–cost ratio of treatments with the kind of products analyzed in this work should be always evaluated before their application as nematode suppressants, because of the high market price of these products which limit their use preferably to high value crops.

Author Contributions: Conceptualization, T.D., M.P., V.C.; data curation, S.L.; formal analysis, T.D., V.C.; investigation, S.L.; methodology, T.D., S.L.; software, S.L.; supervision, M.P.; validation, V.C.; visualization, S.L.; writing—original draft, T.D.; writing—review and editing, T.D., V.C.

Acknowledgments: The authors acknowledge the technical assistance of Fabio Catalano for the arrangement of greenhouse experiment and lab work.

References

1. Nicol, J.M.; Turner, S.J.; Coyne, D.L.; Den Nijs, L.; Hockland, S.; Maafi, Z.T. Current nematode threats to world agriculture. In *Genomics and Molecular Genetics of Plant-Nematode Interactions*; Jones, J., Gheysen, G., Fenoll, C., Eds.; Springer: Dordrecht, NL, USA, 2011; pp. 21–43.
2. Sikora, R.A.; Fernandez, E. Nematode parasites of vegetables. In *Plant Parasitic Nematodes in Subtropical and Tropical Agriculture*; Luc, M., Sikora, R.A., Bridge, J., Eds.; CABI: Wallingford, UK, 2005; pp. 319–392.
3. Le Mire, G.; Nguyen, M.L.; Fassotte, B.; du Jardin, P.; Verheggen, F.; Delaplace, P.; Jijakli, H. Implementing plant biostimulants and biocontrol strategies in the agroecological management of cultivated ecosystems, A review. *Biotechnol. Agron. Soc. Environ.* **2016**, *20*, 299–313.
4. Brown, P.; Saa, S. Biostimulants in agriculture. *Front. Plant Sci.* **2015**, *6*, 671. [CrossRef] [PubMed]
5. Basak, A. Biostimulators. Definitions, classification and legislation. In *Biostimulators in Modern Agriculture: General Aspects*; Gawrońska, H., Ed.; Wie Jutra: Warsaw, Poland, 2008; pp. 7–17.
6. Askary, T.H.; Martinelli, P.R.P. *Biocontrol Agents of Phytonematodes*; CAB International: Wallingford, UK, 2015; p. 470.

7. D'Addabbo, T.; Laquale, S.; Lovelli, S.; Candido, V.; Avato, P. Biocide plants as a sustainable tool for the control of pests and pathogens in vegetable cropping systems. *Ital. J. Agron.* **2014**, *9*, 137–145. [CrossRef]
8. Oka, Y. Mechanisms of nematode suppression by organic soil amendments—A review. *Appl. Soil Ecol.* **2010**, *44*, 101–115. [CrossRef]
9. European Union. Regulation 2019/1009, 2019. (EU) of the European Parliament and of the Council of 5 June 2019 Laying Down Rules on the Making Available on the Market of EU Fertilising Products and Amending Regulations (EC) No 1069/2009 and (EC) No 1107/2009 and Repealing Regulation (EC) No 2003/2003. OJ L 170, 25.6.2019. pp. 1–114. Available online: http://data.europa.eu/eli/reg/2019/1009/oj (accessed on 6 August 2019).
10. D'Addabbo, T.; Radicci, V.; Lucarelli, G.; Carella, A.; Bernad, D.; Martin, E. Effectiveness of a formulation from *Pedaliaceae* plants (Nematon® EC) for the control of the root-knot nematode *Meloidogyne incognita* on greenhouse tomato. *Acta Hortic.* **2011**, *914*, 233–236. [CrossRef]
11. Giannakou, I.O. Efficacy of a formulated product containing *Quillaja saponaria* plant extracts for the control of root-knot nematodes. *Eur. J. Plant Pathol.* **2011**, *130*, 587–596. [CrossRef]
12. D'Addabbo, T.; Curto, G.; Greco, P.; Di Silvestro, D.; Coiro, M.I.; Lamberti, F.; Ferrari, V.; Santi, R.; Carella, A. Preliminary trials with extracts of *Quillaja saponaria* Molina for the control of root-knot nematodes. *Nematol. Medit.* **2005**, *33*, 29–34.
13. Curto, G.; Dallavalle, E.; Matteo, R.; Lazzeri, L. Biofumigant effect of new defatted seed meals against the southern root-knot nematode, *Meloidogyne incognita*. *Ann. Appl. Biol.* **2016**, *169*, 1–10. [CrossRef]
14. Rizvi, R.; Singh, G.; Safiuddin; Ali Ansari, R.; Ali Tiyagi, S.; Mahmood, I. Sustainable management of root-knot disease of tomato by neem cake and *Glomus fasciculatum*. *Cogent Food Agric.* **2015**, *1*, 1008859. [CrossRef]
15. Abbasi, P.A.; Riga, E.; Conn, K.L.; Lazarovits, G. Effect of neem cake soil amendment on reduction of damping-off severity and population densities of plant-parasitic nematodes and soilborne plant pathogens. *Can. J. Plant Pathol.* **2005**, *27*, 38–45. [CrossRef]
16. Khan, S.A.; Abid, M.; Hussain, F. Nematicidal activity of seaweeds against *Meloidogyne javanica*. *Pak. J. Nematol.* **2015**, *33*, 195–203.
17. Rizvi, M.A.; Shameel, M. In vitro nematicidal activities of seaweed extracts from Karachi coast. *Pak. J. Bot.* **2006**, *38*, 1245–1248.
18. Whapham, C.A.; Jenkins, T.; Blunden, G.; Hankins, S. The role of seaweed extracts, *Ascophyllum nodosum*, in the reduction in fecundity of *Meloidogyne javanica*. *Fundam. Appl. Nematol.* **1994**, *17*, 181–183.
19. Wu, Y.; Jenkins, T.; Blunden, G.; Whapham, C.; Hankins, S.D. The role of betaines in alkaline extracts of *Ascophyllum nodosum* in the reduction of *Meloidogyne javanica* and *M. incognita*. *Fundam. Appl. Nematol.* **1997**, *20*, 99–102.
20. Featonby-Smjth, B.C.; Van Staden, J. The effect of seaweed concentrate on the growth of tomato plants in nematode infested soil. *Sci. Hortic.* **1983**, *20*, 137–146. [CrossRef]
21. El-Ansary, M.S.M.; Hamouda, R.A. Biocontrol of root-knot nematode infected banana plants by some marine algae. *Russ. J. Mar. Biol.* **2014**, *40*, 140–146. [CrossRef]
22. Paracer, S.; Armin, C.; Tarjan, A.C.; Hodgson, L.M. Effective use of marine algal products in the management of plant parasitic nematodes. *J. Nematol.* **1987**, *19*, 194–200.
23. Naqvi, B.S.; Khan, A.; Shaikh, D.; Shaikh, M.R. Nematicidal properties of selected marine algae from Karachi coast. *J. Islam. Acad. Sci.* **1992**, *5*, 171–172.
24. De Waele, D.; McDonald, A.H.; De Waele, E. Influence of seaweed concentrate on the reproduction of *Pratylenchus zeae* (Nematoda) on maize. *Nematologica* **1988**, *34*, 71–77.
25. Morgan, K.T.; Tarjan, A.C. Management of sting nematode on centipedegrass with kelp extracts. *Proc. Flor. State Hortic. Soc.* **1980**, *93*, 97–99.
26. Tarjan, A.C. Kelp derivatives for nematode-infected citrus trees. *J. Nematol.* **1977**, *9*, 287.
27. Radwan, M.A.; Farrag, S.A.; Abu-Elamayem, M.M.; Ahmed, N.S. Extraction, characterization, and nematicidal activity of chitin and chitosan derived from shrimp shell wastes. *Biol. Fert. Soils* **2012**, *48*, 463–468. [CrossRef]
28. Escudero, N.; Lopez-Moya, F.; Ghahremani, Z.; Zavala-Gonzalez, E.A.; Alaguero-Cordovilla, A.; Ros-Ibañez, C.; Lacasa, A.; Sorribas, F.J.; Lopez-Llorca, L.V. Chitosan increases tomato root colonization by *Pochonia chlamydosporia* and their combination reduces root-knot nematode damage. *Front. Plant Sci.* **2017**, *8*, 1415. [CrossRef] [PubMed]

29. Asif, M.; Ahmad, F.; Tariq, M.; Khan, A.; Ansari, T.; Khan, F.; Siddiqui, A.M. Potential of chitosan alone and in combination with agricultural wastes against the root-knot nematode, *Meloidogyne incognita* infesting eggplant. *J. Plant Prot. Res.* **2017**, *57*, 288–295. [CrossRef]
30. Mota, L.C.; dos Santos, M.A. Chitin and chitosan on *Meloidogyne javanica* management and on chitinase activity in tomato plants. *Trop. Plant Pathol.* **2016**, *41*, 84–90. [CrossRef]
31. Liang, W.; Yu, A.; Wang, G.; Zheng, F.; Jia, J.; Xu, H. Chitosan-based nanoparticles of avermectin to control pine wood nematodes. *Int. J. Biol. Macromol.* **2018**, *112*, 258–263. [CrossRef]
32. Mwaheb, M.A.; Hussain, M.; Tian, J.; Zhang, X.; Hamid, M.I.; El-Kassim, N.A.; Hassan, G.M.; Xiang, M.; Liu, X. Synergetic suppression of soybean cyst nematodes by chitosan and *Hirsutella minnesotensis* via the assembly of the soybean rhizosphere microbial communities. *Biol. Contr.* **2017**, *115*, 86–94. [CrossRef]
33. Nunes da Silva, M.; Cardoso, A.R.; Ferreira, D.; Brito, M.; Pintado, M.E.; Vasconcelos, M.W. Chitosan as a biocontrol agent against the pinewood nematode (*Bursaphelenchus xylophilus*). *For. Pathol.* **2014**, *44*, 420–423. [CrossRef]
34. Candido, V.; Campanelli, G.; D'Addabbo, T.; Castronuovo, D.; Perniola, M.; Camele, I. Growth and yield promoting effect of artificial mycorrhization on field tomato at different irrigation regimes. *Sci. Hort.* **2015**, *187*, 35–43. [CrossRef]
35. Colella, T.; Candido, V.; Campanelli, G.; Camele, I.; Battaglia, D. Effect of irrigation regimes and artificial mycorrhization on insect pest infestations and yield in tomato crop. *Phytoparasitica* **2014**, *42*, 235–246. [CrossRef]
36. Flor-Peregrín, E.; Verdejo-Lucas, S.; Talavera, M. Combined use of plant extracts and arbuscular mycorrhizal fungi to reduce root-knot nematode damage in tomato. *Biol. Agric. Hortic.* **2017**, *33*, 115–124. [CrossRef]
37. Flor-Peregrín, E.; Azcón, R.; Martos, V.; Verdejo-Lucas, S.; Talavera, M. Effects of dual inoculation of mycorrhiza and endophytic, rhizospheric or parasitic bacteria on the root-knot nematode disease of tomato. *Biocontr. Sci. Technol.* **2014**, *24*, 1122–1136. [CrossRef]
38. Affokpon, A.; Coyne, D.L.; Lawouin, L.; Tossou, C.; Agbèdè, R.D.; Coosemans, J. Effectiveness of native West African arbuscular mycorrhizal fungi in protecting vegetable crops against root-knot nematodes. *Biol. Fertil. Soils* **2011**, *47*, 207–217. [CrossRef]
39. Zhang, L.; Zhang, J.; Christie, P.; Li, X. Pre-inoculation with arbuscular mycorrhizal fungi suppresses root knot nematode (*Meloidogyne incognita*) on cucumber (*Cucumis sativus*). *Biol. Fertil. Soils* **2008**, *45*, 205. [CrossRef]
40. Marro, N.; Lax, P.; Cabello, M.; Doucet, M.E.; Becerra, A.G. Use of the arbuscular mycorrhizal fungus *Glomus intraradices* as biological control agent of the nematode *Nacobbus aberrans* parasitizing tomato. *Braz. Arch. Biol. Technol.* **2014**, *57*, 668–674. [CrossRef]
41. Jonathan, E.I.; Cannayane, I.; Samiyappan, R. Field application of biocontrol agents for the management of spiral nematode, *Helicotylenchus multicinctus*, in banana. *Nematol. Mediterr.* **2004**, *32*, 169–173.
42. Radwan, M.A.; Farrag, S.A.A.; Abu-Elamayem, M.M.; Ahmed, N.S. Biological control of the root-knot nematode, *Meloidogyne incognita* on tomato using bioproducts of microbial origin. *Appl. Soil Ecol.* **2012**, *56*, 58–62. [CrossRef]
43. El-Haddad, M.E.; Mustafa, M.I.; Selim, S.M.; El-Tayeb, T.S.; Mahgoob, A.E.; Aziz, N.H. The nematicidal effect of some bacterial biofertilizers on *Meloidogyne incognita* in sandy soil. *Braz. J. Microbiol.* **2011**, *42*, 105–113. [CrossRef]
44. Ismail, A.E.; Hasabo, S.A. Evaluation of some new Egyptian commercial biofertilizers, plant nutrients and a biocide against *Meloidogyne incognita* root knot nematode infecting sunflower. *Pak. J. Nematol.* **2000**, *18*, 39–49.
45. Biliavska, L.O.; Tsygankova, V.A.; Kozyritska, V.E.; Iutynska, G.O.; Andrusevich, Y.V.; Babich, O.A. Application of new microbial plant resistance/plant growth protection inducers for increasing Chinese cabbage plant tolerance against parasitic nematode *Heterodera schachtii* Schmidt. *Int. J. Res. Biosci.* **2016**, *5*, 64–82.
46. Khan, Z.; Tiyagi, S.A.; Mahmood, I.; Rizvi, R. Effects of N fertilization, organic matter, and biofertilizers on the growth and yield of chilli in relation to management of plant-parasitic nematodes. *Turk. J. Bot.* **2012**, *36*, 73–78.
47. Bridge, J.; Page, S.L.J. Estimation of root-knot nematode infestation levels on roots using a rating chart. *Int. J. Pest Manag.* **1980**, *26*, 296–298. [CrossRef]

48. Hussey, P.S.; Barker, K.R. A comparison of methods of collecting inocula of Meloidogyne spp., including a new technique. *Plant Dis. Rep.* **1973**, *57*, 1025–1028.
49. Chitwood, J.D. Phytochemical based strategies for nematode control. *Ann. Rev. Phytopathol.* **2002**, *40*, 21–49. [CrossRef]
50. Verkleij, F.N. Seaweed extracts in agriculture and horticulture: A review. *Biol. Agric. Hort.* **1992**, *8*, 309–324. [CrossRef]
51. Yakhin, O.I.; Lubyanov, A.A.; Yakhin, I.A.; Brown, P.H. Biostimulants in plant science: A global perspective. *Front. Plant Sci.* **2017**, *7*, 2049. [CrossRef]
52. Oliveira, R.D.; Dhingra, O.D.; Lima, A.O.; Jham, G.N.; Berhow, M.A.; Holloway, R.K.; Vaughn, S.F. Glucosinolate content and nematicidal activity of Brazilian wild mustard tissues against *Meloidogyne incognita* in tomato. *Plant Soil* **2011**, *341*, 155–164. [CrossRef]
53. Pretali, L.; Bernardo, L.; Butterfield, T.S.; Trevisan, M.; Lucini, L. Botanical and biological pesticides elicit a similar induced systemic response in tomato (*Solanum lycopersicum*) secondary metabolism. *Phytochemistry* **2016**, *130*, 56–63. [CrossRef]
54. Bargiacchi, E.; Miele, S.; Romani, A.; Campo, M. Biostimulant activity of hydrolyzable tannins from sweet chestnut (*Castanea sativa* mill.). *Acta Hortic.* **2013**, *1009*, 111–116. [CrossRef]
55. Sofo, A.; Nuzzaci, M.; Vitti, A.; Tataranni, G.; Scopa, A. Control of biotic and abiotic stresses in cultivated plants by the use of biostimulant microorganisms. In *Improvement of Crops in the Era of Climatic Changes*; Ahmad, P., Wani, M.R., Azooz, M.M., PhanTran, L.S., Eds.; Springer Science+Business Media: New York, NY, USA, 2014; pp. 107–117.
56. Vos, C.; Schouteden, N.; Van Tuinen, D.; Chatagnier, O.; Elsen, A.; De Waele, D.; Gianinazzi-Pearson, V. Mycorrhiza-induced resistance against the root–knot nematode *Meloidogyne incognita* involves priming of defense gene responses in tomato. *Soil Biol. Biochem.* **2013**, *60*, 45–54. [CrossRef]
57. Mahanty, T.; Bhattacharjee, S.; Goswami, M.; Bhattacharyya, P.; Das, B.; Ghosh, A.; Tribedi, P. Biofertilizers: A potential approach for sustainable agriculture development. *Env. Sci. Poll. Res.* **2017**, *24*, 3315–3335. [CrossRef] [PubMed]
58. Schouteden, N.; De Waele, D.; Panis, B.; Vos, C.M. Arbuscular mycorrhizal fungi for the biocontrol of plant-parasitic nematodes: A review of the mechanisms involved. *Front. Microbiol.* **2015**, *6*, 1280. [CrossRef] [PubMed]
59. Youssef, M.M.; Eissa, M.F. Biofertilizers and their role in management of plant parasitic nematodes, A review. *J. Biotechnol. Pharm. Res.* **2014**, *5*, 1–6.
60. Vasyukova, N.I.; Zinov'Eva, S.V.; Il'Inskaya, L.I.; Perekhod, E.A.; Chalenko, G.I.; Gerasimova, N.G.; Il'ina, A.V.; Varlamov, V.P.; Ozeretskovskaya, O.L. Modulation of plant resistance to diseases by water-soluble chitosan. *Appl. Biochem. Microbiol.* **2001**, *37*, 103–109. [CrossRef]
61. Reichert, C.L.; Salminen, H.; Weiss, J. Quillaja Saponin Characteristics and Functional Properties. *Ann. Rev. Food Sci. Technol.* **2019**, *10*, 43–73. [CrossRef]
62. Sharma, S.H.S.; Lyons, G.; McRoberts, C.; McCall, D.; Carmichael, E.; Andrews, F.; McCormack, R. Brown seaweed species from Strangford Lough: Compositional analyses of seaweed species and biostimulant formulations by rapid instrumental methods. *J. Appl. Phycol.* **2012**, *24*, 1141–1157.
63. Dong, L.Q.; Zhang, K.Q. Microbial control of plant-parasitic nematodes: A five-party interaction. *Plant Soil* **2006**, *288*, 31–45. [CrossRef]
64. Khalil, M.S.; Badawy, M.E. Nematicidal activity of a biopolymer chitosan at different molecular weights against root-knot nematode, *Meloidogyne incognita*. *Plant Prot. Sci.* **2012**, *48*, 170–178. [CrossRef]

4

Fermented Alfalfa Brown Juice Significantly Stimulates the Growth and Development of Sweet Basil (*Ocimum basilicum* L.) Plants

Szilvia Kisvarga [1], Döme Barna [2,*], Szilvia Kovács [2], Gábor Csatári [2], Ibolya O. Tóth [2], Miklós Gábor Fári [2], Péter Makleit [2], Szilvia Veres [2], Tarek Alshaal [2,3] and Nóra Bákonyi [2]

1 Research Institute for Fruit growing and Ornamentals, National Agricultural Research and Innovation Center (NARIC) Szent-Györgyi Street 4, 2100 Gödöllő, Hungary; fullerina@gmail.com

2 Department of Agricultural Botany, Plant Physiology and Biotechnology, Institute of Crop Sciences (MEK), Debrecen University, Böszörményi Street 138, 4032 Debrecen, Hungary; szkovacs@agr.unideb.hu (S.K.); gaborcsat96@gmail.com (G.C.); olahne@agr.unideb.hu (I.O.T.); fari@agr.unideb.hu (M.G.F.); pmakleit@agr.unideb.hu (P.M.); szveres@agr.unideb.hu (S.V.); alshaaltarek@gmail.com (T.A.); nbakonyi@agr.unideb.hu (N.B.)

3 Soil and Water Department, Faculty of Agriculture, Kafrelsheikh University, Kafr El-Sheikh 33516, Egypt

* Correspondence: barna.dome@agr.unideb.hu

Abstract: Fertilization management is a key issue in plant nutrition to produce plants with good quality and quantity. Deproteinized leaf juice or brown juice (BJ) is a by-product during the isolation of leaf protein from biomass crops such as alfalfa. The idea of using BJ as a biostimulant fits well in the aspect of circular economy since BJ is currently a problematic issue of the leaf protein production approach. Fractionation of one-kilogram fresh biomass results in approximately 500 cm^3 BJ. Due to fast spoil of fresh BJ, if left on room temperature, it is found that fermentation of fresh BJ using lactic acid bacteria and reducing its pH increases its stability and storage on room temperature. In the present study, we examined the effect of fermented alfalfa BJ on vegetative, physiological, and anatomical properties of the versatile sweet basil (*Ocimum basilicum* L. 'Bíborfelhő') plants. Sweet basil seedlings were sprayed at different doses of fermented alfalfa BJ (i.e., 0.5%, 1.0%, and 2.5%) and tap water served as a control (0.0% BJ). The results revealed that foliar application of fermented alfalfa BJ significantly improved the biometrical features of sweet basil plants. Plants treated with fermented BJ showed significantly higher values of all the measured parameters compared to the control (0.0%), except for the number of leaves per plants where control plants (0.0%) had more leaves. However, the leaves of control plants (0.0%) were smaller than treated plants as data of leaf area showed. Fermented alfalfa BJ significantly increased the content of photosynthetic pigments (chl a and chl b), relative chlorophyll (SPAD value), lengths of stem and root, fresh masses of stem, root, and leaves, volumes of stem and root, and leaf area. Despite all rates of fermented BJ displayed higher values over control plants (0.0%), the rate of 0.5% was the best one supported by results. Application of fermented alfalfa BJ influenced the anatomical parameters as well. These findings demonstrate the possible use of fermented alfalfa BJ as a promising novel plant biostimulant.

Keywords: sweet basil; alfalfa brown juice; fermentation; biostimulation; chlorophyll pigments; histological changes

1. Introduction

The increase in the world's population leads to several issues that we already have to face, and we must find sustainable solutions for them to save our life on the planet. One of these concerns is the sustainable supply of high-quality protein. To solve the protein issue, several approaches were proposed to identify novel protein sources or alternatives [1–3]. Alfalfa (*Medicago sativa* L.) leaf protein concentrate (LPC) is a promising element either in human or animal diet as the human population of the Earth (7.2 billion) is growing rapidly causing a high demand for animal protein [4,5]. The isolation of leaf protein resulted in four products, i.e., green juice, fiber, leaf protein concentrate, and deproteinized juice or referred to as brown juice (BJ). The amount of produced BJ during the isolation of leaf protein and the utilization of this product is a huge obstacle for the wider recognition of LPC production. Fractionation of one-kilogram fresh biomass results in 450–550 mL BJ [6–8]. Therefore, disposal of BJ is a big challenge that the leaf protein isolate approach faces. Another reason making disposal of BJ risky is the high content of several bioactive components such as sugars, free amino acids, minerals and vitamins that BJ contains [7,9,10]. Therefore, finding an alternative use of BJ is urgent due to environmental concerns, besides maximizing the benefit from this waste that is very rich in several valuable compounds and nutrients. Instead of dumping it we could make a valuable product, thus making a step towards the circular economy concept [11,12]. The BJ contains about 40% carbohydrates (mainly monosaccharides, like glucose and fructose) and 3% nitrogen-based on dry mass [12]; additionally, numerous biologically active components like phenols, amino acids, macro- and microelements and biostimulators, etc. [7,11,13]. However, fresh BJ cannot be stored at room temperature, after a few days it gets spoiled at pH 5–6 due to its high sugar content. In our previous study, inoculation of BJ with lactic acid bacteria was not only found to increase BJ stability at room temperature but also substantially improved the nutritional characteristics of BJ [7] through converting sugars into organic acids decreasing the pH to almost 4.5. Therefore, BJ seems to be an ideal component in animal feeding programs as well as plant nutrition and soil stabilization. Several authors have previously suggested BJ as a plant fertilizer, fodder, and growth medium for microbes [10,14–16]. Fermented BJ can be applied as a very effective foliar biostimulant. We have observed remarkable effects of fermented alfalfa BJ on the growth dynamic of plumed cockscomb [7]. Additionally, lactic acid bacteria as a plant growth-promoting bacteria represents an additional benefit of fermentation of BJ as it promotes plant growth [12,17].

Sweet basil (*Ocimum basilicum* L.) is a well-known annual herb, member of the Lamiaceae family. It is one of the 150 species of the Ocimum genus, a very important medicinal, spice and fresh vegetable, culinary herb and industrial plant, cultivated for aromatic and medicinal use on large areas in many countries [18–23]. It is native to India, Asia and Africa, but grows in many regions of the world, including Italy, Thailand, Vietnam and Laos [24]. Sweet basil is popularly used in traditional Chinese medicine, because of its selected purified components and antiviral activity [25]. The latest scientific developments revealed the strong pharmacological action and nutritional aspects because of antioxidant content [25], additionally, it is a rich source of acylated and glycosylated anthocyanins being a valuable source for the food industry [26].

The aim of this study was to examine the impact of the fermented BJ on the biometrical, physiological and anatomical features of sweet basil plants.

2. Materials and Methods

2.1. Sources of Brown Juice and Plant Materials

The fresh alfalfa BJ was obtained from the Proteomill Green Protein Biorefinery Factory (Tedej Ltd., Hajdúnánás, Hungary). The fresh BJ was fermented using lactic acid bacteria to avoid its fast spoiling. The physicochemical properties of fresh and fermented BJ as well as the fermentation process were described by Bákonyi et al. [7]. The seeds of sweet basil (*Ocimum basilicum* 'Bíborfelhő') were obtained from the National Agricultural Research and Innovation Center (NARIC, Budapest, Hungary).

2.2. *Experimental Study*

A pot experiment under greenhouse conditions was carried out at the NARIC to assess the possible growth stimulation effect of fermented BJ using the multipurpose sweet basil as a plant model. The experimental design was the Randomized Complete Block design (RCB) with 15 replicates. A polyethylene pot ($7 \times 7 \times 8$ cm) was filled with white peat for young plants (Klassman-Deilmann TS 3 FINE type, Geeste, Germany). The characteristics of growth medium are as follows: fine structure, pH (H_2O) 6, N 140 mg L^{-1}, P (P_2O_5) 100 mg L^{-1}, K (K_2O) 180 mg L^{-1}, Mg 100 mg L^{-1}, S 150 mg L^{-1}. The sweet basil seeds were sown in the nursery substrate on 16 July 2018. The germinated seedlings were treated once a week with fermented BJ at different rates (i.e., 0.5, 1, and 2.5%) in the early stage of development (stage 1 BBCH) [27]. On 1 August 2018, the seedlings turned to stage 2 BBCH, all identical and healthy, were transferred to the pots. One pot contained one seedling and each treatment contained 15 pots. Fermented BJ was sprayed on the plants twice a week (on Tuesdays and Fridays) from 15 August till the end of the experiment (11 September) at rates of 0.5, 1.0, and 2.5%. Final application volume of 250 mL BJ was equally shared among all replicates of the same treatment (15 plants). The control plants (0.0%) were sprayed with the equivalent amounts of tap water. The experiment was terminated on 11 September (stage 5 BBCH) and plants were harvested, and samples were collected for the further biometric, physiological and anatomical analyses.

2.2.1. Analysis of Biometric Features of Basil Plants

At the end of the experiment, before flowering (BBCH stage 5), all plants in each treatment were harvested and the following vegetative parameters were measured: root and stem length (cm), root and stem volume (mL), root and stem fresh mass (g plant^{-1}), the number of leaves (pcs plant^{-1}) and leaf area. Roots were carefully removed from the pots and washed by tap water on a sieve to remove the adhered growth medium particles. Length was measured with a measuring ruler, while volume was determined by a graduated cylinder, fresh mass by OHAUS Pioneer PA214 analytical balance and the number of leaves by counting. The leaf area (cm^2) of sweet basil plants were measured by AreaMeter 350 (Opti-Sciences, Hudson, NY, USA). Six plants were analyzed and represented as a mean ± SD ($n = 6$). In the case of leaf area, we used the data of nine plants. All leaves of the examined plants were measured.

2.2.2. Determination of Chlorophyll Contents

At the end of the experimental period, the relative chlorophyll values were measured by SPAD 502 chlorophyll meter (Minolta, Japan) using the last fully developed leaves. The chlorophyll-a and -b contents were extracted based on the method of Moran and Porath [28] and determined by the method of Wellburn [29], Vidician and Cachita-Cosma [30]. We used the following formulas: "Chlorophyll a (mg·g^{-1}) = (11.65 a664–2.69 a647)" and "Chlorophyll b (mg·g^{-1}) = (20.81 a647–4.53 a664)". The samples were taken from the last fully expanded leaves and chlorophyll pigments were extracted by 5 mL *N,N*-dimethylformamide (DMF) added to 0.05 g leaf disc. The samples were soaked in this solvent for 48 h at room temperature in the dark. After 48 h, the discs were removed and the contents of chlorophyll-a and -b were measured by METEREKSP-830 spectrophotometer. All the measurements were repeated three times for each plant making up to nine measurements for each of the treatments.

2.2.3. Anatomical Analysis

We used three individual plants per treatment for the stem's histological examination. 15 different cross-sections per plant internodes ($n = 45$) were prepared as described in the following: each plant was cut into smaller pieces and the third internodes (from beneath) fixed separately in Strasburger-Flemming's solution [31], which is a mixture of glycerin:alcohol:water (1:1:1) for a week. Then, several cross-sections were prepared using blades, after clarification, they were stained with Toluidin-blue. Each analysis was performed under a light microscope (Zeiss Axioscope

2+; Zeiss International, Oberkochen, Ostalbkreis, Germany) with a compatible camera, and Scope Photo software (Scopetek, München, Germany) was used for processing the images. The measured parameters were: thickness of the epidermis, primary cortex, pith including primary and secondary vascular tissues.

2.3. *Statistical Analysis*

Results of the experiments were subjected to one-way ANOVA by SigmaPlot 12.0 and IBM SPSS Statistics 24 software and the means were compared by Tukey Test [32] at $p \leq 0.05$. Before the ANOVA test, in SPSS the Levene's Test for Equality of Variances was performed in the case of anatomical data. SigmaPlot 12.0 automatically ran a check test for Equality of Variance. The Equality of Variance test for different variables at the four treatments of BJ (i.e., 0.0, 0.5, 1.0, and 2.5%) were negative, $p \leq 0.05$, and the variances showed homogeneity.

3. Results

3.1. *Plant Biometric Features*

Results of root and stem length of sweet basil are presented in Figure 1. The treated plants with fermented BJ had taller stems and roots compared to the control plants (0.0%). Most importantly, fermented BJ enhanced plant development resulting in higher values of both root and stem. Despite all treated plants showed a taller stem than control plants (0.0%), increasing the rate of applied fermented BJ slightly reduced the stem length. However, the differences in stem length between treatments of fermented BJ were not significant. Treated plants with 0.5% fermented BJ showed an increase of 50.4% in stem length compared to control (0.0%) as the highest measured increase. Similarly, root length showed a relationship as root length gradually increased with increasing the rate of the applied fermented BJ. The application of fermented BJ significantly increased root length compared to the control (0.0%); however, no significant differences in root length were reported between different rates of fermented BJ (Figure 1).

Stem volume of sweet basil plants follows a dose-relationship response to fermented BJ; increasing rates of fermented BJ significantly and gradually caused an increase in stem volume. The highest increase in stem volume (38.6%) compared to untreated control plants (0.0%) was recorded by 2.5% BJ (Figure 2). Similar effect was noticed for root volume, as it increased upon spraying plants with increased rates of fermented BJ. Nevertheless, the highest rate of fermented BJ (2.5%) showed the same value as for treatment of 0.5% fermented BJ, which was two-fold higher than the control plants (0.0%). The highest root volume was 1.16 mL and was found when plants received 1.0% fermented BJ.

Results of stem and root fresh mass are presented in Figure 3. The fresh mass of stem gradually increased as a result of increased rates of fermented BJ. Significantly, all fermented BJ increased the stem fresh mass. While control plants (0.0%) possessed a stem fresh mass of 0.73 g plant^{-1}, treated plants with 2.5% fermented BJ showed a stem fresh mass of 1.52 g plant^{-1} as the highest recorded value. The differences between fermented BJ treatments were not significant. In contrast to stem fresh mass, root fresh mass increased as rate of fermented BJ increased up to 1.0%, then decreased at 2.5% fermented BJ. However, all treatments showed significant increases in the fresh mass of root system in comparison to control plants (0.0%) (Figure 3). Plants received 1.0% of fermented BJ, which was almost three-fold higher in their root fresh mass than control plants (0.0%).

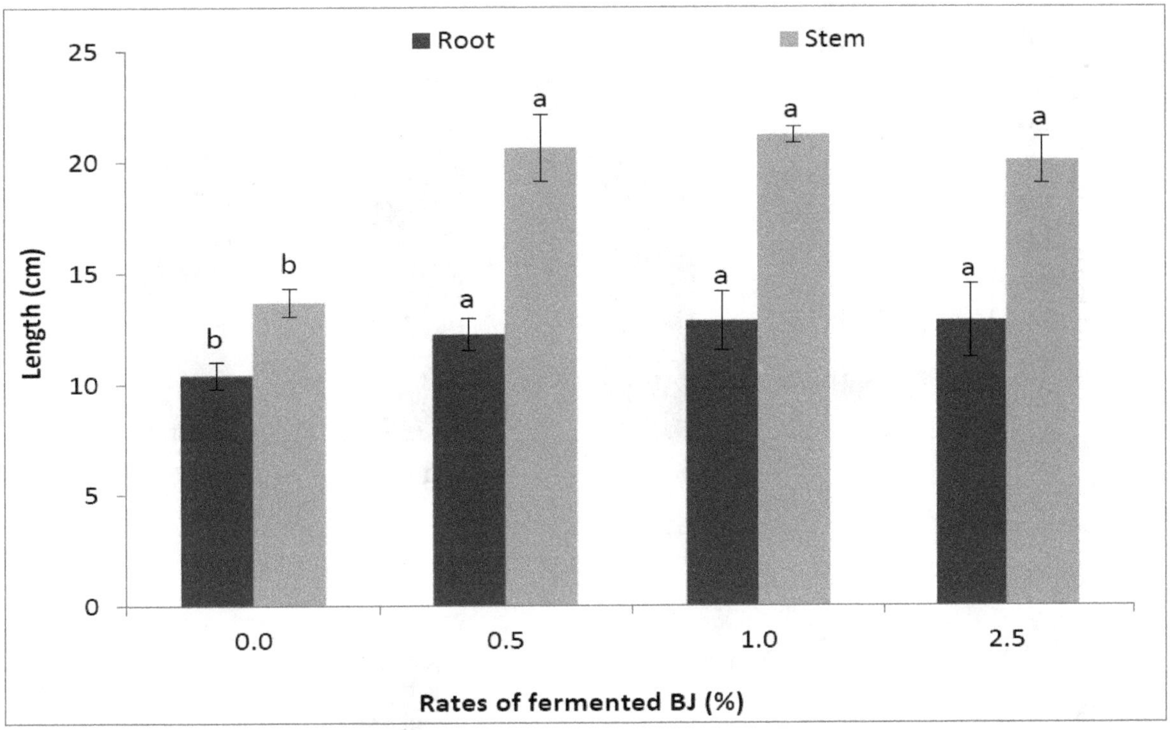

Figure 1. Root and stem length of sweet basil plants treated with different rates of fermented BJ. Sample size $n = 6$ (mean ± SD). Different letters above the same columns show significant differences at the level of $p \leq 0.05$.

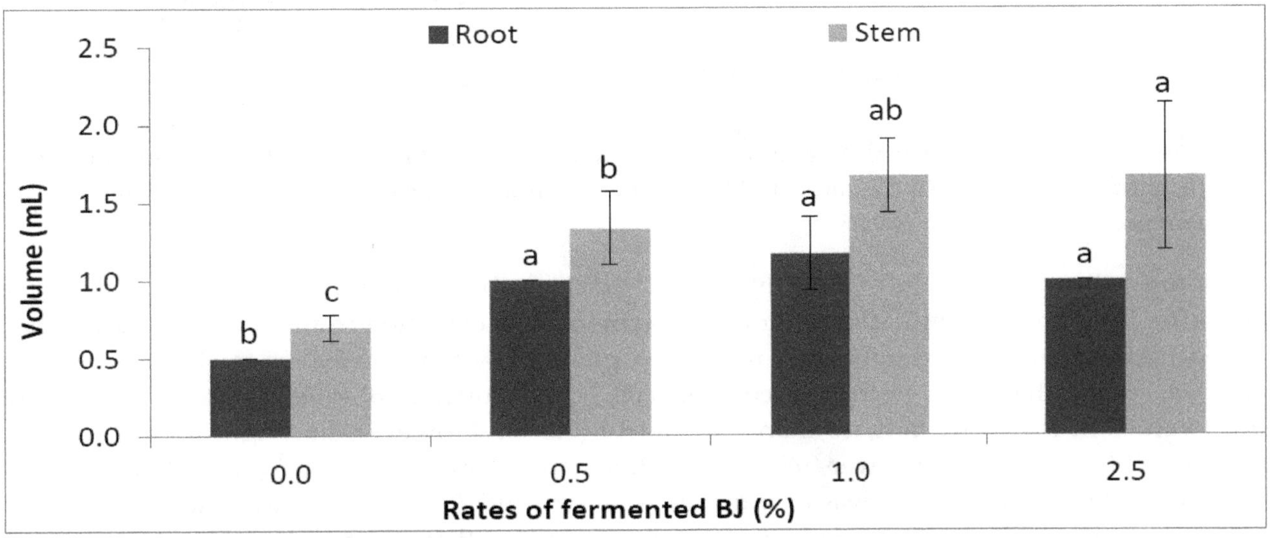

Figure 2. Stem and root volume of sweet basil plants treated with different rates of fermented BJ. Sample size $n = 6$ (mean ± SD). Different letters above the same columns show significant differences at the level of $p \leq 0.05$.

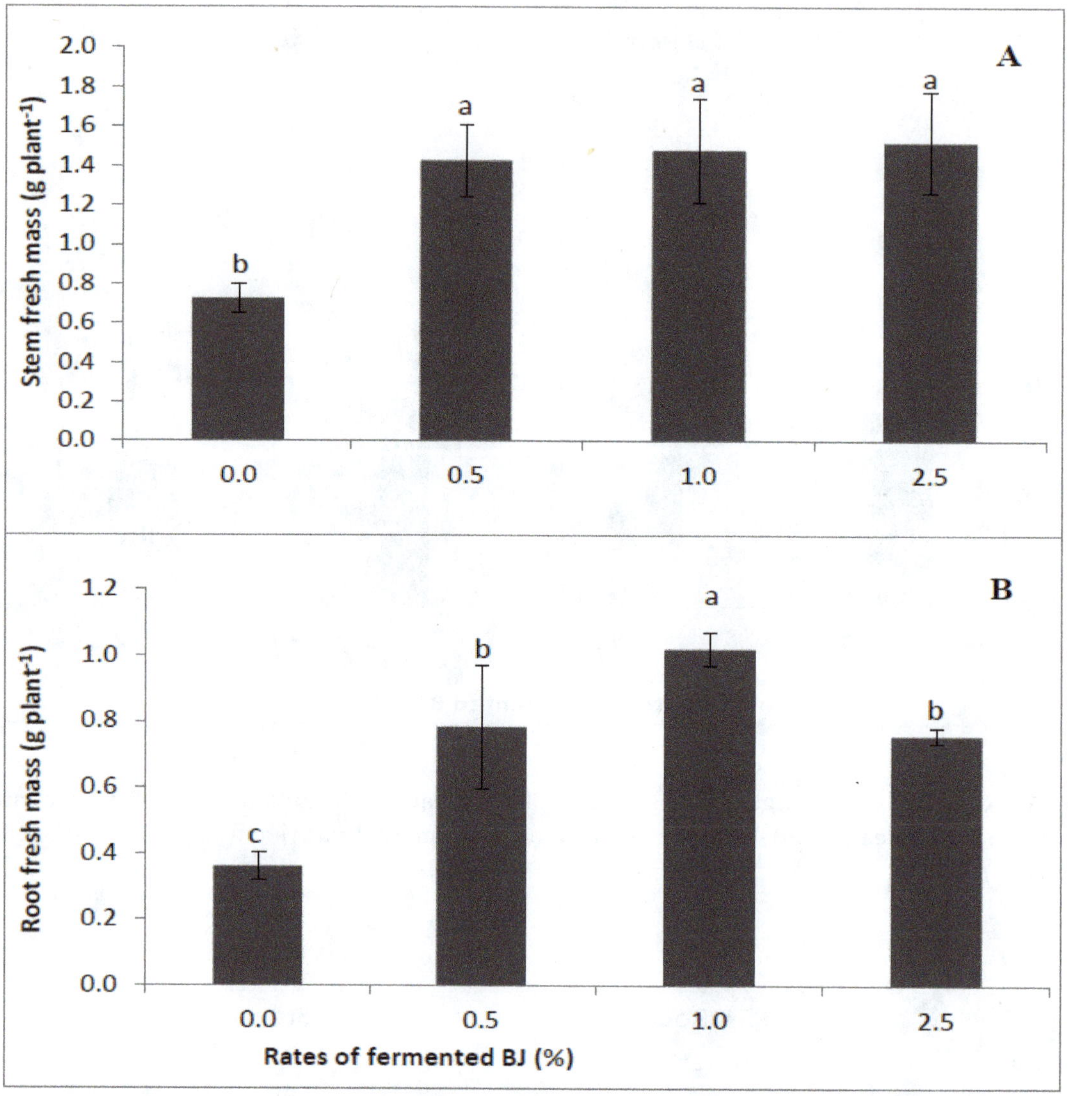

Figure 3. Fresh mass of stem (**A**) and root (**B**) of sweet basil plants treated with different rates of fermented BJ. Sample size $n = 6$ (mean ± SD). Different letters above the same columns show significant differences at the level of $p \leq 0.05$.

Figure 4 shows the results of the leaves fresh mass, number of leaves and leaf area of sweet basil plants after treating them with different rates of fermented BJ extracted from alfalfa biomass. Results showed that fermented BJ significantly improved plant growth as the leaves fresh mass of treated plants was higher than those untreated (control, 0.0%). All fermented BJ rates resulted in significantly higher values of leaves fresh mass than the control (0.0%). However, the highest applied rate of fermented BJ (2.5%) showed a lower fresh mass value than that measured at 0.5 and 1.0% of fermented BJ. The highest leaves fresh mass was recorded at treatment with 0.5 and 1.0% fermented BJ (Figure 4A). Contrarily, the number of leaves per plant gradually reduced with increasing rate of applied fermented BJ (Figure 4B). Control plants (0.0%) possessed the highest number of leaves (22 leaves per plant), while the plants sprayed with 2.5% fermented BJ had 11.7 leaves per plant. The differences among all treatments including the control (0.0%) were statistically significant. Despite control plants (0.0%) possessing a higher number of leaves than the treated plants with fermented BJ, the leaf area of plants sprayed with different rates of fermented BJ was larger than the control plants (0.0%). While the control plants (0.0%) had a leaf area of 71.4 cm^2, the plants of the treatment of 0.5% fermented BJ displayed a leaf area of 131.5 cm^2 (Figure 4C). Moreover, other treatments exhibited significantly higher leaf areas than control plants (0.0%) as well.

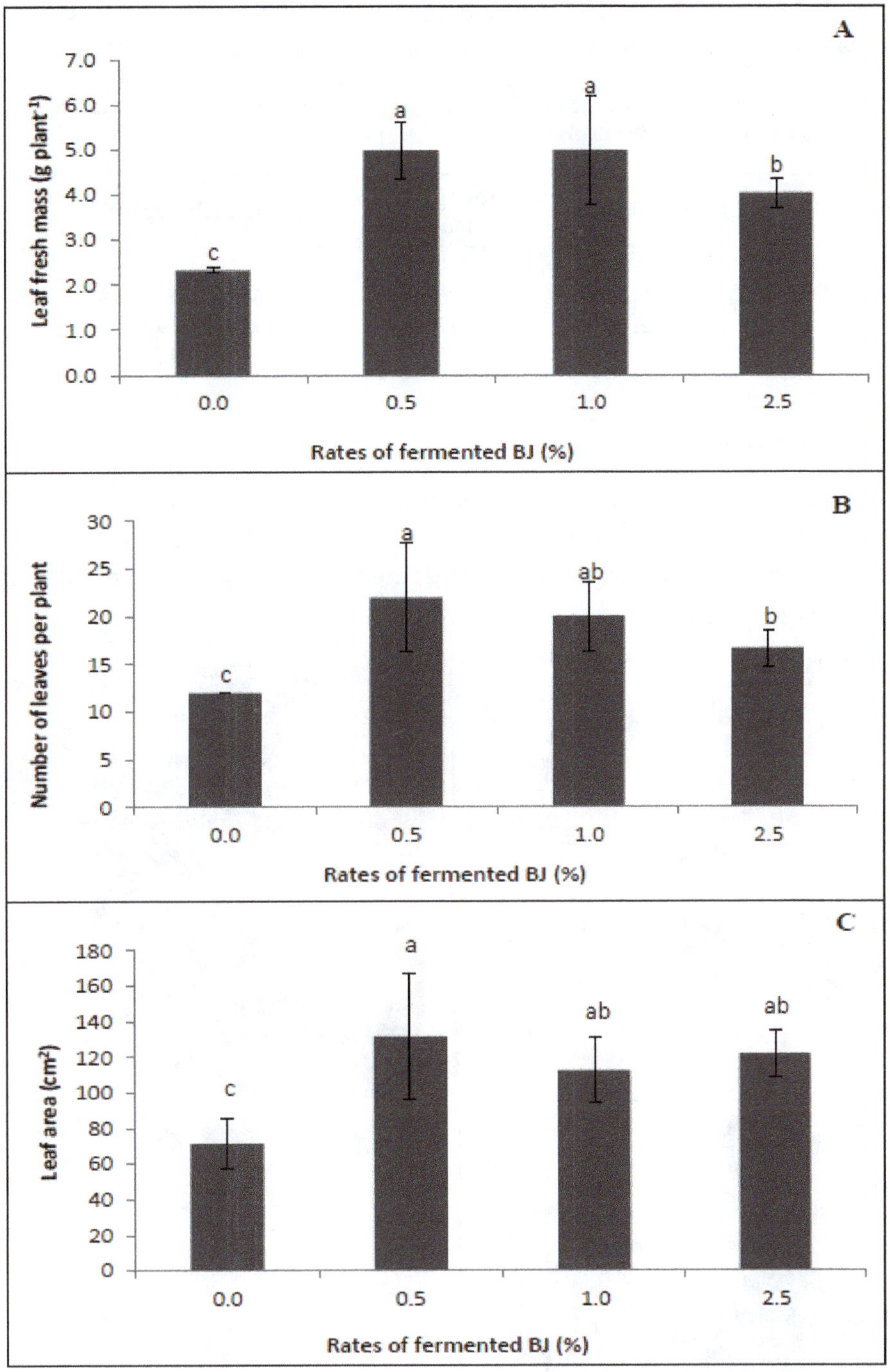

Figure 4. (**A**) Leaves fresh mass; (**B**) number of leaves; (**C**) leaf area of sweet basil plants treated with different rates of fermented BJ. Sample size $n = 6$ (mean ± SD). Different letters above the same columns show significant differences at the level of $p \leq 0.05$.

3.2. *Contents of Photosynthetic Pigment*

The results showed that the content of chlorophylls increased upon spraying sweet basil plants with fermented BJ (Table 1). The contents of chlorophyll-a (chl a) and chlorophyll-b (chl b) in the control (0.0%) leaves were 6.90 and 1.95 mg g^{-1}, respectively, and significantly increased to 8.04 (chl a) and 2.66 (chl b) mg g^{-1} after treating plants with 0.5% (chl a) and 2.5% (chl b) fermented BJ, respectively. The total content of chlorophylls ranged from 8.86 to 10.60 mg g^{-1}, 2.5% BJ application increased the chlorophyll content by 19% compared to the control (0.0%). Chlorophyll a/b ratio in treated plants with fermented BJ ranged from 2.97 to 3.50; while the control plants (0.0%) possessed a chlorophyll a/b ratio of 3.52 (Table 1). The SPAD value showed a significant, 28% increase when leaves were treated with 0.5% concentration of fermented alfalfa BJ in comparison to the control (0.0%).

Table 1. Content of chlorophyll in sweet basil leaves (mg g^{-1}) and their relative changes (%) compared to the control plants (0.0%) ($n = 9$).

	Chlorophyll-a	Chlorophyll-b	Chlorophyll-a/b Ratio	Total Chlorophyll	SPAD Value
0.0%	6.91 b	1.96 b	3.52 a	8.86 a	27.00 b
0.5%	8.05 a (16.52%)	2.46 ab (25.78%)	3.26 ab	10.51 ab (18.57%)	34.60 a (28.15%)
1.0%	6.94 ab (0.56%)	2.00 ab (2.26%)	3.47 ab	8.95 ab (0.93%)	32.40 ab (20.00%)
2.5%	7.93 ab (14.86%)	2.67 a (36.12%)	2.97 b	10.60 b (19.56%)	31.60 ab (17.04%)

Different letters (a, b, ab) in the same columns show significant differences at the level of $p \leq 0.05$.

3.3. *Anatomical Traits of Sweet Basil*

The cortex is made up of angular collenchyma (two to four cells thick) and typical parenchymatous cells. The pith consists of mainly parenchymatous cells, but the ratio of primary and secondary vascular tissues is dependent on the treatments. Treatments 0.5% and 1.0% resulted in well-developed secondary vascular tissues (Figure 5).

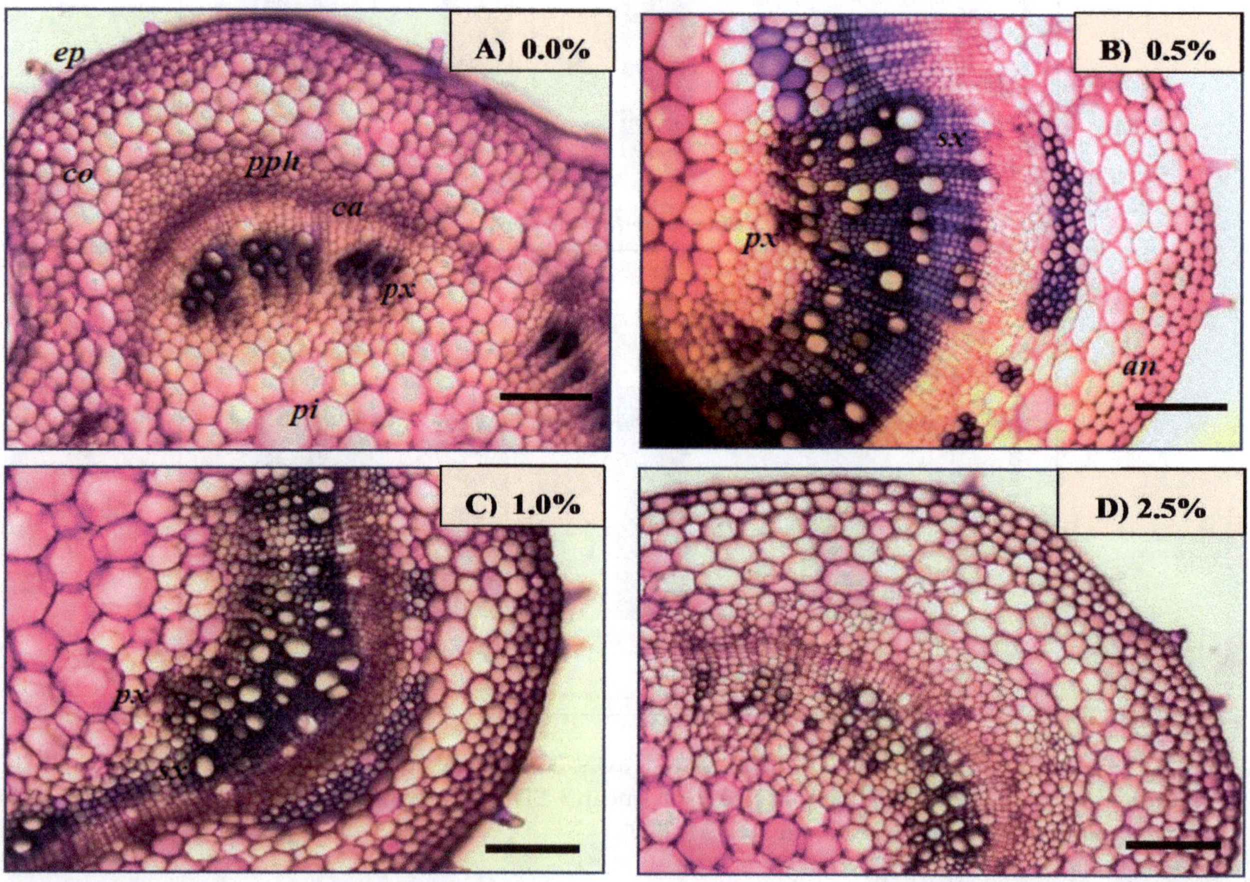

Figure 5. Anatomical sections of sweet basil stem. Subfigures show samples treated with different rates of fermented BJ such as (**A**) 0.0%, (**B**) 0.5%, (**C**) 1.0%, (**D**) 2.5%. ***ep*** epidermis, ***co*** cortex, ***ang*** angular collenchyma, ***pi*** pith, ***pph*** primary phloem ***ca*** cambium ***px*** primary xylem ***sx*** secondary xylem, after treating with different rates of fermented BJ (i.e., control (0.0%), 0.5%, 1.0%, 2.5%). Scale bar is 200 μm.

Differences which are visible in the tissues caused by the treatments were proven true by our measurements. All levels of concentration increased the thickness of the epidermis, however only two tests were significant. The impact of all treatments was a decrease in the thickness of the primary cortex.

Treatments 0.5% and 1.0% increased both the extension of the pith and the proportion of vascular tissue in it, which were proven statistically significant. Growth of the secondary vascular tissue was the highest at the 0.5% treatment. It can be concluded that a greater brown juice concentration, which was 2.5% in this analysis, may block the development of secondary vascular tissues (Table 2).

Table 2. Anatomical traits of basil stem tissue (epidermis, cortex, pith involved primary and secondary vascular tissues) (μm) after treatment with fermented BJ. Sample size $n = 45$ (Mean ± SD).

BJ Rate	Epidermis	Cortex	Pith	Vascular Tissue
0.0%	23.75 ± 5.55‡ b	273.55 ± 48.18 a	2202.89 ± 239.13 c	234.41 ± 53.75 c
0.5%	27.92 ± 7.81 a	259.39 ± 73.72 a	2529.03 ± 280.56 b	548.68 ± 107.39 a
1.0%	27.37 ± 8.22 ab	260.97 ± 51.36 a	2684.10 ± 259.04 a	539.41 ± 96.54 a
2.5%	28.88 ± 7.54 a	242.16 ± 56.36 a	2101.83 ± 162.30 c	290.38 ± 59.46 b

Different letters (a, b, ab) in the same columns show significant differences at the level of $p \leq 0.05$.

Consequently, the use of an incremental concentration of brown juice can increase the thickness of the stem and the vascular tissue in it but only to a certain extent.

4. Discussion

The BJ is largely produced during the isolation of leaf protein from several green leafy crops. The storage of fresh BJ in room temperature is the main concern due to the high content of carbohydrates particularly in the form of monosaccharaides. Lactic acid bacteria showed a considerable effect on reducing BJ pH as a result of organic acids production under an anaerobic condition and consequently increased the stability and handling of BJ at a pH of 4.5–4.8 [7].

Fermented BJ can be exploited as an organic fertilizer or growth stimulator due to its richness in free amino acids, soluble sugars, vitamins, organic acids and other nutrients. In addition, lactic acid bacteria—as plant growth promoting bacteria—represent an additional benefit for using fermented BJ as a fertilizer [7,33]. In the present study, fermented BJ proved its efficiency as a promising plant growth stimulator as it caused a significant increase in photosynthetic pigments including chl a, chl b, total chlorophyll and relative chlorophyll (SPAD value). Similar findings have been previously reported by Bergstrand et al. [34] who proved in their study on nitrogen speciation in pot experiment of sweet basil fertilized by different organic manures, i.e., blood meal + Baralith®Enslow and poultry manure, that the plant-based organic fertilizer treatment induced the chlorophyll content. Moreover, applying biofertilizers, i.e., Nitrajin (including Azotobacter, Azospirillum and Pseudomonase), increased the amount of photosynthetic pigments and the leaf area of sweet basil [35]. Additionally, Ertani et al. [36] mentioned that the extract prepared from alfalfa biomass using enzymes contained growth stimulant compounds like triacontanol and indole-3-acetic acid, which significantly improved the relative chlorophyll and growth of maize plants under salt-stress conditions.

Noticeably, all rates of BJ resulted in higher values of stem and root length (Figure 1), stem and root volume, stem and root fresh mass, number of leaves and leaf area. This was in agreement with those documented earlier by El-Ziat et al. [37]. They cited that organic fertilization and humic acid application improved growth parameters of 'Red Rubin' basil plant, i.e., fresh weight, plant height and leaf area, in a greenhouse experiment. Organic fertilization of basil using organic NPK fertilizer (4–3–4) (Organic Fertilizer, Mighty Grow®Fruitdale, AL) resulted in changes in both fresh and dry weight, and in nutrient uptake as well [24]. Onofrei et al. [18] stated that different organic foliar applications, i.e., Fylo®, Geolino Plants&Flowers®, Cropmax®, Fitokondi®, stimulated the content of total phenolic compounds contributing to healthier vegetable production of *Ocimum basilicum* L.

Unique positive changes in secondary vascular tissues observed as a result of 0.5% and 1.0% BJ treatments (Figure 5 and Table 2), which according to our best knowledge have not been published before. However, similar effects were reported on plumed cockscomb (*Celosia argantea* var. plumosa 'Arrabona') plants by Bákonyi et al. [7] who cited significant changes in histological parameters after treating plants with fermented BJ. The tissue structure of the stem of sweet basil analyzed is a typical structure of a plant at the age of 10–12 weeks. Stems were covered by the epidermis (single row) with a thin layer of cuticle. Contrary to Venkateshappa and Sreenath [38], more types of the trichomes are identified on the surface of the stem, e.g., non-glandular uniseriate hair (composed of three cells) and glandular capitate hairs, which supports the findings of Werker et al. [39] and Nasssar et al. [40].

In a few cases, the application of fermented BJ led to a slight reduction in some measured parameters. However, all rates of fermented BJ were better than the control plants (0.0%, untreated). This beneficial role of fermented BJ is owed to its high content of phytoavailable nutrients such as N, P, K, Ca, Mg, S, Mn, Fe, Cu, Zn, and Mo [7]. Similar findings were reported earlier by Ream et al. [41]. They revealed that BJ at the rate of 1.25 cm enhanced the yield and growth of corn, alfalfa and bromegrass. However, they also stated that higher rates of BJ (2.5 cm) caused a reduction in development and yields of all crops. Another study cited similar results where the application of BJ obtained from alfalfa biomass at low rates positively improved the germination of many crops, i.e., cowpea, mung bean and groundnut. Negative impacts were reported when BJ was used at higher rates, above 10% [33]. The detrimental effect of high rates of BJ may be attributed to the existence of some phytotoxic organic compounds in BJ [6]. Moreover, the reduction in plant growth at a high applied rate of fermented BJ could be due to the high electrical conductivity of BJ [7]. Additionally, spraying plants with acidic solutions (low pH as at the high rate of fermented BJ) is known to lessen the stomatal conductance [42]; therefore, a reduction in plant growth is expected. This hypothesis was supported by the earlier findings of Long et al. [43].

The results obtained from the present study alongside with our previous published work [7] strongly support the hypothesis about prospective studies on BJ either fermented or fresh to improve soil properties through the soil application technique. In alkaline, salt-affected soils, and sandy soils the expected benefits of using fermented BJ would be more due to the high content of macro and microelements, free amino acids [7,42,44]. Additionally, the soluble sugars will facilitate and improve the growth of rhizosphere bacteria supporting them against the unflavored growth conditions [45]. Another substantial result for using fermented BJ via soil application is increasing the soil aggregates due to a high sugar content in BJ. Moreover, the low pH of fermented BJ is an advantage of using fermented BJ as a soil conditioner since it will partially modify the local soil pH around the root system facilitating the uptake of especially microelements by plant roots [43]. However, in our next studies, we are going to work on these hypotheses.

5. Conclusions

This study demonstrates the potentiality of fermented alfalfa BJ to enhance both quantity and quality of one of the most important multipurpose crops, i.e., sweet basil, which is cultivated on large areas in many countries. Our results showed a significant increase in photosynthetic pigments leading to better vegetative growth; and consequently, enhancement in important medical phytochemicals in sweet basil could occur. For a plant such as sweet basil (used as spice and fresh vegetable, and culinary herb), larger leaf area, number of leaves, and thickness of stem (including vascular tissues) as well as fresh weight are very important decisive traits. The results also indicated that concentrations of 0.5 and 1.0% fermented BJ were the most valuable and have a considerable biostimulator effect on leaves as these treatments were more effective improving leaf parameters in comparison to control plants (0.0%).

Additionally, we came to the conclusion based on the results that fermented alfalfa BJ has great potential as a biofertilizer and plant growth promoter. Future experiments are needed to justify more of these findings with other horticultural and agricultural crops.

Author Contributions: Conceptualization, N.B., S.K. (Szilvia Kisvarga) and M.G.F.; methodology, S.K. (Szilvia Kisvarga), D.B. and I.O.T.; software, D.B.; validation, D.B.; formal analysis, G.C. and S.K. (Szilvia Kovács); investigation, S.V. and P.M.; resources, N.B. data curation, T.A.; writing—original draft preparation, N.B., T.A. and D.B.; writing—review and editing, N.B., T.A., S.V. and D.B.; visualization, G.C. and S.K. (Szilvia Kovács); supervision, M.G.F.; project administration, N.B.; funding acquisition, N.B. and M.G.F. All authors have read and agreed to the published version of the manuscript.

Acknowledgments: We acknowledge the financial support of the followings: "Complex Rural Economic and Sustainable Development, Elaboration of its Service Networks in the Carpathian Basin (Project ID: EFOP-3.6.2-16-2017-00001, Hungary)" research project and Széchenyi 2020 under the GINOP-2.2.1-15-2017-00051.

References

1. de Boer, J.; Aiking, H. Strategies towards healthy and sustainable protein consumption: A transition framework at the levels of diets, dishes, and dish ingredients. *Food Qual. Prefer.* **2019**, *73*, 171–181. [CrossRef]
2. Tenerio, A.T.; Kyriakopoulou, K.E.; Suarez-Garcia, E.; van den Berg, C.; van der Goot, A.J. Understanding differences in protein fractionation from conventional crops, and herbaceous and aquatic biomass—Consequences for industrial use. *Trends Food Sci. Technol.* **2018**, *71*, 235–245. [CrossRef]
3. Fári, M.G.; Domokos-Szabolcsy, E. Method for Producing Plant Protein Coagulum. Hun Patent WO2019150144, 8 August 2019.
4. Gaweł, E.; Grzelak, M.; Janyszek, M. Lucerne (*Medicago sativa* L.) in the human diet—Case reports and short reports. *J. Herb. Med.* **2017**, *10*, 8–16. [CrossRef]
5. Tripathi, A.D.; Mishra, R.; Maurya, K.K.; Singh, R.B.; Wilson, D.W. Estimates for world population and global food availability for global health. *Role Funct. Food Secur. Glob. Health* **2019**, 3–24. [CrossRef]
6. Pirie, N.W. *Leaf Protein and Other Aspects of Fodder Fractionation*; Cambridge University Press: Cambridge, UK, 1987; p. 209.
7. Bákonyi, N.; Kisvarga, S.; Barna, D.; Tóth, I.O.; El-Ramady, H.; Abdalla, N.; Kovács, S.; Rozbach, M.; Fehér, C.; Elhawat, N.; et al. Chemical Traits of Fermented Alfalfa Brown Juice: Its Implications on Physiological, Biochemical, Anatomical, and Growth Parameters of Celosia. *Agronomy* **2020**, *10*, 247. [CrossRef]
8. Jadhav, R.K.; Chhaya, G. Influence of deproteinised foliage fluid liquid biofertilizer on nitrate reductase activity of Eleusine coracana plants. *Curr. Bot.* **2018**, 33–36. [CrossRef]
9. Thomsen, M.H.; Bech, D.; Kiel, P. Manufacturing of Stabilised Brown Juice for L-lysine Production—From University Lab Scale over Pilot Scale to Industrial Production. *Chem. Biochem. Eng. Q.* **2003**, *18*, 37–46.
10. Bákonyi, N.; Tóth, I.O.; Barna, D.; Fári, M.G. *Preliminary Experiments on Preserving Alfalfa Brown Juice for Bio-Industrial Use (Előkísérletek a Lucerna Barnalé Tartósítására Bioipari Hasznosítás Céljából)*; XXXVII, Óvári Tudományos Napok; Széchenyi István University: Mosonmagyaróvár, Hungary, 2018.
11. Manwatkar, W.G.; Gogle, D.P. The Effect of Deproteinised Juice (DPJ) on Seed Germination and Seedling Growth of Different Plants (by Paper Towel Method). *Int. J. Life Sci.* **2014**, *A2*, 65–68.
12. Lamont, J.R.; Wilkins, O.; Bywater-Ekegärd, M.; Smith, D.L. From yogurt to yield: Potential applications of lactic acid bacteria in plant production. *Soil Biol. Biochem.* **2017**, *111*, 1–9. [CrossRef]
13. Cheynier, V.; Comte, G.; Davies, K.M.; Lattanzio, V.; Martens, S. Plant phenolics: Recent advances on their biosynthesis, genetics, and ecophysiology. *Plant Physiol. Biochem.* **2013**, *72*, 1–20. [CrossRef]
14. Shende, G.C.; Gogle, D.P. To study the effect of various concentration of deproteinized leaf juice (DPJ) of selected plants on growth of Aspergillus niger. *Int. J. Life Sci.* **2016**, *A6*, 186–188.
15. Jadhav, R.K.; Kadam, R.; Joshi, D. Physiology of Gramineae Crops by Deproteinised Foliage Extract and its Influence on Photosynthetic Chemistry. *J. Phytochem.* **2019**, *I*, 149–163.
16. Zanin, V. *A New Nutritional Idea for Man: Lucerne Leaf Concentrate*; APEF, Association Pour la Promotion des Extraits Foliaires en Nutrition: Paris, France, 1998.
17. Smith, D.L.; Subramanian, S.; Lamont, J.R.; Bywater-Ekegärd, M. Signaling in the phytomicrobiome: Breadth and potential. *Front. Plant Sci.* **2015**, *6*, 921. [CrossRef] [PubMed]
18. Onofrei, V.; Burducea, M.; Lobiuc, A.; Teliban, G.-C.; Ranghiuc, G.; Robu, T. Influence of Organic Foliar Fertilization on Antioxidant Activity and Content of Polyphenols in *Ocimum Basilicum* L. *Acta Pol. Pharm.* **2017**, *74*, 611–615.
19. Alishah, H.M.; Heidari, R.; Hassani, H.; Dizaji, A.A. Effect of Water Stress on Some Morphological and Biochemical Characteristics of Purple Basil (*Ocimum basilicum*). *J. Biol. Sci.* **2006**, *6*, 763–767. [CrossRef]
20. Nurzyńska-Wierdak, R.; Rożek, E.; Dzida, K.; Borowski, B. Growth response to nitrogen and potassium fertilization of common basil (*Ocimum basilicum* L.) plants. *Hortorum Cultus* **2012**, *11*, 275–288.
21. Gülçin, I.; Elmastaş, M.; Aboul-Enein, H.Y. Determination of antioxidant and radical scavenging activity of Basil (*Ocimum basilicum* L. Family Lamiaceae) assayed by different methodologies. *Phytother. Res.* **2007**, *21*, 354–361. [CrossRef]

22. Labra, M.; Miele, M.; Ledda, B.; Grassi, F.; Mazzei, M.; Sala, F. Morphological characterization, essential oil composition and DNA genotyping of *Ocimum basilicum* L. cultivars. *Plant Sci.* **2004**, *167*, 725–731. [CrossRef]
23. Hikmasari, I. *Handbook of Herbs and Spices*, 2nd ed.; Woodhead Publishing Limited: Cambridge, UK, 2012; Volume 1.
24. Bufalo, J.; Cantrell, C.; Astatkie, T.; Jeliazkov, V.; Gawde, A.; Boaro, C.S.F. Organic versus conventional fertilization effects on sweet basil (*Ocimum basilicum* L.) growth in a greenhouse system. *Ind. Crop. Prod.* **2015**, *74*, 249–254. [CrossRef]
25. Chiang, L.-C.; Ng, L.-T.; Cheng, P.-W.; Chiang, W.; Lin, C.-C. Antiviral activities of extracts and selected pure constituents of *Ocimum basilicum*. *Clin. Exp. Pharm. Physiol.* **2005**, *32*, 811–816. [CrossRef]
26. Phippen, W.B.; Simon, J.E. Anthocyanins in Basil (*Ocimum basilicum* L.). *J. Agric. Food Chem.* **1998**, *46*, 1734–1738. [CrossRef]
27. Hess, M.; Barralis, G.; Bleiholder, H.; Buhr, L.; Eggers, T.; Hack, H.; Stauss, R. Use of the extended BBCH scale—General for the descriptions of the growth stages of mono; and dicotyledonous weed species. *Weed Res.* **1997**, *37*, 433–441. [CrossRef]
28. Moran, R.; Porath, D. Chlorophyll Determination in Intact Tissues Using *N,N*-Dimethylformamide. *Plant Physiol.* **1980**, *65*, 478–479. [CrossRef]
29. Wellburn, A.R. The Spectral Determination of Chlorophylls a and b, as well as Total Carotenoids, Using Various Solvents with Spectrophotometers of Different Resolution. *J. Plant Physiol.* **1994**, *144*, 307–313. [CrossRef]
30. Vidican, T.I.; Cachita-Cosma, D. Determination of assimilator pigment content in cladodes of Opuntia fragilis var. Fragilis exposed to light of different colors emitted by LEDs. *Studia Univ. Vasile Goldis Arad* **2010**, *20*, 49–54.
31. Kálmán, K.; Nyakas, A.; Mihalik, E.; Nagy, E. Plantanatomical practicality. In *Növényanatómiai Praktikum*; JATE Press: Budapest, Hungary, 2019.
32. Tukey, J.W. Comparing Individual Means in the Analysis of Variance. *Biometrics* **1949**, *5*, 99–114. [CrossRef]
33. Reddy, G.U.; Deshmukh, V.R.; Joshi, R.N.; Kayama, R. Utilization of Alfalfa (*Mediacago sativa* L.) Whey as a Fertilizer in Irrigation. *Jap. J. Grassl. Sci.* **1987**, *33*, 32–37. [CrossRef]
34. Bergstrand, K.-J.; Löfkvist, K.; Asp, H. Dynamics of nitrogen availability in pot grown crops with organic fertilization. *Biol. Agric. Hortic.* **2018**, *35*, 143–150. [CrossRef]
35. Larimi, S.B.; Shakiba, M.; Mohammadinasab, A.D.; Vahed, M.M. Changes in nitrogen and chlorophyll density and leaf area of sweet basil (*Ocimum basilicum* L.) affected by biofertilizer and nitrogen application. *Int. J. Biosci.* **2014**, *5*, 256–265. [CrossRef]
36. Ertani, A.; Schiavon, M.; Muscolo, A.; Nardi, S. Alfalfa plant-derived biostimulant stimulate short-term growth of salt stressed *Zea mays* L. plants. *Plant Soil* **2013**, *364*, 145–158. [CrossRef]
37. El-Ziat, R.A.; Swaefy, H.M.; Esmail, S.E.A. The Response of Red Rubin Basil Plant to Organic Fertilizer and Humic Acid versus Chemical Fertilizers. *Middle East J. Agric. Res.* **2018**, *7*, 740–751.
38. Venkateshappa, S.M.; Sreenath, K.P. Some Species of Lamiaceae- Comparative Anatomical Studies. *Indo Am. J. Pharm. Res.* **2013**, *3*, 6.
39. Werker, E.; Putievsky, E.; Ravid, U.; Dudai, N.; Katzir, I. Glandular hairs and essential oil in developing leaves of *Ocimum basilicum* L. (Lamiaceae). *Ann. Bot.* **1993**, *71*, 43–50. [CrossRef]
40. Nassar, M.A.; El-Sagai, M.U.; Azoz, S.N. Anatomical and Phytochemical studies on *Ocimum basilicum* L. plant (Lamiaceae). *Int. J. Adv. Res.* **2014**, *2*, 204–226.
41. Ream, H.W.; Smith, D.; Walgenbach, R.P. Effects of Deproteinized Alfalfa Juice Applied to Alfalfa—Bromegrass, Bromegrass, and Corn1. *Agron. J.* **1977**, *69*, 685–689. [CrossRef]
42. Anugoolprasert, O.; Kinoshita, S.; Naito, H.; Shimizu, M.; Ehara, H. Effect of low pH on the growth, physiological characteristics and nutrient absorption of sago palm in a hydroponic system. *Plant Prod. Sci.* **2012**, *15*, 125–131. [CrossRef]
43. Long, A.; Zhang, J.; Yang, L.T.; Ye, X.; Lai, N.W.; Tan, L.L.; Lin, D.; Chen, L.S. Effects of low pH on photosynthesis, related physiological parameters, and nutrient profiles of citrus. *Front. Plant Sci.* **2017**, *8*, 185. [CrossRef]
44. Porra, R.J.; Thompson, W.A.; Kriedemann, P.E. Determination of accurate extinction coefficients and simultaneous equations for assaying chlorophylls a and b extracted with four different solvents: Verification of the concentration of chlorophyll standards by atomic absorption spectroscopy. *Biochim. Biophys. Acta* **1989**, *975*, 384–394. [CrossRef]
45. Ferreira, A.S.; Santos, M.A.D.; Corrêa, G.F. Soil microbial response to glucose and phosphorus addition under agricultural systems in the Brazilian Cerrado. *An. Acad. Bras. Ciênc.* **2013**, *85*, 395–403. [CrossRef]

5

Biostimulants Application Alleviates Water Stress Effects on Yield and Chemical Composition of Greenhouse Green Bean (*Phaseolus vulgaris* L.)

Spyridon A. Petropoulos [1,*], Ângela Fernandes [2], Sofia Plexida [1], Antonios Chrysargyris [3], Nikos Tzortzakis [3], João C. M. Barreira [2], Lillian Barros [2] and Isabel C. F. R. Ferreira [2,*]

1 Department of Agriculture, Crop Production and Rural Environment, University of Thessaly, 38446 N. Ionia, Magnissia, Greece; sofiaplexida@uth.gr

2 Centro de Investigação de Montanha (CIMO), Instituto Politécnico de Bragança, Campus de Santa Apolónia, 5300-253 Bragança, Portugal; afeitor@ipb.pt (Â.F.); jbarreira@ipb.pt (J.C.M.B.); lillian@ipb.pt (L.B.)

3 Department of Agricultural Sciences, Biotechnology and Food Science, Cyprus University of Technology, 3036, Lemesos, Cyprus; a.chrysargyris@cut.ac.cy (A.C.); nikolaos.tzortzakis@cut.ac.cy (N.T.)

* Correspondence: spetropoulos@uth.gr (S.A.P.); iferreira@ipb.pt (I.C.F.R.F.)

Abstract: The increasing scarcity of water demands proper water management practices to ensure crop sustainability. In this study, the effect of drought stress and biostimulants application on the yield and chemical composition of green pods and seeds of common bean (*Phaseolus vulgaris* L.) was evaluated. For this purpose, four commercially available biostimulant products, namely Nomoren (G), EKOprop (EK), Veramin Ca (V), and Twin-Antistress (TW), were tested under two irrigation regimes: normal irrigation (W+) and water-holding (W-) conditions. The highest increase (20.8%) of pods total yield was observed in EKW+ treatment due to the formation of more pods of bigger size compared to control treatment (CW+). In addition, the highest yield under drought stress conditions was recorded for the GW- treatment (5691 ± 139 kg/ha). Regarding the effects of biostimulants on the protein and ash content of pods, the application of VW+ treatment (first harvest of pods; 201 ± 1 and 79 ± 1 g/kg dw for proteins and ash content, respectively) and GW+ (second harvest of pods; 207.1 ± 0.1 and 68.4 ± 0.5 g/kg dw for proteins and ash content, respectively) showed the best results. For seeds, the application of GW+ treatment resulted in the highest content for fat, protein, and ash content (52.7 ± 0.1, 337 ± 1, 56 ± 1 g/kg dw) and energetic value (5474 ± 3 kcal/kg dw). γ-tocopherol was the main detected tocopherol in pods and seeds, and it was significantly increased by the application of TWW- (first harvest of pods; 6410 ± 40 μg/kg dw), VW- (second harvest of pods; 3500 ± 20 μg/kg dw), and VW+ (seeds; 39.8 ± 0.1 g/kg dw) treatments. EKW- treatment resulted in the lowest oxalic acid content for both pod harvests (26.3 ± 0.1 g/kg dw and 22.7 ± 0.2 g/kg dw for the first and second harvest of pods, respectively) when compared with the rest of the treatments where biostimulants were applied, although in all the cases, the oxalic acid content was considerably low. Fructose and sucrose were the main sugars detected in pods and seeds, respectively, while the highest content was recorded for the TWW- (first harvest of pods) and GW- (second harvest of pods and seeds) treatments. The main detected fatty acids in pods and seeds were α-linolenic, linoleic, and palmitic acid, with a variable effect of the tested treatments being observed. In conclusion, the application of biostimulants could be considered as an eco-friendly and sustainable means to increase the pod yield and the quality of common bean green pods and seeds under normal irrigation conditions. Promising results were also recorded regarding the alleviation of negative effects of drought stress, especially for the application of arbuscular mycorrhizal fungi (AMF; G treatment), which increased the total yield of green pods. Moreover, the nutritional value and chemical composition of pods and seeds was positively affected by biostimulants application, although a product specific effect was recorded depending on the irrigation regime and harvesting time (pods and seeds).

Keywords: arbuscular mycorrhizal fungi; biofertilizers; common bean; *Glomus* spp.; organic acids; pod quality; seaweed extracts; seed quality; tocopherols; total sugars

---

1. Introduction

The increasing concerns for food security in a rapidly growing world population has rendered the necessary intensification of agricultural production for the achievement of higher crop yield and total production. Protected vegetables cultivation is the most intensified cropping system and requires high amounts of fertilizers and pesticides [1,2]. However, despite the fact that higher fertilizer rates result in increased total yield, this practice is not always favorable when the quality of the final product is also considered. On the contrary, it is very common for excessive fertilization to stimulate vegetative growth and increase susceptibility to pathogens, which may result in increased product losses, as well as high nutrient losses due to leaching [3].

In addition, the increasing scarcity of water availability for human activities and irrigation in particular is a worldwide phenomenon and demands appropriate water management practices to ensure crop sustainability and economic activities related to water, especially in semi-arid and arid regions [4]. The use of biostimulants can diminish effects of environmental abiotic stress factors such as water stress, improve soil water-holding capacity and root conformation, and increase root growth with beneficial effects on nutrient and water use efficiency and yield; hence, the past decade has witnessed tremendous growth in the use of biostimulants in the farming sector [5–7]. The use of biostimulants containing arbuscular mycorrhizal fungi (AMF), saprophytic fungi, or algae extracts is considered an environmental friendly technique for the alleviation of adverse impact of osmotic stress, by increasing wáter and the nutrient uptake of crops and tolerance to biotic and abiotic stress [8–10].

Plant biostimulants usually consist of amino acids and peptide mixtures [11]. They also contain a wide number of bioactive compounds that are able to improve various physiological processes that stimulate plant growth and increase nutrient use efficiency without adverse effects on crop yield and final product quality, while at the same time reduce chemical fertilizers inputs [5,12]. However, the effect of biostimulants may differ from species to species, while it greatly depends on environmental factors during and after application, as well as on the dose and time of application [13,14]. For example, the application of saprophytic fungi (*Trichoderma harzianum* ALL-42) was associated with increased shoot biomass production and the number of lateral shoots in *Phaseolus vulgaris* plants due to the beneficial effects of root colonization by fungi on plant root growth [15]. On the other hand, seaweed extracts (*Ascophyllum nodosum*) increased the plant growth and overall yield of leafy vegetables such as spinach and lettuce [16–18], while in bean plants, the application of extracts enhanced root growth and plant development, especially when water stress conditions were imposed [19]. The biostimulatory activity of symbiotic bacteria such as *Bacillus* sp. is mostly associated with adaptation mechanisms for improved water retention though alterations in plant cell wall composition and hormones production (e.g., indole-3-acetic acid (IAA)) [6]. Therefore, environmental friendly methods such as applying biostimulants for stimulating early growth in vegetable crops and ensuring high yields are innovative agricultural practices that have to be further investigated in order to improve our understanding of their functions and the involved mechanisms of action [20].

Common bean (*Phaseolus vulgaris* L.) is a drought-sensitive vegetable crop, and water stress may have a detrimental effect on crop yield [21] and the chemical composition of pods and seeds [22,23]. So far, there is limited literature regarding the use of biostimulants on common bean plants, whereas various studies have tested the effects of biostimulants on other legume species, especially under drought stress conditions. In particular, the application of *Pseudomonas aeruginosa* GGRJ21 strain on mung bean (*Vigna radiata* (L.) R. Wilczek) under greenhouse and field conditions up-regulated the expression of drought stress-responsive genes, which resulted in better plant growth and development under water stress conditions [24]. Foliar application of amino acids on faba beans (*Vicia faba* L.)

subjected to salt stress showed significant ameliorative effects that were mainly associated with the use of amino acids as carbon and nitrogen pools, which further increased photosynthetic apparatus efficiency [25]. Kumar et al. [26] reported the synergistic effects of *Pseudomonas putida* and *Bacillus amyloliquefaciens* on chickpea plants subjected to water stress through the evaluation of several traits, including the activation of plant defense and soil enzymes and plant growth parameters. Moreover, the inoculation of common bean plants with *Azospirillum brasilense* altered root morphology, which allowed plants to overcome water stress without increasing plant biomass compared to non-inoculated plants [27]. In another study, Klimek-Kopyra et al. [28] suggested that biostimulants application on seeds of seven winter pea cultivars (*Pisum sativum* L.) may increase frost tolerance through the increased germination percentage and growth rate of seedlings, although a varied response depending on biostimulant x cultivar combination was observed. In contrast, Galvão et al. [19] suggested that the application of *Bacillus amyloliquefaciens BV 03* and/or the combination of *B amyloliquefaciens BV 03* with *A. nodosum* extracts did not alleviate water deficit effects on common bean plants. According to Dourado-Neto et al. [29] the use of hormones with biostimulant activity (combination of kinetin, indole butyric acid, and gibberellic acid) on common bean plants through seed treatment, sowing, or foliar spraying may increase the number of grains per pod and grains yield. Moreover, licorice root extracts may have a beneficial role on mitigating the negative effect of salt stress on *P. vulgaris* plants' growth and yield, as well as on the total soluble carbohydrates, soluble sugars, and nutrients content [30]. The combined application of salicylic acid and *Moringa oleifera* leaves extracts has been also reported to mitigate salinity stress effects on common bean plants through the improvement of green pods and seeds yield and the physicochemical characteristics of pods and seeds [31]. Other examples of biostimulants use on common beans crop include the application of *Lolium perenne* foliage extracts as potent cell defense elicitors [32] and the positive effect of aqueous extracts of moringa leaves and garlic cloves on the yield and chemical composition of snap beans [33,34].

Most of the studies regarding the mitigating effects of biostimulants to abiotic stressors refer to high salinity or nutrient deficiency stress. The main goal of this study was to record the effects of natural biostimulants on the yield, nutritional value, and chemical composition under drought conditions. For this purpose, a drought-sensitive species, namely the common bean (*P. vulgaris*), was selected and grown in a protected environment under water stress conditions, and the use of commercially available biostimulants products was evaluated as an environmentally friendly and sustainable method for increasing the yield and quality of end-products through the improvement of the chemical composition of the final products (pods and/or seeds) without compromising yield.

2. Materials and Methods

2.1. Plant Material and Growing Conditions

The experiment was carried out during the growing period of summer–autumn 2017. Sowing took place on 11 August 2017 and seeds of bean (*Phaseolus vulgaris* L.) were sown directly in soil within the unheated plastic greenhouse at the experimental farm of the University of Thessaly, Greece. Seeds were sown in double rows with a spacing of 50 cm between the rows, and the plant density was 2 plants/m^2 (20,000 plants/ha), while each treatment consisted of 6 plants and was replicated three times (180 plants in total). The soil at 0–30 cm depth was clay (26% sand, 32% silt, and 42% clay); pH: 8.0 (1:1 soil/H2O); organic matter content: 3.1%; CaCO3: 10.8%; available P (Olsen method): 70.9 mg/kg; total N (Kjeldahl method): 1.8 g/kg; exchangeable K2O (ammonium acetate method): 195 mg/kg; electrical conductivity (ECe): 0.95 dS/m. The growth conditions throughout the experimental period are presented in Figure 1. Two factors were applied in a split-plot factorial design, namely water stress and biostimulants. Biostimulants treatments included: (1) Control (C: no biostimulants added), (2) Nomoren (G; Anthis S.A., Greece) (3) Twin-Antistress (TW; Microspore Hellas–Sacom Hellas, Greece), (4) Veramin Ca (V; Microspore Hellas–Sacom Hellas, Greece), and (5) EKOprop (EK; Anthesis S.A., Greece). Regarding the detailed composition of each product, Nomoren contains 20% arbuscular mycorrhizal fungi (AMF) (*Glomus* spp.). Twin-Antistress contains natural microorganisms based on *Bacillus subtilis*, as well as yeast and *Ascophyllum nodosum* extracts, as well as N (organic): 1%, organic carbon: 10%, and organic matter (<50 kDa): 30%. Veramin Ca contains an amino acid complex of vegetable origin with *Aloe vera* extract, and CaO: 15.6%. EKOprop contains a mixture of arbuscular mycorrhizal fungi (*Glomus* spp: 1%), rhizospere symbiotic bacteria (*Bacillus* spp., *Streptomyces* spp., *Pseudomonas* spp.,: 1.6 × 109 CFU/g in total), and saprophytic fungi (*Trichoderma* spp.: 5 × 105 CFU/g) (Table 1).

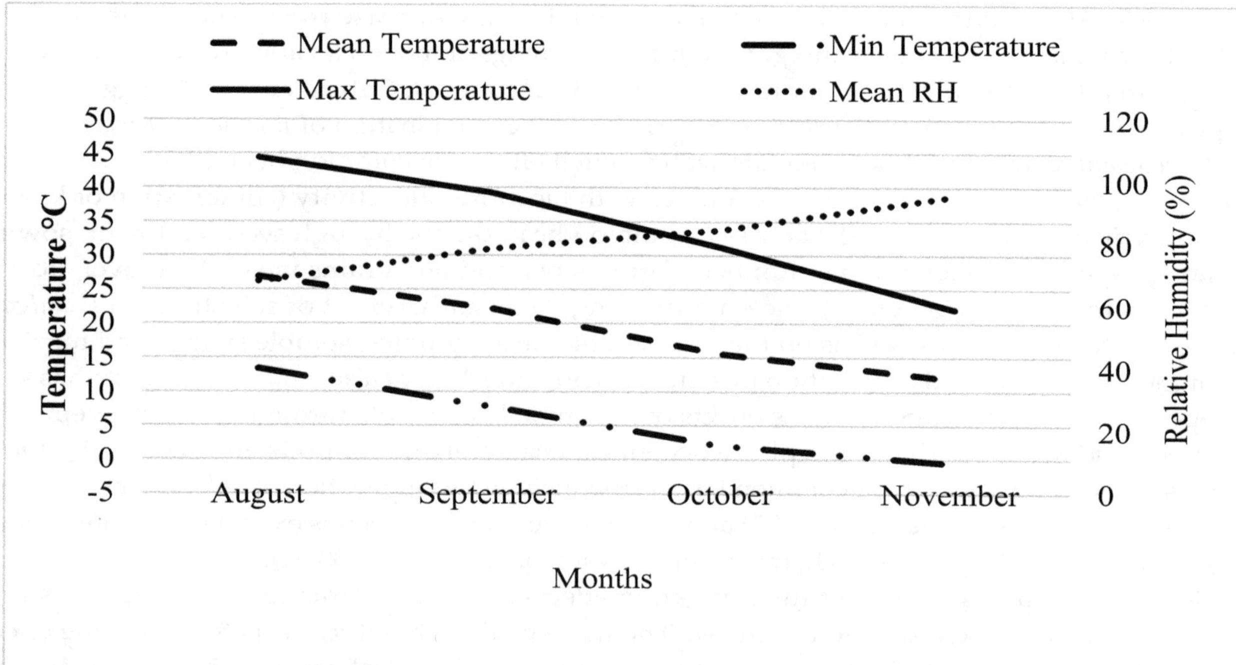

Figure 1. Environmental conditions (mean, max, and min temperature and mean relative humidity (RH)) throughout the experimental period.

Water stress treatments were previously described by the authors and scheduled with the use of tensiometers (Irrometer-Moisture Indicator, Irrometer, Riverside, CA) including: (a) normally irrigated plants (W+) where irrigation was applied approximately twice a week and when tensiometer readings were between 10% and 15%, and (b) water-stressed plants (W-) where water holding was applied with irrigation being implemented approximately once a week and when tensiometer readings were between 40% and 50% [35,36]. Tensiometer readings are percent levels that correspond to soil moisture content ranging from field capacity (0%) to dry soil (100%). Irrigation was applied through a drip irrigation system with one dripper per plant and a water flow rate of 4.0 L/h for each dripper. The total amount of irrigation water was 350 m3/ha (17.5 L per plant) for normally irrigated plants and 210 m^3/ha (10.5 L per plant) for water-stressed plants. Biostimulants were applied according to the directions for use of each product at 10, 20 and 30 days after sowing (DAS) as following: (G) was applied with irrigation water at 5 L/ha for each dose; (TW) was applied with irrigation water 5 L/ha for each dose; (V) was applied with foliar spraying at 500 g/100 L H2O for each dose; and (EK) was applied with irrigation water at 1 kg/ha for each dose. Water holding started after the second application of biostimulants (21 DAS). The harvest of green pods took place at marketable maturity at 60 DAS (first harvest) and 70 DAS (second harvest), while seeds were collected from fully mature green pods at 103 DAS. All harvests were carried out on the same plants. After harvest, the fresh and dry weight of pods, as well the fresh and dry weight of seeds, number of seeds per pod, and 100 seeds weight were recorded. The number of seeds per pod and the weight of 100 seeds was evaluated from 10 pods for each plot (30 pods per treatment). Batch samples of pods and seeds were put in deep-freezing conditions, lyophilized, ground with a mortar and pestle, and stored at freezing conditions (−80 °C) until further analyses.

Table 1. Detailed composition and application guides for the tested biostimulants.

Product	Composition	Application Method	Dose
Nomoren (G)	20% of arbuscular mycorrhizal fungi (AMF) (*Glomus* spp.)	Irrigation water	5 L/ha
Twin-Antistress (TW)	Natural microorganisms based on *Bacillus subtilis*, and yeast and *Ascophyllum nodosum* extracts, as well as N (organic): 1%, organic carbon: 10%, and organic matter (<50 kDa): 30%	Irrigation water	5 L/ha
Veramin Ca (V)	Amino acid complex of vegetable origin with *Aloe vera* extract, and CaO: 15.6%	Foliar spraying	500 g/100 L H2O
EKOprop (EK)	Mixture of arbuscular mycorrhizal fungi (*Glomus* spp: 1%), rhizospere symbiotic bacteria (*Bacillus* spp., *Streptomyces* spp., *Pseudomonas* spp.; 1.6×10^9 CFU/g in total), and saprophytic fungi (*Trichoderma* spp.: 5×10^5 CFU/g)	Irrigation water	1 kg/ha

2.2. Chemical Analyses

2.2.1. Nutritional Value

Sample were analyzed in terms of nutritional compounds (moisture, fat, ash, proteins, and carbohydrates) according to the Association of Analytical Communities (AOAC) methods [37]. Moisture was determined by pods and seeds drying at 105 ± 5 °C until constant weight. Crude protein was evaluated by the macro-Kjeldahl method (N × 6.25) using an automatic distillation and titration unit (model Pro-Nitro-A, JP Selecta, Barcelona, Spain), ash content was determined by incineration at 550 ± 15 °C, and the crude fat was determined by extraction with petroleum ether using a Soxhlet apparatus (Behr Labor Technik, Dusseldorf, Germany). Total carbohydrates (g/kg dw) were determined by difference according to the equation: 1000 – (g moisture + g fat + g ash + g proteins), and energy (kcal/kg dw) was determined according to the equation: 4 × (g proteins + g carbohydrates) + 9 × (g fat).

2.2.2. Minerals Composition

Mineral composition analysis was performed in forced-air dried (at 72 °C) and ground to powder pods and seeds, after dry ashing and extraction with 2 N HCl according to the method described by Chrysargyris et al. [38]. Atomic absorption spectrophotometry (PG Instruments AA500FG, Leicestershire, UK) was used for Ca, Mg, Mn, Zn, and Cu content determination, while flame photometry (Lasany Model 1832, Lasany International, Haryana, India) was used for and Na and K content determination. Nitrogen and phosphorus content were determined by Kjeldahl (Digest Automat K-439 and Distillation Kjelflex K-360, BÜCHI, Flawil, Switzerland) and spectrophotometry methods (Multiskan GO, Thermo Fisher Scientific, Waltham, Massachusetts, USA), respectively. The determination of minerals composition was performed only in pods of the second harvest and seeds. Pods of the first harvest were not evaluated due to insufficient amounts of samples for specific treatments, which did not allow a complete set of data. Results are expressed on a dry weight basis.

2.2.3. Tocopherols

Tocopherols were determined in the lyophilized samples by HPLC fluorescence, following a procedure previously described using tocol (Matreya, Pleasant Gap, Pensylvania, USA) as internal standard [39]. Tocopherols standards (α-, β-, γ-, and δ-isoforms, Sigma-Aldirch, St. Louis, MO, USA) were used for compounds identification, and quantification was assessed by the internal standard method. Results were obtained using the Clarity 2.4 software (DataApex, Prague, Czech Republic) and expressed in μg/kg dw and mg/kg dw for pods and seeds, respectively.

2.2.4. Organic Acids

Organic acids were determined in the lyophilized sample and determined by a high-performance liquid chromatography system equipped with a diode array detector (HPLC-DAD), following a procedure previously described [40]. Compounds were identified and quantified by comparison of the retention time, spectra, and peak area recorded at 245 nm and 215 nm (for ascorbic acid and remaining acids, respectively), with those obtained from commercial standards (oxalic, malic, fumaric, and ascorbic acids, Sigma-Aldrich, St. Louis, MO, USA). The results were recorded and processed using LabSolutions Multi LC-PDA software (Shimadzu Corporation, Kyoto, Japan) and were expressed in g/kg dw and mg/kg dw for pods and seeds, respectively.

2.2.5. Free Sugars

Free sugars were determined by HPLC coupled to a refractive index (RI) detector (Knauer, Smartline system 1000, Berlin, Germany) using the internal standard (IS; melezitose). The lyophilized sample was extracted using a methodology previously described [40]. Compounds were identified by comparison with standards (Sigma-Aldrich, St. Louis, MO, USA), and quantification was performed by the IS method. Results were processed using the Clarity 2.4 software (DataApex, Prague, Czech Republic) and expressed in g per kg dw.

2.2.6. Fatty Acids

Fatty acids profile was characterized after a transesterification procedure and according to the method previously described [41]. The analysis was carried out with gas-liquid chromatography with flame ionization detection (GC-FID; DANI1000, Contone, Switzerland). Fatty acids identification and quantification (Clarity DataApex 4.0 Software, Prague, Czech Republic) were performed by comparing the relative retention times of fatty acid methyl ester (FAME) peaks from samples with standards (reference standard mixture 47885-U, Sigma, St. Louis, MO, USA). Results were expressed in the percentage of each fatty acid.

2.3. Statistical Analysis

2.3.1. Experimental Layout and Statistical Treatment

The experimental design was laid out in a split plot arrangement with each main plot consisting of water stress treatments (W+ or W-), while fully randomized sub-plots comprised the biostimulants treatments. Each subplot contained 6 plants and each main plot contained 30 plants. Pod yield components were evaluated in 18 plants for each treatment ($n = 18$), whereas for seed yield, 30 randomly selected pods from each treatment were measured. In order to constitute a representative and adequate sample of the tested treatments, batches of several samples of pods and seeds were taken at random from each plot in order to obtain three different samples. Then, these batches were powdered to obtain homogenous samples. For each methodology, three extractions were carried out, and the analyses were performed in triplicate. Statistical analysis was conducted with the aid of Statgraphics 5.1.plus (Statpoint Technologies, Inc., Warrenton, VA, USA). Data were evaluated by a two-way analysis of variance (two-way ANOVA), and significant interactions of the tested factors (water regime and biostimulant treatment) were observed. Therefore, all the means for each pod harvest and seeds were compared separately by using the Tukey's honestly significant difference (HSD) test ($p = 0.05$).

2.3.2. Linear Discriminant Analysis (LDA)

Linear discriminant analysis (LDA) was applied to evaluate the overall effects of different biostimulants, independently of water level, in each phenological stage (first and second harvest of pods and seeds). The stepwise technique, considering the Wilk's Λ test with the usual probabilities of F

(3.84 to enter and 2.71 to be removed) for variable selection, was employed. With this procedure, it was aimed to estimate the association between the single categorical dependent variables (biostimulant treatments: C, G, TW, V, EK) and the quantitative independent variables (analyzed parameters: proximate composition, organic acids, tocopherols, sugars, fatty acids). In all cases, a leaving-one-out cross-validation procedure was carried out to assess the model performance.

3. Results and Discussion

3.1. Yield and Growth Parameters

The yield and growth characteristics of pods and seeds are presented in Table 2. Yield was positively affected by the application of EK treatment in normally irrigated plants (EKW+) where higher yields compared to the control and the rest of biostimulant treatments were observed (5284 ± 120 kg/ha, 3701 ± 88 kg/ha, and 8985 ± 196 kg/ha, for the first harvest, second harvest, and total yield, respectively). This increase in pod yield was attributed to the higher number of pods harvested from both harvests in plants treated with the specific biostimulants, while the mean pod weight was also the highest in EKW+ treatment only for aggregated results.

On the contrary, the application of V treatment in water-stressed plants (VW-) resulted in the lowest yields for the first and second harvest and consequently in the total yield of green pods (2213 ± 90 kg/ha, 1749 ± 59 kg/ha, and 3962 ± 147 kg/ha for the first harvest, second harvest, and total yield, respectively). Additionally, the number of pods per plant and consequently total yield was higher in normally irrigated plants comparing to water-stressed ones, while GW- and TWW- treatments were the most effective at alleviating the negative effects of stress conditions. Similar results have been reported by Aimo et al. [42], who suggested that the *Crocus sativus* yield was increased after AMF application due to the higher number of flowers.

This was also the case in our study, where the products containing AMF (G) or a mixture of AMF, saprophytic fungi, and rizosphere bacteria (EK) resulted in higher yields in water-stressed and normally irrigated plants, respectively. According to German et al., the inoculation of common bean plants with *Azospirillum brasilense* increased tap root length as well as the proportion of long and thin roots at the early growth stages, which are critical for plant adaptation to water stress conditions [27]. Moreover, Weber et al. reported an increase of fruit setting and total yield in strawberry plants as the result of *Ascophyllum nodosum* extracts application [43].

In the present study, the application of products containing *A. nodosum* extracts had a positive effect on total yield under water stress conditions (TWW-) when compared to non-biostimulant treated plants (CW-), although the product containing AMFs (GW-) was shown to be more effective. According to Arthur et al. [44], biostimulants such as seaweed extracts may contain plant hormones (cytokins and auxins) that induce flower formation. Seaweed extracts (*Ascophyllum nodosum*) have been also reported to have a positive effect on the plant growth of lettuce, carrot, and strawberry through increased hormone activity and K uptake [18,45,46].

Moreover, the increased yield for plants treated with biostimulants is associated with improved plant tolerance to abiotic stress conditions, which according to Battacharyya et al. [47] could be attributed to various protective mechanisms such as the regulation of related genes, the accumulation of osmolytes, the improvement in water-use efficiency, and other effects on the plant rhizosphere. Moreover, Ahmad et al. [10] reported that the inoculation of Indian mustard plants with *Trichoderma harzianum* alleviated osmotic stress effects through the activation of plant antioxidant mechanisms.

Table 2. Yield and growth characteristics of bean plants in relation to water stress and biostimulants application (mean ± SD).

Treatment	1st Harvest			2nd Harvest			Total				
	Number of Pods/Plant	Mean Pod Weight (g)	Yield (kg/ha)	Number of Pods/Plant	Mean Pod Weight (g)	Yield (kg/ha)	Number of Pods/Plant	Mean Pod Weight (g)	Yield (kg/ha)	100 Seeds Weight (g)	Seeds per Pot
CW+ [¥]	24 ± 1c	9 ± 1b	4296 ± 105d	17 ± 4c	9 ± 1b	3138 ± 43c	41 ± 4d	9 ± 1c	7434 ± 120d	101 ± 2f	5.4 ± 0.5a
VW+	26 ± 2b	9.7 ± 0.6a	4933 ± 154b	18 ± 1b	9 ± 1b	3318 ± 89b	44 ± 4b	10 ± 1b	8311 ± 233b	112 ± 3d	4.8 ± 0.9b
EKW+	27 ± 1a	9.7 ± 0.8a	5284 ± 120a	18.6 ± 0.8a	10 ± 1a	3701 ± 88a	46 ± 3a	10 ± 2a	8985 ± 196a	108 ± 2e	5.0 ± 0.8b
GW+	26 ± 2b	9 ± 1b	4621 ± 85.0c	16.4 ± 0.8c	10 ± 2a	3318 ± 81b	42 ± 2c	9 ± 1c	7939 ± 160c	116 ± 2b	5.5 ± 0.9a
TWW+	24 ± 2c	8 ± 1d	3799 ± 109e	17.5 ± 0.9c	10 ± 2a	3357 ± 96b	42 ± 3c	9 ± 1c	7156 ± 187e	116 ± 2b	5.5 ± 0.9a
CW-	14 ± 3f	8.8 ± 0.9c	2456 ± 75h	13 ± 3f	9 ± 2b	2435 ± 77e	27 ± 2h	9 ± 1c	4891 ± 137h	114 ± 3bc	5.0 ± 0.9b
VW-	14 ± 2f	8 ± 1d	2213 ± 90i	12.0 ± 0.6g	7.3 ± 0.8d	1749 ± 59f	26 ± 2i	7.8 ± 0.8f	3962 ± 147j	115 ± 3b	4.4 ± 0.5c
EKW-	16.7 ± 0.8e	8 ± 1d	2515 ± 50h	14.1 ± 0.9e	8 ± 2c	2147 ± 62c	31 ± 2g	7.6 ± 0.7g	4662 ± 80i	122 ± 2a	5.0 ± 0.6b
GW-	20.2 ± 0.9d	8 ± 1d	3172 ± 79f	15 ± 1d	8 ± 2c	2519 ± 81de	36 ± 2e	8 ± 1e	5691 ± 139f	113 ± 3cd	5.1 ± 0.5b
TWW-	16.7 ± 0.8e	8.2 ± 0.8d	2748 ± 64g	15.4 ± 0.8d	8 ± 2c	2603 ± 71d	32 ± 2f	8.3 ± 0.9d	5351 ± 127g	114 ± 2bc	5.0 ± 0.7b

[¥] W+: indicates normal irrigation regime; W-: indicates water-holding irrigation regime; C: Control; V: Veramin Ca; EK: EKOprop; G: Nomoren; TW: Twin-Antistress. Means in the same column followed by different Latin letters are significantly different according to Tukey's honestly significant difference (HSD) test ($p = 0.05$).

Regarding the weight of 100 seeds, the highest value was observed for the EK treatment under water stress conditions (EKW-), while normally irrigated plants with no added biostimulants (CW+) presented the lowest values (122 ± 2 and 101 ± 2 for EKW- and CW+ treatments, respectively) (Table 2). Moreover, the weight of 100 seeds and number of seeds per pod were higher and lower, respectively, for water-stressed plants regardless of biostimulants treatment when compared to the control treatment of normally irrigated plants, indicating that water stress may affect the fertilization process and consequently the number of seeds per pod. The beneficial use of plant-growth promoting rhizobacteria under water stress conditions has been also reported by Sarma and Saikia [24], who suggested that the inoculation of mung bean plants with *Pseudomonas aeruginosa* strains alleviated water stress effects through the scavenging of oxidative enzymes. Moreover, Korir et al. [48] reported a synergistic effect of plant growth promoting bacteria with common bean rhizobia that enhanced plant growth and development, while Farouk and Abdul Qados [49] suggested that folic acid application increased the seed yield and chemical composition of pea plants (*Pisum sativum*).

3.2. Nutritional Value

The nutritional value of the pods and seeds is presented in Table 3. The application of biostimulants did not have a beneficial effect on the moisture content of pods in the first harvest when compared with the control treatment for either normally irrigated (CW+) or water-stressed plants (CW-). Similar trends were observed under prolonged water stress (second harvest), while for normally irrigated plants, the VW+ treatment resulted in the highest moisture content values. For both harvests, the lowest values were recorded for pods harvested from water-stressed plants treated with the V treatment (VW-). These findings could be probably attributed to a functional allocation equilibrium where under biostimulant treatments, plants allocated resources and biomass in fruit; thus, a reduction in moisture content (or similarly an increased dry matter content) was observed [50]. Moreover, plants under stress tend to accumulate minerals and metabolites as a means to maintain high water potential [5]. However, considering that these trends were observed both in water-stressed and normally irrigated plants highlights the need for further investigation. On the other hand, the moisture content of seeds was beneficially affected by the various biostimulants in plants grown under water stress conditions, especially for G treatment (GW-), where the highest values were observed. Regarding the rest of the nutritional value parameters, a varied effect of biostimulants and irrigation treatments was observed in terms of fat, protein, ash, and carbohydrates content and the energetic value of pods and seeds. In particular and for the first harvest of pods, protein, ash content, and energetic value were beneficially affected under the normal irrigation regime, while carbohydrates content was the highest for water-stressed plants with no added biostimulants. Similarly, a beneficial effect of GW+ treatment on protein and ash content was observed for the pods of the second harvest, whereas fat and carbohydrates content were the highest for water-stressed plants that received no biostimulants (CW-) or the G treatment (GW-). The energetic value was the highest for normally irrigated plants that received no biostimulants. Seeds' nutritional value was beneficially affected by the application of G treatment under normal irrigation conditions (GW+) when proteins, fat, and ash content were considered, whereas EK and G treatments increased energetic value under the same irrigation treatments (5468 ± 6 kcal/kg dw and 5474 ± 3 kcal/kg dw, respectively). The highest carbohydrates content values were recorded for normally irrigated plants where no biostimulant or the V treatment was applied (CW+ and VW+, respectively), as well as for water-stressed plants that received the EK treatment (EKW-). Significant differences were also observed between normally irrigated and water-stressed plants in a biostimulant treatment-specific manner, although no direct comparisons between the corresponding treatments were performed due to the presence of significant interactions among the tested factors. Nevertheless, it is worth highlighting the beneficial effects of biostimulants on the protein content of green pods under water stress conditions compared to the corresponding control treatment (CW-). The effect of biostimulants on plant metabolism and the quality of end-products has been previously reported by Colla et al. [51], while Przygocka-Cyna and Grzebisz [52] have associated the use of biostimulants with the improvement in plant nutrient uptake and therefore with the better nutritional value of the end products. Moreover, according to Elsheikh et al., the inoculation of faba bean (*Vicia faba*) plants with arbuscular mycorrhiza increased the protein content in the seeds, regardless of irrigation conditions, suggesting the improved nutritional status of plants as the main reason for this increase [53]. Similarly, Farouk and Abdul Qados [49] suggested that folic acid and hydrogen peroxide application may improve the nutritional value of pea seeds through the increase of protein and carbohydrates content, while Elsheikh and Mohamedzein reported that the inoculation of groundnut with *Glomus* sp. and *Bradyrhizobium* sp. increased the protein content of seeds. Regarding the biostimulatory activity of seaweed extracts, Kocira et al. [54] and Castellanos-Barriga et al. [55] reported contrasting effects of *Ecklonia maxima* and *Ulva lactuca* extracts on the nutritional value of common bean (*P. vulgaris*) and mung bean (*Vigna radiata*) seeds, respectively. These differences in the literature reports highlight the variable biostimulatory effects of seaweed extracts, which contain a wide range of compounds associated with improved plant nutrient uptake, phytohormone-like activities, tolerance to abiotic stressors, and the modulatory effects of plant metabolism and physiology [47].

Table 3. Nutritional (g/kg dw) and energetic value (kcal/kg dw) of the studied pods and seeds of beans in relation to irrigation regime (mean ± SD).

Treatment	Moisture (%)	Fat	Proteins	Ash	Carbohydrates	Energy
			1st Harvest of Pods			
CW+ ¥	96.4 ± 0.7a	48.2 ± 0.8e	165 ± 6h	59.6 ± 0.4h	728 ± 3b	4630 ± 20f
VW+	94.2 ± 0.4c	45.6 ± 0.6g	201 ± 1a	79 ± 1a	674 ± 2i	4821 ± 3a
EKW+	95.0 ± 0.6b	48 ± 2e	197 ± 4b	73.7 ± 0.2b	681 ± 1h	4790 ± 20b
GW+	92.5 ± 0.7de	53.7 ± 0.2b	174.5 ± 0.1e	71.6 ± 0.5c	700 ± 1f	4658 ± 1e
TWW+	93 ± 1d	50.1 ± 0.5c	193 ± 3c	70 ± 3d	688 ± 1g	4760 ± 10c
CW-	93 ± 2d	49.4 ± 0.8cd	150 ± 02i	66 ± 2f	735 ± 2a	4553 ± 8h
VW-	88 ± 1g	48.7 ± 0.8de	181.5 ± 0.5d	67 ± 1e	703 ± 2e	4713 ± 1d
EKW-	93 ± 1d	45.5 ± 0.3g	169.5 ± 0.4g	61 ± 1g	724 ± 1c	4665 ± 1e
GW-	89.4 ± 0.8f	47 ± 3f	171.6 ± 0.3f	66 ± 2f	715 ± 1d	4671 ± 9e
TWW-	92 ± 1e	67 ± 1a	173.2 ± 0.1e	69.5 ± 0.1d	690 ± 1g	4599 ± 3g
			2nd Harvest of pods			
CW+	91 ± 1b	27.1 ± 0.2e	204.6 ± 0.3b	49.6 ± 0.7cd	719 ± 1f	4915 ± 2a
VW+	92 ± 2a	32 ± 2d	174 ± 1e	56.9 ± 0.8b	737 ± 1e	4739 ± 2f
EKW+	91.1 ± 0.5b	37.1 ± 0.2b	195.9 ± 0.3c	51 ± 1c	716 ± 1f	4831 ± 2d
GW+	90.7 ± 0.4c	34 ± 2c	207.1 ± 0.1a	68.4 ± 0.5a	690 ± 1g	4898 ± 5b
TWW+	90 ± 2d	27.3 ± 0.2e	175 ± 1e	49 ± 2d	748 ± 1d	4768 ± 2e
CW-	91.1 ± 0.5b	43 ± 2a	150.2 ± 0.1h	49 ± 1d	758 ± 1c	4580 ± 6g
VW-	87.6 ± 0.8f	32 ± 2d	174.3 ± 0.2e	52 ± 2c	741 ± 3e	4742 ± 4f
EKW-	89 ± 1e	21.4 ± 0.7f	189.2 ± 0.4d	45 ± 1e	745 ± 1d	4861 ± 1c
GW-	90 ± 2d	18 ± 1h	164.0 ± 0.4g	49 ± 2d	769 ± 1a	4747 ± 5f
TWW-	90 ± 1d	19.8 ± 0.8g	171 ± 1f	45 ± 2e	764 ± 1b	4776 ± 2e
			Seeds			
CW+	69.8 ± 0.3a	40.0 ± 0.1c	305 ± 6e	51.4 ± 0.5cd	604 ± 4a	5360 ± 20g
VW+	66.7 ± 0.7d	37 ± 1d	310.2 ± 0.2d	53 ± 2b	600 ± 2a	5403 ± 5f
EKW+	67.7 ± 0.9b	33.2 ± 0.1g	320 ± 2b	53 ± 2b	593 ± 3c	5468 ± 6ab
GW+	67 ± 2c	52.7 ± 0.1a	337 ± 1a	56 ± 1a	554 ± 1e	5474 ± 3a
TWW+	66.3 ± 0.7e	45 ± 3b	317.6 ± 0.1c	49.7 ± 0.1f	588 ± 2d	5410 ± 7e
CW-	62 ± 1g	40.2 ± 0.1c	322.7 ± 0.1b	51 ± 1de	587 ± 1d	5453 ± 1c
VW-	65.8 ± 0.9f	35 ± 2e	320.2 ± 0.1b	51.7 ± 0.3c	593 ± 2c	5461 ± 6b
EKW-	67.3 ± 0.6c	34 ± 1f	317 ± 1c	48.1 ± 0.8g	601 ± 1a	5448 ± 6c
GW-	69.8 ± 0.5a	34 ± 1f	317.1 ± 0.4c	50.9 ± 0.4e	598 ± 1b	5450 ± 2c
TWW-	66.2 ± 0.4e	35.1 ± 0.6e	316.5 ± 0.4c	50 ± 1f	598 ± 1b	5442 ± 3d

¥ W+: indicates normal irrigation regime; W-: indicates water-holding irrigation regime; C: Control; V: Veramin Ca; EK: EKOprop; G: Nomoren; TW: Twin-Antistress. Means in the same column and the same harvest (first and second harvest of pods and seeds) followed by different Latin letters are significantly different according to Tukey's honestly significant difference (HSD) test (p = 0.05).

3.3. Mineral Composition

The mineral composition of pods and seeds is presented in Table 4. The combination of biostimulants application and irrigation regime had a varied effect on mineral content of pods and seeds with no specific trends being observed. In particular, the application of V treatment under water stress conditions (VW-) increased the nitrogen content of pods without being significantly different from normally irrigated control plants (CW+).

Similarly, the highest values of nitrogen content in seeds were recorded for GW- and TWW-. Positive effects were observed for K content in the pods of normally irrigated plants that received G treatment (GW+), whereas no significant differences were observed in the K content of seeds for the tested treatments. EKW+ and GW+ resulted in the highest values for P content in pods and seeds, respectively, whereas contrasting effects of the irrigation regime × TW treatment combination were observed on Ca content where TWW- and TWW+ increased its content in pods and seed, respectively.

The Na content of pods was the highest when EK treatment was applied regardless of the irrigation regime, while similar results were observed in seeds for the control and V treatment. The Mg content of pods and seeds increased when no biostimulants or the V treatment were applied on water-stressed plants, respectively. Moreover, the Cu content of pods was beneficially affected by the EK treatment, regardless of the irrigation regime, whereas for seeds, the highest values were recorded for the TWW+ treatment.

Regarding the Zn content, the highest values in pods and seeds were observed for the EKW+ and VW- treatments, respectively. Finally, the Mn content of pods increased when V and G treatments were applied, regardless of the irrigation regime, whereas for seeds, the application of biostimulants had a negative effect on Mn content when compared to control treatments (CW- and CW+).

The impact of biostimulants on the nutrient content of agricultural products could be attributed to the fact that they usually contain various minerals in their composition [5]. Mineral uptake from plants may help in maintaining high stomatal conductance and leaf water potential; therefore, biostimulants may improve the nutritional status of plants and induce tolerance to abiotic stress factors such as drought stress [56].

In addition, according to Chrysargyris et al. [18], the application of *Ascophyllum nodosum* seaweed extracts alleviated the negative effects of K deficiency on lettuce plants, while the beneficial effect of AMFs as biostimulants has been associated with higher P uptake from plants [57]. This was the case in our study under normal irrigation conditions where the P content of pods was the highest for the EK treatment.

The results from the study of Colla et al. [58] confirm the beneficial effect of biostimulant application on the Ca content of tomato fruit for the seaweed extracts treatment, which was also observed in our study for TW treatment, regardless of the irrigation regime. Moreover, it has been reported in several studies that the inoculation of plants with mixtures of bacteria has better results in nutrient mobilization and uptake compared to inoculation with a single bacterium [26], which was also the case in our study where EK treatment increased the P, Na, Cu, and Zn content in the pods of normally irrigated plants.

Table 4. Mineral composition of the studied pods (second harvest) and seeds of beans in relation to the irrigation regime, expressed on a dry weight basis (mean ± SD).

Treatment	N (g/kg)	K (g/kg)	P (g/kg)	Na (g/kg)	Ca (g/kg)	Mg (mg/kg)	Cu (mg/kg)	Zn (mg/kg)	Mn (mg/kg)
					Pods				
CW+ [¥]	26.9 ± 0.3a	17.7 ± 0.3c	6.3 ± 0.4b	0.05 ± 0.03c	6.6 ± 1.6e	104 ± 5h	138 ± 17c	24 ± 1d	4.1 ± 0.4b
VW+	24.6 ± 0.2b	18.7 ± 0.4b	4.2 ± 0.4e	0.07 ± 0.06c	4.3 ± 0.9h	235 ± 17c	78 ± 18e	29 ± 2c	4.5 ± 0.7a
EKW+	24.6 ± 0.3b	19.1 ± 0.3b	7.3 ± 0.2a	0.16 ± 0.02a	5.4 ± 0.7g	131 ± 21g	205 ± 18a	38 ± 4a	3.8 ± 0.4c
GW+	23.0 ± 0.5cd	20.0 ± 0.4a	6.2 ± 0.4b	0.07 ± 0.03c	6.1 ± 1f	219 ± 46d	170 ± 15b	21 ± 1e	4.5 ± 0.1a
TWW+	23.2 ± 0.8c	19.0 ± 0.5b	4.2 ± 0.4e	0.17 ± 0.03a	10.7 ± 3.2c	314 ± 6b	162 ± 19b	33 ± 3b	4.0 ± 0.5bc
CW-	22.4 ± 0.3e	16.7 ± 0.3d	5.2 ± 0.7d	0.03 ± 0.01d	4 ± 1.3h	341 ± 16a	98 ± 18d	29 ± 2c	3.8 ± 0.2c
VW-	26.5 ± 0.5a	15.1 ± 0.2f	6.0 ± 0.2c	0.14 ± 0.02b	11.5 ± 3.5b	171 ± 12e	46 ± 16f	28 ± 2c	4.2 ± 0.8ab
EKW-	23.2 ± 0.2c	17.9 ± 0.6c	5.5 ± 0.1d	0.16 ± 0.08a	8.5 ± 2.6d	147 ± 8f	200 ± 27a	28 ± 3c	3.8 ± 0.5c
GW-	21.5 ± 0.1f	17.1 ± 0.8d	5.3 ± 0.7d	0.07 ± 0.01c	10.2 ± 2.3c	311 ± 14b	169 ± 21b	21.8 ± 0.3e	4.4 ± 0.3a
TWW-	22.8 ± 0.5d	16.1 ± 0.3e	5.9 ± 0.2c	0.06 ± 0.01c	13.2 ± 4.7a	311 ± 35b	145 ± 12c	20 ± 3e	4.1 ± 0.2b
					Seeds				
CW+	42.9 ± 0.3b	1.6 ± 0.2a	13.3 ± 1.4b	0.18 ± 0.08a	11.1 ± 2.3e	7 ± 1f	158 ± 45c	13 ± 2g	5 ± 1a
VW+	41.3 ± 0.4d	1.6 ± 0.3a	9.4 ± 1.8g	0.16 ± 0.05a	6.8 ± 1.7f	11 ± 3cd	99 ± 38e	21 ± 1b	0.5 ± 0.2g
EKW+	41.9 ± 0.8c	1.6 ± 0.2a	12.7 ± 2.3c	0.05 ± 0.01b	10.9 ± 1.4e	6 ± 3fg	193 ± 11b	13 ± 2g	4.4 ± 0.9cd
GW+	41.9 ± 0.1c	1.7 ± 0.3a	14.0 ± 6.6a	0.05 ± 0.01b	12.6 ± 2.5c	5 ± 4g	165 ± 15c	18 ± 3cd	3.5 ± 0.4e
TWW+	41.9 ± 0.1c	1.6 ± 0.3a	8.0 ± 0.9h	0.03 ± 0.01b	15.2 ± 1.5a	13 ± 1b	199 ± 39a	17 ± 3de	4.6 ± 0.4bc
CW-	39.5 ± 0.3e	1.6 ± 0.4a	10.0 ± 2.5f	0.15 ± 0.04a	10.9 ± 2.4e	10.0 ± 0.6de	120 ± 50d	13 ± 1g	5.1 ± 0.3a
VW-	42.8 ± 0.5b	1.6 ± 0.1a	7.2 ± 0.5i	0.15 ± 0.06a	3.7 ± 1.3g	15 ± 3a	90 ± 8f	22.6 ± 0.3a	2 ± 1f
EKW-	42.2 ± 0.8c	1.6 ± 0.9a	11.9 ± 2.6d	0.03 ± 0.01b	11.7 ± 1.7d	2.9 ± 0.3h	190 ± 49b	16 ± 2ef	4.2 ± 0.3d
GW-	44.7 ± 0.2a	1.6 ± 0.6a	11.0 ± 1.2e	0.04 ± 0.01b	14.1 ± 1.3b	9.2 ± 0.6e	168 ± 30c	19.2 ± 0.9c	4.2 ± 0.4d
TWW-	44.7 ± 0.2a	1.6 ± 0.1a	7.5 ± 1.2hi	0.04 ± 0.01b	12.3 ± 3.7c	12 ± 2bc	160 ± 43c	15.1 ± 0.2f	4.8 ± 0.5b

[¥] W+: indicates normal irrigation regime; W-: indicates water-holding irrigation regime; C: Control; V: Veramin Ca; EK: EKOprop; G: Nomoren; TW: Twin-Antistress. Means in the same column and the same plant part (pods and seeds) followed by different Latin letters are significantly different according to Tukey's honestly significant difference (HSD) test ($p = 0.05$).

3.4. Tocopherols

The main detected tocopherols in pods were γ-tocopherol, followed by α-tocopherol, while seeds contained mainly γ-tocopherol and less amounts of δ- and α-tocopherol (Table 5). The application of TW treatment resulted in a significant increase of γ-tocopherol (104% and 18.3% for the TWW+ and TWW- treatments, respectively) and total tocopherols (82.3% and 19.4% for the TWW+ and TWW- treatments, respectively) in the pods of the first harvest compared to the control treatments (CW+ and CW-). In contrast, the application of G treatment had a negative effect on the γ-tocopherol content in pods of the first harvest under water stress conditions (reduced by 45.7%), while EK treatment resulted in the lowest content of α-tocopherol for the same irrigation regime (1800 ± 40 μg/kg dw). In the second harvest, the highest values for γ- and total tocopherols were recorded for the VW- treatment (3500 ± 20 μg/kg dw and 5250 ± 10 μg/kg dw, respectively), while pods from the control treatment (CW-) contained the highest amounts of α-tocopherol (1810 ± 20 μg/kg dw). In addition, drought stress increased the individual and total tocopherols content for all the biostimulant treatments, except for G treatment, where normally irrigated plants had a higher content of tocopherols compared to water-stressed ones. The observed increase of tocopherols in pods under prolonged water stress conditions could be attributed to the induction of self-defense mechanisms by biostimulants application and the production of antioxidant compounds such as tocopherols [59]. However, the variable effects of the tested biostimulants indicate a diverse plants x biostimulant interaction, as well as the induction of different mechanisms in each combination depending on the biostimulant composition and the severity of stress [60]. Therefore, although in the first harvest TWW- treatment induced tocopherols biosynthesis as a non-enzymatic antioxidant mechanism, under prolonged stress conditions, VW- (second harvest) and VW+ (seeds) treatments were beneficial to tocopherols content. Regarding seeds, the presence of α- and γ-tocopherol has been previously reported in common bean seeds by Kan et al. [61]. In our study, γ-, δ-, and the total tocopherols content was the highest under normal irrigation conditions and for those plants that did not receive biostimulants or the V treatment was applied. In contrast, EK treatment had a negative effect on tocopherols content under water stress conditions. According to the literature, Ca and amino acids supplementation (as in the case of V treatment in our study) may induce the biosynthesis of non-enzymatic antioxidants such as tocopherols and increase tolerance against drought stress [62–65].

Table 5. Composition in tocopherols of the studied pods (μg/kg dw) and seeds (mg/kg dw) of beans in relation to the irrigation regime (mean ± SD).

Treatment	α-Tocopherol	γ-Tocopherol	Total Tocopherols
		1st Harvest of Pods	
CW+ [¥]	770 ± 20f	2970 ± 20i	3740 ± 10i
VW+	880 ± 40c	4880 ± 30f	5760 ± 10f
EKW+	930 ± 30b	5230 ± 50d	6160 ± 20d
GW+	306 ± 3g	4020 ± 60g	4330 ± 50g
TWW+	760 ± 20f	6060 ± 90b	6820 ± 70b
CW-	798 ± 3e	5416 ± 1c	6214 ± 3c
VW-	810 ± 30d	4990 ± 50e	5800 ± 90e
EKW-	278 ± 3i	3910 ± 30h	4190 ± 20h
GW-	290 ± 20h	2940 ± 60j	3240 ± 80j
TWW-	1010 ± 10a	6410 ± 40a	7420 ± 30a

Table 5. *Cont.*

Treatment	α-Tocopherol	γ-Tocopherol		Total Tocopherols
	1st Harvest of Pods			
	2nd Harvest of pods			
CW+	288 ± 6g	1560 ± 60i		1840 ± 60i
VW+	236 ± 6h	1670 ± 10h		1900 ± 10h
EKW+	266 ± 3g	1530 ± 30j		1800 ± 40j
GW+	508 ± 6e	3280 ± 20b		3790 ± 30d
TWW+	160 ± 10i	1940 ± 40g		2100 ± 30g
CW-	1810 ± 20a	2820 ± 30d		4630 ± 50b
VW-	1750 ± 20b	3500 ± 20a		5250 ± 10a
EKW-	720 ± 20d	2440 ± 20e		3160 ± 10e
GW-	326 ± 8f	2000 ± 40f		2320 ± 50f
TWW-	1210 ± 10c	3090 ± 20c		4300 ± 40c
	Seeds			
	α-Tocopherol	γ-Tocopherol	δ-Tocopherol	Total Tocopherols
CW+	0.96 ± 0.01a	39.1 ± 0.1b	2.48 ± 0.03a	42.5 ± 0.1a
VW+	0.52 ± 0.01c	39.8 ± 0.1a	2.28 ± 0.01b	42.7 ± 0.1a
EKW+	0.50 ± 0.01c	38.5 ± 0.1c	1.95 ± 0.02f	40.9 ± 0.1c
GW+	0.52 ± 0.01c	36.7 ± 0.1d	2.21 ± 0.03c	39.5 ± 0.1d
TWW+	0.53 ± 0.02c	38.4 ± 0.1c	2.01 ± 0.03e	41.0 ± 0.1c
CW-	0.46 ± 0.01d	36.1 ± 0.1e	2.16 ± 0.05d	38.7 ± 0.1e
VW-	0.57 ± 0.01b	35.4 ± 0.1f	1.91 ± 0.02f	38.0 ± 0.1f
EKW-	0.96 ± 0.05a	32.0 ± 0.1h	1.48 ± 0.02g	34.5 ± 0.1h
GW-	0.59 ± 0.02b	39.0 ± 0.1b	1.91 ± 0.02f	41.5 ± 0.1b
TWW-	0.45 ± 0.02d	34.7 ± 0.1g	2.14 ± 0.08d	37.3 ± 0.1g

¥ W+: indicates normal irrigation regime; W-: indicates water-holding irrigation regime; C: Control; V: Veramin Ca; EK: EKOprop; G: Nomoren; TW: Twin-Antistress. Means in the same column and the same harvest (1st and 2nd pod harvests and seeds) followed by different Latin letters are significantly different according to Tukey's honestly significant difference (HSD) test ($p = 0.05$).

3.5. Organic Acids

The main detected organic acids in pods were malic and oxalic acid, while ascorbic acid was detected in less amounts in specific treatments of the first harvest (Table 6). On the other hand, malic and oxalic acid were the main organic acids detected in seeds, followed by ascorbic acid and traces of fumaric acid. The application of the VW+ treatment resulted in the highest content of oxalic and malic acid and total organic acids in pods of the first harvest (26.3 ± 0.1 g/kg dw, 23.1 ± 0.1 g/kg/dw, and 49.4 ± 0.1 g/kg/dw, respectively). Similarly, in the second harvest, the highest content of oxalic and malic acid was recorded for the GW+ and VW- treatments (22.7 ± 0.2 g/kg/dw and 24.6 ± 0.2 g/kg/dw, respectively), while total organic acids content was most abundant in the GW+ treatment (46.9 ± 0.1 g/kg/dw). Regarding seeds, VW- treatment resulted in the highest content of oxalic and malic acid, and total organic acids (752 ± 1 mg/kg/dw, 1440 ± 40 mg/kg/dw, and 2780 ± 30mg/kg/dw, respectively), whereas the highest amounts of ascorbic acid were detected in GW+ treatment (715 ± 4 mg/kg dw). Considering the antinutritional properties of oxalic acid, it is worth mentioning that EKW- treatment resulted in the lowest content for both pod harvests when compared with the rest of the treatments where biostimulants were applied, although in all the cases, the oxalic acid content was considerably low. On the other hand, in the case of seeds, the application of EKW+ and TWW+ treatments significantly reduced the oxalic acid content compared to the control and the rest of the biostimulant treatments. Although there are reports in the literature that suggest that organic acids increase under stress conditions, according to Zushi and Matsuzoe [66], this increase could be attributed only to a concentration effect due to the increase in dry matter content under stress conditions. According to other studies, the composition of biostimulants may significantly affect the organic acids composition, especially those biostimulant products that contain microorganisms such as Twin-Antistress and EKOprop in our study [36,67]. However, although the total organic acids content of pods harvested from water-stressed plants was in general higher in biostimulant-treated plants compared to the control treatment (CW-), the application of the EKW- treatment resulted in a significant reduction of organic acids content. This finding is reflected to the reduced total pod yield for this treatment (see Table 2), suggesting a non-effective alleviating mechanism against water stress related to biostimulant product composition. Moreover, the effect of V treatment on the oxalic acid content of seeds under water stress conditions (VW-) could be attributed to Ca addition, which is associated with calcium oxalate formation for the removal of excessive calcium or oxalic acid [67].

3.6. Free Sugars

The sugars composition of pods and seeds is presented in Table 6. The main detected sugar in the pods of both harvests was fructose, followed by glucose and sucrose, whereas in seeds, only sucrose was detected. Similarly with our study, Kan et al. [61] detected sucrose as the main sugar in common bean seeds, while they also detected the presence of glucose; this difference could be attributed to the different harvesting stages (fully dried seeds comparing to fully developed green seeds in our study), which may affect hydrolysis and the transformation of sugars after harvest [68].

In the first harvest, the highest content of individual and total sugars were recorded in TWW- (fructose: 198 ± 1 g /kg dw), CW- (glucose: 135 ± 5 g /kg dw; total sugars: 333 ± 7 g /kg dw), and VW- (sucrose: 7.6 ± 1 g /kg dw) treatments. Under prolonged water stress (second harvest), the application of G treatment (GW-) resulted in the highest content of fructose, glucose, and total sugars (232 ± 7 g /kg dw, 140 ± 7 g /kg dw, and 380 ± 10 g/kg dw, respectively), while the sucrose content was the highest for the TWW- treatment (15.3 ± 0.8 g/kg/dw). Similarly, the highest content of sucrose in seeds was recorded for the GW-treatment (22.9 ± 0.7 g/kg dw).

The low levels of sucrose in pods could be attributed to the inhibitory activity of hexose sugars (fructose and glucose) to sucrose synthase activity [68]. Moreover, for most of the tested biostimulants and control treatments, the total and individual sugars content was higher in water-stressed plants than normally irrigated plants, especially for G treatment—that resulted in the highest total sugars content, which could be associated with osmoprotective effects against water stress [69].

Considering the involvement of soluble sugars in plant defense mechanisms as well as in the regulation of stress and growth-related genes, the findings of our study suggest an efficient defense mechanism against water stress for the AMF-containing biostimulant product (G treatment), as already justified by the increased pods yield under water stress conditions for the same treatment (see Table 2).

According to the literature, inoculation with AMF is associated with increased soluble sugars content in *Ipomea batatas* and *Vigna subterranea* under drought stress, since sugars may serve as organic carbon pools to be used for photosynthates and biomass production [70]. Apart from the osmoregulatory role of sugars in plant defense mechanisms against stress, sucrose content is also related with secondary metabolites biosynthesis, which may also contribute to the overall non-enzymatic tolerance of plants under stress [71].

Table 6. Composition in organic acids and sugars of the studied pods and seeds of beans in relation to the irrigation regime (mean ± SD).

Treatment	Oxalic Acid (g/kg dw)	Malic Acid (g/kg dw)	Ascorbic Acid (g/kg dw)	Total Organic Acids (g/kg dw)	Fructose (g/kg dw)	Glucose (g/kg dw)	Sucrose (g/kg dw)	Total Sugars (g/kg dw)
				1st Harvest of pods				
CW+ ¥	14.0 ± 0.1h	18.4 ± 0.2d	tr	32.5 ± 0.1h	151 ± 4f	84.5 ± 0.1g	5.8 ± 0.4b	241 ± 4h
VW+	26.3 ± 0.1a	23.1 ± 0.1a	tr	49.4 ± 0.1a	168 ± 5e	90 ± 2f	nd	257 ± 7f
EKW+	21.8 ± 0.2c	20.4 ± 0.5c	0.10 ± 0.01b	42.4 ± 0.4b	194 ± 3b	107 ± 3c	nd	301 ± 6c
GW+	19.3 ± 0.1e	21.7 ± 0.4b	tr	41.0 ± 0.3d	173 ± 0d	85 ± 3g	4.7 ± 0.3d	263 ± 4e
TWW+	23.2 ± 0.1b	18.1 ± 0.2de	0.5 ± 0.1a	41.8 ± 0.3c	173 ± 1d	101 ± 2d	3.5 ± 0.1f	277 ± 2d
CW-	18.5 ± 0.1f	16.5 ± 0.2f	0.50 ± 0.03a	35.5 ± 0.1f	194 ± 2b	135 ± 5a	4.3 ± 0.1e	333 ± 7a
VW-	17.0 ± 0.1g	20.6 ± 0.4c	0.10 ± 0.01b	37.7 ± 0.4e	153 ± 8f	84 ± 6g	7.6 ± 0.1a	250 ± 10g
EKW-	16.4 ± 0.1g	17.9 ± 0.3e	tr	34.3 ± 0.3g	171 ± 4d	106 ± 3c	3.1 ± 0.3g	280 ± 8d
GW-	20.9 ± 0.1d	16.7 ± 0.1f	tr	37.7 ± 0.2e	184 ± 2c	94 ± 6e	3.1 ± 0.3g	281 ± 8d
TWW-	21.2 ± 0.1cd	20.4 ± 0.3c	0.50 ± 0.01a	42.1 ± 0.4bc	198 ± 1a	110 ± 3b	5.2 ± 0.4c	314 ± 3b
				2nd Harvest of pods				
CW+ ¥	10.9 ± 0.1d	15.0 ± 0.4h	tr	25.9 ± 0.4h	170 ± 2e	82 ± 4g	3.5 ± 0.4e	256 ± 6h
VW+	10.7 ± 0.1e	17.8 ± 0.1d	tr	28.5 ± 0.1f	180 ± 2d	91 ± 2f	2.4 ± 0.2f	273 ± 4f
EKW+	10.5 ± 0.1f	17.4 ± 0.1e	tr	28.0 ± 0.1g	181 ± 4d	101 ± 3d	4.3 ± 0.2d	287 ± 6d
GW+	22.7 ± 0.2a	24.1 ± 0.3b	tr	46.9 ± 0.1a	195 ± 1b	105 ± 1c	1.76 ± 0.01g	301 ± 1c
TWW+	8.9 ± 0.2h	16.4 ± 0.3g	tr	25.3 ± 0.4j	170 ± 1e	106 ± 3c	3.5 ± 0.7e	280 ± 3e
CW-	10.9 ± 0.2d	20.6 ± 0.3c	tr	31.5 ± 0.1d	180 ± 5d	96 ± 2e	14 ± 2b	290 ± 9d
VW-	9.9 ± 0.2g	24.6 ± 0.2a	tr	34.5 ± 0.4c	162 ± 6f	90 ± 3f	13.1 ± 0.5c	265 ± 9g
EKW-	8.3 ± 0.1i	17.1 ± 0.2f	tr	25.5 ± 0.3i	191 ± 3c	110 ± 10b	13.3 ± 0.9c	318 ± 14b
GW-	16.3 ± 0.1b	20.4 ± 0.2c	tr	36.8 ± 0.2b	232 ± 7a	140 ± 7a	4.2 ± 0.4d	380 ± 10a
TWW-	13.2 ± 0.1c	17.3 ± 0.3e	tr	30.6 ± 0.4e	163 ± 4f	77 ± 3h	15.3 ± 0.8a	256 ± 8h

Table 6. *Cont.*

	Seeds						
	Oxalic acid (mg/kg dw)	Malic acid (mg/kg dw)	Ascorbic acid (mg/kg dw)	Fumaric acid (mg/kg dw)	Total organic acids (mg/kg dw)	Sucrose (g/kg dw)	Total sugars (g/kg dw)
CW+ [¥]	224 ± 3g	1170 ± 20c	657 ± 6b	tr	2050 ± 30d	19.5 ± 0.3c	19.5 ± 0.3c
VW+	293 ± 1f	1383 ± 7b	612 ± 3c	tr	2290 ± 10b	19.3 ± 0.9c	19.3 ± 0.9c
EKW+	185 ± 8h	970 ± 30d	524 ± 2e	tr	1680 ± 40f	22.3 ± 0.5b	22.3 ± 0.5b
GW+	600 ± 10b	910 ± 60e	715 ± 4a	tr	2220 ± 60c	16.6 ± 0.5f	16.6 ± 0.5f
TWW+	185 ± 3h	316 ± 3h	488 ± 9g	tr	990 ± 20i	18.9 ± 0.4d	18.9 ± 0.4d
CW-	530 ± 10d	tr	530 ± 10e	tr	1040 ± 40h	19.2 ± 0.6c	19.2 ± 0.6c
VW-	752 ± 1a	1440 ± 40a	580 ± 10d	tr	2780 ± 30a	15.5 ± 0.7g	15.5 ± 0.7g
EKW-	579 ± 8c	690 ± 30f	515 ± 7f	tr	1780 ± 40e	17.1 ± 0.5e	17.1 ± 0.5e
GW-	530 ± 10d	tr	396 ± 3i	tr	964 ± 1j	22.9 ± 0.7a	22.9 ± 0.7a
TWW-	440 ± 10e	490 ± 20g	449 ± 9h	tr	1380 ± 20g	15 ± 1h	15 ± 1h

[¥] W+: indicates normal irrigation regime; W-: indicates water-holding irrigation regime; C: Control; V: Veramin Ca; EK: EKOprop; G: Nomoren; TW: Twin-Antistress. Means in the same column and the same harvest (1st and 2nd pod harvests and seeds) followed by different Latin letters are significantly different according to Tukey's honestly significant difference (HSD) test ($p = 0.05$). Tr: traces.

3.7. Fatty Acids

The main fatty acids composition is presented in Table 7. Seventeen individual fatty acids were detected in pods and seeds regardless of the irrigation treatment and harvest (data not shown). Pods were abundant in α-linolenic (C18:3n3), linoleic (C18:2n6c), and palmitic acid (C16:0) followed by stearic (C18:0), oleic (C18:1n9c), behenic (C22:0), and lignoceric acid (C24:0), which were detected in lower amounts.

Similarly, in seeds, the most abundant fatty acids were α-linolenic, linoleic, and palmitic acid, followed by stearic and oleic acid. In the first harvest of pods, GW- and EKW- treatments had a beneficial effect on palmitic and linoleic acid content in water-stressed plants (25.6 ± 0.3% and 35.02 ± 0.01%, respectively), whereas in normally irrigated plants, TWW+ treatment resulted in the highest content of α-linolenic acid (42.76 ± 0.06%).

In the second harvest, fatty acids composition showed a varied response, with control and TW treatment resulting in the highest content of linoleic and palmitic acid for normally irrigated plants (43.3 ± 0.1% and 28.90 ± 0.06, respectively), whereas α-linolenic acid content was the highest for the GW- treatment (34.33 ± 0.09%). For seeds, the highest amounts of α-linolenic, linoleic and palmitic acid were recorded in the treatments of GW-, CW+, and VW+ (59.08 ± 0.02%, 29.54 ± 0.02%, and 11.86 ± 0.03%, respectively).

Polyunsaturated fatty acids (PUFA) were the most abundant fatty acids class, followed by saturated (SFA) and monounsaturated fatty acids (MUFA) in both seeds and pods due to the high amounts of α-linolenic and linoleic acids.

Overall, the ratios of PUFA/SFA and n-6/n-3 fatty acids were higher than 0.45 and lower than 4.0 for all the tested treatments, respectively, which according to Petropoulos et al. [72] is indicative for the good nutritional value of a food product.

Moreover, the increase of PUFAs under water stress conditions comparing to the control treatment (CW-) for all the biostimulant treatments except for GW-(first harvest) and EKW- (second harvest) treatments indicates the stimulation of plant antioxidant mechanisms which effectively quenched the reactive oxygen species (ROS) that appear after stress initiation and induce lipid peroxidation and decrease fatty acids content [10,73].

Fatty acids may also serve as organic carbon pools to be used for photosynthates and biomass production [70]. Therefore, considering that inoculation with AMFs and bacteria induces synergistic effects between plants and symbionts that may improve plant nutrient and water uptake, this could be the reason for the increased content of fatty acids.

Table 7. The main fatty acids composition (%) of the studied pods and seeds of common bean (mean ± SD).

	CW+ [¥]	VW+	EKW+	GW+	TWW+	CW-	VW-	EKW-	GW-	TWW-
					1st Harvest of pods					
C16:0	21.72 ± 0.04d	20.87 ± 0.07f	19.35 ± 0.05i	24.1 ± 0.1b	18.4 ± 0.1j	23.55 ± 0.06c	19.9 ± 0.1h	20.26 ± 0.05g	25.6 ± 0.3a	21.18 ± 0.09e
C18:0	5.30 ± 0.01c	4.87 ± 0.01d	3.95 ± 0.01g	9.81 ± 0.02a	3.65 ± 0.01h	5.35 ± 0.03c	4.23 ± 0.02f	4.27 ± 0.01f	5.59 ± 0.03b	4.48 ± 0.01e
C18:1n9c	1.75 ± 0.02d	1.34 ± 0.01g	1.51 ± 0.01f	2.08 ± 0.01b	1.60 ± 0.01e	1.79 ± 0.01d	1.72 ± 0.01d	2.18 ± 0.01a	1.93 ± 0.01c	1.53 ± 0.01f
C18:2n6c	27.60 ± 0.03f	28.69 ± 0.04d	28.65 ± 0.01d	26.98 ± 0.01g	29.94 ± 0.03b	28.52 ± 0.05d	28.60 ± 0.01d	32.02 ± 0.01a	27.74 ± 0.08e	29.33 ± 0.04c
C18:3n3	38.50 ± 0.01f	39.53 ± 0.04d	42.21 ± 0.04b	32.62 ± 0.01j	42.76 ± 0.06a	35.51 ± 0.02h	41.36 ± 0.06c	36.83 ± 0.05g	33.1 ± 0.1i	38.7 ± 0.1e
C22:0	1.27 ± 0.03b	0.96 ± 0.09d	0.82 ± 0.01f	1.06 ± 0.06c	0.77 ± 0.01g	1.47 ± 0.05a	0.86 ± 0.03e	0.88 ± 0.05e	1.46 ± 0.05a	0.99 ± 0.02d
C24:0	1.09 ± 0.02f	1.29 ± 0.02c	1.15 ± 0.01e	1.03 ± 0.02g	1.04 ± 0.01g	1.32 ± 0.04c	1.24 ± 0.02d	1.12 ± 0.02ef	1.56 ± 0.02a	1.39 ± 0.01b
SFA	31.06 ± 0.03d	29.33 ± 0.02e	26.80 ± 0.07h	37.64 ± 0.03a	25.02 ± 0.07h	33.35 ± 0.08c	27.62 ± 0.06g	27.87 ± 0.06f	35.9 ± 0.2b	29.42 ± 0.06e
MUFA	2.75 ± 0.06c	2.30 ± 0.05f	2.18 ± 0.02g	2.62 ± 0.01d	2.17 ± 0.01g	2.43 ± 0.01e	2.26 ± 0.02f	3.12 ± 0.01a	2.99 ± 0.02b	2.25 ± 0.01f
PUFA	66.19 ± 0.03f	68.37 ± 0.07e	71.02 ± 0.05b	59.74 ± 0.02i	72.81 ± 0.08a	64.22 ± 0.09g	70.12 ± 0.07c	69.01 ± 0.07d	61.1 ± 0.2h	68.33 ± 0.06e
					2nd Harvest of pods					
C16:0	18.8 ± 0.2g	26.81 ± 0.08b	22.4 ± 0.1d	21.4 ± 0.1e	28.90 ± 0.06a	23.6 ± 0.1c	17.09 ± 0.06i	23.6 ± 0.1c	19.1 ± 0.2f	17.7 ± 0.1h
C18:0	4.38 ± 0.03f	9.68 ± 0.02a	4.81 ± 0.01d	4.00 ± 0.01g	6.78 ± 0.01b	5.35 ± 0.02c	3.69 ± 0.01h	5.30 ± 0.03c	4.68 ± 0.03e	4.38 ± 0.01f
C18:1n9c	1.48 ± 0.01h	1.71 ± 0.01f	1.75 ± 0.01f	1.46 ± 0.01h	2.29 ± 0.01b	1.68 ± 0.01g	1.95 ± 0.01e	2.13 ± 0.01d	2.21 ± 0.02c	2.82 ± 0.01a
C18:2n6c	28.46 ± 0.02d	24.80 ± 0.01g	28.46 ± 0.04d	28.89 ± 0.05c	26.04 ± 0.02f	28.39 ± 0.01d	32.46 ± 0.02b	28.03 ± 0.04e	34.33 ± 0.09a	28.41 ± 0.01d
C18:3n3	43.3 ± 0.1a	31.42 ± 0.01h	35.0 ± 0.1g	40.02 ± 0.07d	29.67 ± 0.01i	36.12 ± 0.03e	40.43 ± 0.07c	35.08 ± 0.03g	35.63 ± 0.07f	42.78 ± 0.02b
C22:0	0.64 ± 0.02h	1.1 ± 0.1c	1.7 ± 0.1a	1.04 ± 0.04d	0.927 ± 0.005f	0.95 ± 0.02e	1.18 ± 0.06b	1.68 ± 0.06a	0.46 ± 0.01i	0.720 ± 0.008g
C24:0	0.901 ± 0.008f	1.32 ± 0.01a	1.03 ± 0.01e	1.17 ± 0.02d	1.27 ± 0.05b	1.22 ± 0.03c	0.91 ± 0.02f	1.20 ± 0.04cd	0.930 ± 0.006f	0.84 ± 0.01g
SFA	26.1 ± 0.1h	40.95 ± 0.03a	31.75 ± 0.02e	29.1 ± 0.1f	40.23 ± 0.09b	32.94 ± 0.05d	24.3 ± 0.1j	33.6 ± 0.1c	26.8 ± 0.2g	25.50 ± 0.07i
MUFA	1.93 ± 0.01h	2.63 ± 0.04e	4.6 ± 0.2a	1.78 ± 0.01i	3.90 ± 0.07b	2.37 ± 0.03g	2.52 ± 0.01f	2.99 ± 0.02d	3.08 ± 0.02c	3.06 ± 0.02c
PUFA	71.9 ± 0.1b	56.42 ± 0.01i	63.6 ± 0.1g	69.1 ± 0.1e	55.87 ± 0.02j	64.68 ± 0.02f	73.1 ± 0.1a	63.43 ± 0.09h	70.2 ± 0.2d	71.44 ± 0.05c
					Seeds					
C16:0	11.60 ± 0.04b	11.86 ± 0.03a	11.16 ± 0.06c	10.46 ± 0.04i	10.73 ± 0.05g	10.93 ± 0.04d	10.58 ± 0.01h	10.83 ± 0.05f	10.89 ± 0.01e	10.38 ± 0.02j
C18:0	2.61 ± 0.01b	2.69 ± 0.01a	2.19 ± 0.01h	2.25 ± 0.01g	2.10 ± 0.01i	2.54 ± 0.01c	2.38 ± 0.01e	2.43 ± 0.01d	2.33 ± 0.01f	2.27 ± 0.01g
C18:1n9c	2.35 ± 0.01b	1.80 ± 0.01d	1.95 ± 0.02c	1.54 ± 0.01h	1.58 ± 0.02g	2.56 ± 0.01a	1.70 ± 0.01e	1.63 ± 0.01f	1.46 ± 0.01i	1.79 ± 0.01d
C18:2n6c	29.54 ± 0.02a	25.28 ± 0.01e	27.36 ± 0.03b	26.00 ± 0.01c	24.89 ± 0.03h	24.97 ± 0.01f	25.84 ± 0.01d	24.31 ± 0.03j	24.52 ± 0.01i	24.93 ± 0.02g
C18:3n3	51.48 ± 0.03i	55.96 ± 0.02g	55.54 ± 0.02h	57.75 ± 0.05d	58.41 ± 0.05c	56.88 ± 0.01f	57.46 ± 0.04e	58.72 ± 0.03b	59.08 ± 0.02a	58.77 ± 0.03b
SFA	16.15 ± 0.06b	16.66 ± 0.01a	14.63 ± 0.05f	14.32 ± 0.03h	14.45 ± 0.01g	15.30 ± 0.02c	14.66 ± 0.03f	15.07 ± 0.07d	14.78 ± 0.01e	14.15 ± 0.02i
MUFA	2.72 ± 0.01a	2.02 ± 0.02e	2.39 ± 0.03b	1.84 ± 0.01f	2.17 ± 0.06c	2.77 ± 0.01a	1.97 ± 0.02e	1.82 ± 0.01f	1.59 ± 0.01g	2.08 ± 0.02d
PUFA	81.13 ± 0.05i	81.32 ± 0.03h	82.98 ± 0.02f	83.84 ± 0.05a	83.39 ± 0.07d	81.93 ± 0.01g	83.37 ± 0.05d	83.11 ± 0.06e	83.63 ± 0.01c	83.77 ± 0.04b

C16:0 palmitic acid, C18:0 stearic acid, C18:1n9c oleic acid, C18:2n6c linoleic acid, C18:3n3 c, C22:0 behenic acid, C24:0 lignoceric acid; SFA: saturated fatty acids; MUFA: monounsaturated fatty acids; PUFA: polyunsaturated fatty acids; n6/n3: ratio of omega-6/omega-3 fatty acids. [¥] W+: indicates normal irrigation regime; W-: indicates water-holding irrigation regime; C: Control; V: Veramin Ca; EK: EKOprop; G: Nomoren; TW: Twin-Antistress. Means in the same row followed by different Latin letters are significantly different according to Tukey's honestly significant difference (HSD) test ($p = 0.05$).

3.8. LDA Analysis

3.8.1. First Harvest

In the first harvest, the linear discriminant analysis (LDA) selected PUFA, C17:0, ascorbic acid, glucose, α-tocopherol, C16:1, C14:0, C16:0, carbohydrates, and C22:1 as variables with discriminant ability, which is equivalent to say that these were the parameters showing the most profound changes in result of using different biostimulants (Figure 2). Function 1 separated primarily samples treated with G, which was placed in the farthest position in the negative side of the axis; in turn, the biostimulants with the most similar effect according to this function (which was the most important, as it included 85.1% of the observed variance) were V and TW. In turn, Function 2 separated mainly C samples, which was mostly due to their higher contents in carbohydrates and glucose (in this case, specifically in W-samples). The most noticeable effect of Function 3 was the individualization of markers corresponding to EK samples, which was mostly due to the levels of PUFA, C16:0, and C14:0.

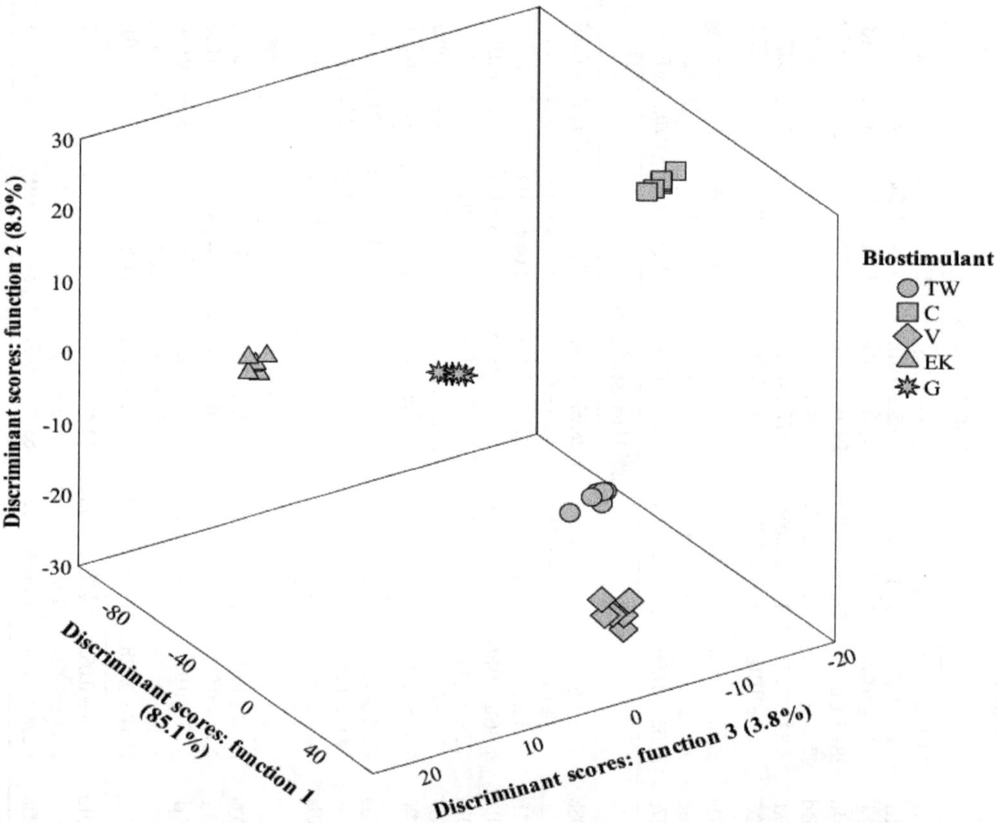

Figure 2. Canonical discriminant functions coefficients defined from the evaluated parameters plotted to show the effect of biostimulants treatments on *Phaseolus vulgaris* green pods of the first harvest under different irrigation regimes (normal irrigation and water stress).

3.8.2. Second Harvest

Concerning the second harvest, the LDA selected C22:0, fructose, organic acids, C21:0, proteins, C16:1, C24:0, C20:1, C18:0, C18:1n9c, MUFA, and sugars as the variables with the highest differences as a result of using different biostimulants (highest discriminant ability) (Figure 3). Function 1 was especially effective in separating samples treated with TW or EK, which was mostly due to their higher C18:1n9c contents (regardless of the irrigation treatment). On the other hand, Function 2 separated samples treated with G, which was placed in the farthest position in comparison to control samples. Considering all the functions together, it was possible to conclude that the biostimulant treatment V was the one that induced the least differences in comparison to the control.

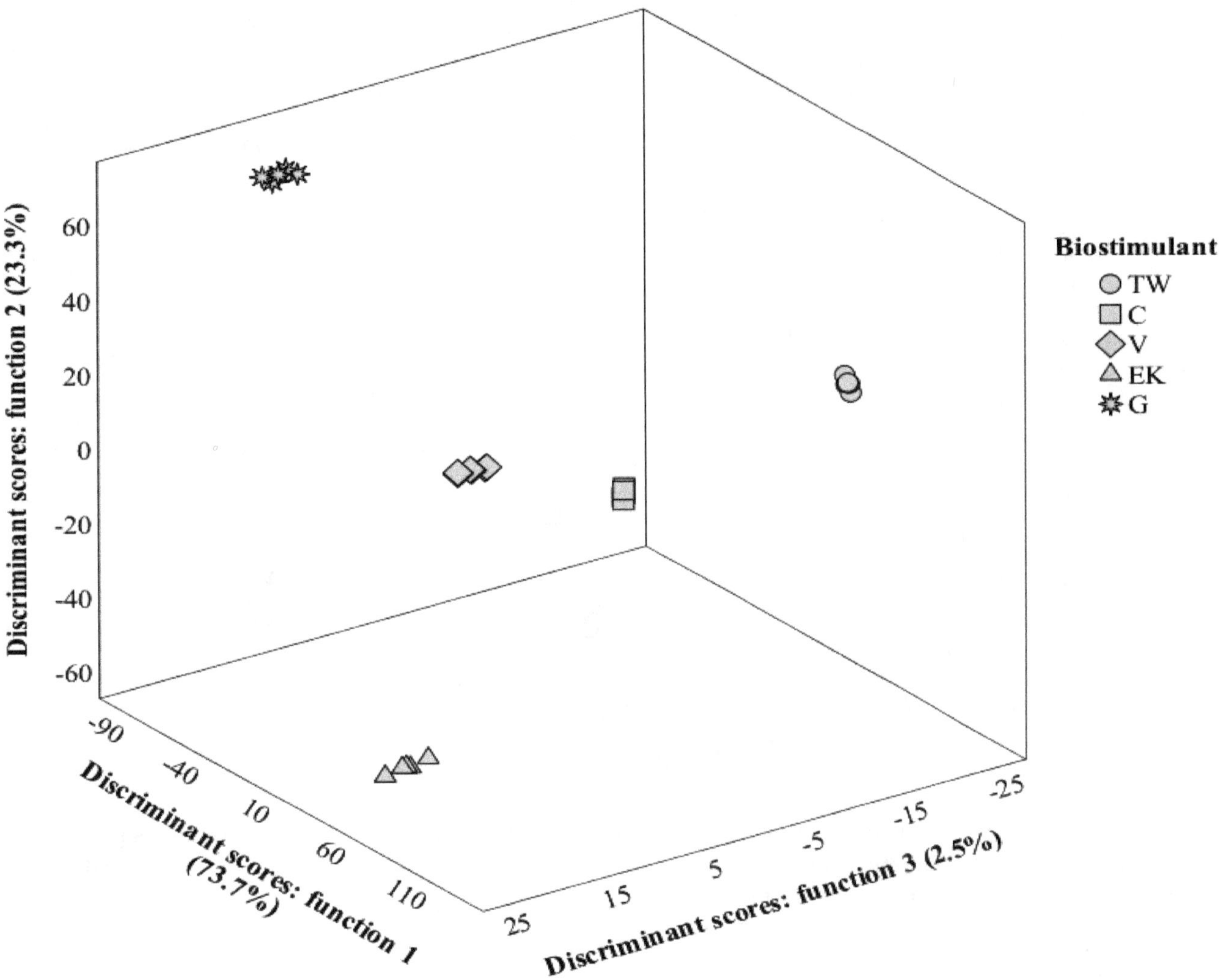

Figure 3. Canonical discriminant functions coefficients defined from the evaluated parameters plotted to show the effect of biostimulants treatments on *Phaseolus vulgaris* green pods of the second harvest under different irrigation regimes (normal irrigation and water stress).

3.8.3. Seeds

The effects of the tested biostimulants on seeds were also more pronounced for fatty acids, as indicated by the variables classified as being discriminant: C20:1, C20:0, sucrose, C22:1, lipids, C17:0, C15:0, organic acids, PUFA, C16:0, C18:2n6c, and C16:1 (Figure 4). According to Function 1, all biostimulants had similar effects (markers are almost vertically aligned), while untreated samples (C) were completely individualized (negative side of the axis); among the selected variables, the one showing the highest correlation with this function was C20:1, which showed higher percentages in C samples, independently of water level. Function 2, in turn, was mostly correlated to C16:1 and sucrose, contributing mainly to separate samples treated with EK, while Function 3 was more highly correlated with lipids content, contributing to separate samples treated with G.

In all the former LDAs, the classification performance was 100% accurate both for originally grouped cases as well as for cross-validated ones.

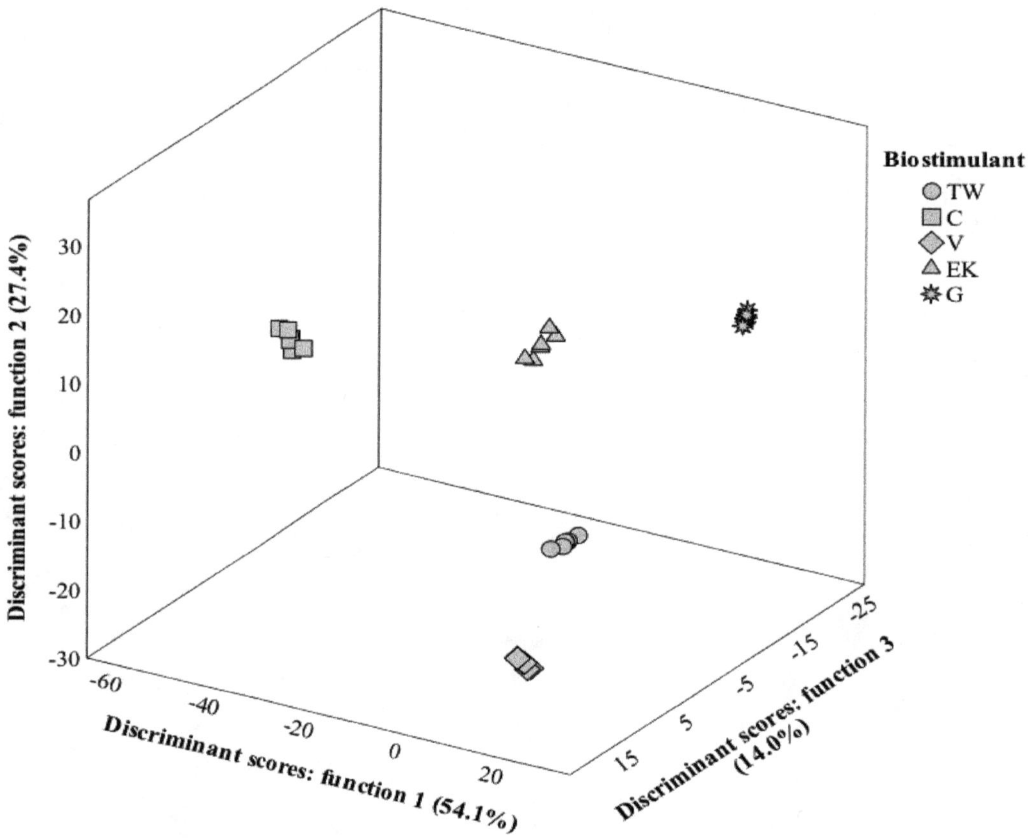

Figure 4. Canonical discriminant functions coefficients defined from the evaluated parameters plotted to show the effect of biostimulants treatments on *Phaseolus vulgaris* seeds under different irrigation regimes (normal irrigation and water stress).

4. Conclusions

The results of the present study showed a varied effect of biostimulants and water treatments on pod yield and the quality of common bean green pods and seeds, while significant differences were also observed between normally irrigated and water-stressed plants in a biostimulants treatment-specific manner. Promising results were also recorded regarding the alleviation of negative effects of drought stress where the application of arbuscular mycorrhizal fungi (AMF; G treatment) increased the crop yield of green beans. Moreover, the nutritional value and chemical composition of pods and seeds was positively affected by biostimulants application, although a product specific effect was recorded depending on the irrigation regime and harvesting time (pods and/or seeds). In conclusion, the application of biostimulants could be considered as an eco-friendly and sustainable tool to increase the pod yield and quality of common bean green pods and seeds under normal irrigation and/or drought stress conditions. Considering that the tested biostimulants contain beneficial microorganisms such as AMF, symbiotic rizosphere bacteria, and saprophytic fungi, its application not only could benefit crops but it could also improve soil properties and preserve soil quality. However, future research is needed to investigate in depth the mechanisms of action of biostimulant product, the application dose efficiency, as well as the most effective application regime and the possible effect of genotype x biostimulants interactions.

Author Contributions: S.A.P. conceived and designed the research, administered and supervised the project, carried out the cultivation, wrote the original draft, and reviewed and edited the final manuscript; A.F. performed chemical analyses, data curation, and methodology; S.P. carried out the cultivation and prepared the original draft; A.C. performed chemical analyses, data curation, and methodology; N.T. performed chemical analyses, data curation, prepared the original draft, and edited the final manuscript; J.C.M.B. performed the LDA analysis of the data and the interpretation of the statistical analysis results; L.B. performed chemical analyses, data curation, and methodology, wrote the original draft, and reviewed and edited the final manuscript; I.C.F.R.F. obtained funding, administered and supervised the project, and reviewed and edited the final manuscript. All authors have read and agreed to the published version of the manuscript.

References

1. Huang, Z.A.; Jiang, D.A.; Yang, Y.; Sun, J.W.; Jin, S.H. Effects of nitrogen deficiency on gas exchange, chlorophyll fluorescence, and antioxidant enzymes in leaves of rice plants. *Photosynthetica* **2004**, *42*, 357–364. [CrossRef]
2. Zhao, D.; Reddy, K.R.; Kakani, V.G.; Reddy, V.R. Nitrogen deficiency effects on plant growth, leaf photosynthesis, and hyperspectral reflectance properties of sorghum. *Eur. J. Agron.* **2005**, *22*, 391–403. [CrossRef]
3. Liebman, M.; Davis, A.S. Integration of soil, crop and weed management in low-external-input farming systems. *Weed Res.* **2000**, *40*, 27–47. [CrossRef]
4. Postel, S.L. Entering an era of water scarcity: The challenges ahead. *Ecol. Appl.* **2000**, *10*, 941–948. [CrossRef]
5. Bulgari, R.; Cocetta, G.; Trivellini, A.; Vernieri, P.; Ferrante, A. Biostimulants and crop responses: A review. *Biol. Agric. Hortic.* **2015**, *31*, 1–17. [CrossRef]
6. Van Oosten, M.J.; Pepe, O.; De Pascale, S.; Silletti, S.; Maggio, A. The role of biostimulants and bioeffectors as alleviators of abiotic stress in crop plants. *Chem. Biol. Technol. Agric.* **2017**, *4*, 1–12. [CrossRef]
7. Ziosi, V.; Zandoli, R.; Di Nardo, A.; Biondi, S.; Antognoni, F.; Calandriello, F. Biological activity of different botanical extracts as evaluated by means of an array of in vitro and in vivo bioassays. *Acta Hortic.* **2012**, *1009*, 61–66. [CrossRef]
8. Abdel Latef, A.A.H.; Chaoxing, H. Does inoculation with *Glomus mosseae* improve salt tolerance in pepper plants? *J. Plant Growth Regul.* **2014**, *33*, 644–653. [CrossRef]
9. Abu Seif, Y.I.; El-Miniawy, S.E.D.M.; Abu El-Azm, N.A.I.; Hegazi, A.Z. Response of snap bean growth and seed yield to seed size, plant density and foliar application with algae extract. *Ann. Agric. Sci.* **2016**, *61*, 187–199. [CrossRef]
10. Ahmad, P.; Hashem, A.; Abd-Allah, E.F.; Alqarawi, A.A.; John, R.; Egamberdieva, D.; Gucel, S. Role of Trichoderma harzianum in mitigating NaCl stress in Indian mustard (*Brassica juncea* L) through antioxidative defense system. *Front. Plant Sci.* **2015**, *6*, 1–15. [CrossRef]
11. Halpern, M.; Bar-Tal, A.; Ofek, M.; Minz, D.; Muller, T.; Yermiyahu, U. The use of biostimulants for enhancing nutrient uptake. *Adv. Agron.* **2015**, *130*, 141–174.
12. Koleška, I.; Hasanagić, D.; Todorović, V.; Murtić, S.; Klokić, I.; Paradiković, N.; Kukavica, B. Biostimulant prevents yield loss and reduces oxidative damage in tomato plants grown on reduced NPK nutrition. *J. Plant. Interact.* **2017**, *12*, 209–218. [CrossRef]
13. Rouphael, Y.; Franken, P.; Schneider, C.; Schwarz, D.; Giovannetti, M.; Agnolucci, M.; De Pascale, S.; Bonini, P.; Colla, G. Arbuscular mycorrhizal fungi act as biostimulants in horticultural crops. *Sci. Hortic. (Amsterdam)* **2015**, *196*, 91–108. [CrossRef]
14. Kunicki, E.; Grabowska, A.; Sękara, A.; Wojciechowska, R. The effect of cultivar type, time of cultivation, and biostimulant treatment on the yield of spinach (*Spinacia oleracea* L.). *Folia Hortic.* **2010**, *22*, 9–13. [CrossRef]
15. Pereira, J.L.; Queiroz, R.M.L.; Charneau, S.O.; Felix, C.R.; Ricart, C.A.O.; Lopes Da Silva, F.; Steindorff, A.S.; Ulhoa, C.J.; Noronha, E.F. Analysis of *Phaseolus vulgaris* response to its association with *Trichoderma harzianum* (ALL-42) in the presence or absence of the phytopathogenic fungi Rhizoctonia solani and Fusarium solani. *PLoS ONE* **2014**, *9*, 1–23. [CrossRef]
16. Rouphael, Y.; Giordano, M.; Cardarelli, M.; Cozzolino, E.; Mori, M.; Kyriacou, M.C.; Bonini, P.; Colla, G. Plant-and seaweed-based extracts increase yield but differentially modulate nutritional quality of greenhouse spinach through biostimulant action. *Agronomy* **2018**, *8*, 126. [CrossRef]
17. Xu, C.; Leskovar, D.I. Effects of *A. nodosum* seaweed extracts on spinach growth, physiology and nutrition value under drought stress. *Sci. Hortic. (Amsterdam)* **2015**, *183*, 39–47. [CrossRef]
18. Chrysargyris, A.; Xylia, P.; Anastasiou, M.; Pantelides, I.; Tzortzakis, N. Effects of *Ascophyllum nodosum* seaweed extracts on lettuce growth, physiology and fresh-cut salad storage under potassium deficiency. *J. Sci. Food Agric.* **2018**, *98*, 5861–5872. [CrossRef]
19. Galvão, Í.M.; dos Santos, O.F.; de Souza, M.L.C.; de Jesus Guimarães, J.; Kühn, I.E.; Broetto, F. Biostimulants action in common bean crop submitted to water deficit. *Agric. Water Manag.* **2019**, *225*, 105762. [CrossRef]

20. De Pascale, S.; Rouphael, Y.; Colla, G. Plant biostimulants: Innovative tool for enhancing plant nutrition in organic farming. *Eur. J. Hortic. Sci.* **2018**, *82*, 277–285. [CrossRef]
21. Karimzadeh Soureshjani, H.; Nezami, A.; Kafi, M.; Tadayon, M. Responses of two common bean (*Phaseolus vulgaris* L.) genotypes to deficit irrigation. *Agric. Water Manag.* **2019**, *213*, 270–279. [CrossRef]
22. Hummel, M.; Hallahan, B.F.; Brychkova, G.; Ramirez-Villegas, J.; Guwela, V.; Chataika, B.; Curley, E.; McKeown, P.C.; Morrison, L.; Talsma, E.F.; et al. Reduction in nutritional quality and growing area suitability of common bean under climate change induced drought stress in Africa. *Sci. Rep.* **2018**, *8*, 1–11. [CrossRef] [PubMed]
23. Arruda, I.M.; Moda-Cirino, V.; Koltun, A.; dos Santos, O.J.A.P.; Moreira, R.S.; Moreira, A.F.P.; Gonçalves, L.S.A. Physiological, biochemical and morphoagronomic characterization of drought-tolerant and drought-sensitive bean genotypes under water stress. *Physiol. Mol. Biol. Plants* **2018**, *24*, 1059–1067. [CrossRef] [PubMed]
24. Sarma, R.K.; Saikia, R. Alleviation of drought stress in mung bean by strain *Pseudomonas aeruginosa* GGRJ21. *Plant. Soil* **2014**, *377*, 111–126. [CrossRef]
25. Sadak, M.S.; Abdelhamid, M.T.; Schmidhalter, U. Effect of foliar application of aminoacids on plant yield and some physiological parameters in bean plants irrigated with seawater. *Acta Biol. Colomb.* **2015**, *20*, 141–152.
26. Kumar, M.; Mishra, S.; Dixit, V.; Kumar, M.; Agarwal, L.; Chauhan, P.S.; Nautiyal, C.S. Synergistic effect of *Pseudomonas putida* and *Bacillus amyloliquefaciens* ameliorates drought stress in chickpea (*Cicer arietinum* L.). *Plant. Signal. Behav.* **2016**, *11*, 1–9. [CrossRef]
27. German, M.A.; Burdman, S.; Okon, Y.; Kigel, J. Effects of *Azospirillum brasilense* on root morphology of common bean (*Phaseolus vulgaris* L.) under different water regimes. *Biol. Fertil. Soils* **2000**, *32*, 259–264. [CrossRef]
28. Klimek-Kopyra, A.; Kliszcz, A.; Ślizowska, A.; Kot, D. Application of biostimulants influences shoot and root characteristics of seedlings of winter pea (Pisum sativum L.). *Acta Agrobot.* **2019**, *72*, 1–8. [CrossRef]
29. Dourado-Neto, D.; Dario, G.J.A.; Barbieri, A.P.P.; Martin, T.N. Biostimulant action on agronomic efficiency of corn and common beans. *Biosci. J.* **2014**, *30*, 371–379.
30. Rady, M.M.; Desoky, E.S.M.; Elrys, A.S.; Boghdady, M.S. Can licorice root extract be used as an effective natural biostimulant for salt-stressed common bean plants? *S. Afr. J. Bot.* **2019**, *121*, 294–305. [CrossRef]
31. Rady, M.M.; Mohamed, G.F. Modulation of salt stress effects on the growth, physio-chemical attributes and yields of *Phaseolus vulgaris* L. plants by the combined application of salicylic acid and *Moringa oleifera* leaf extract. *Sci. Hortic. (Amsterdam)* **2015**, *193*, 105–113. [CrossRef]
32. Selby, C.; Carmichael, E.; Sharma, H.S.S. Bio-refining of perennial ryegrass (*Lolium perenne*): Evaluation of aqueous extracts for plant defence elicitor activity using French bean cell suspension cultures. *Chem. Biol. Technol. Agric.* **2016**, *3*, 1–7. [CrossRef]
33. Elzaawely, A.A.; Ahmed, M.E.; Maswada, H.F.; Xuan, T.D. Enhancing growth, yield, biochemical, and hormonal contents of snap bean (*Phaseolus vulgaris* L.) sprayed with moringa leaf extract. *Arch. Agron. Soil Sci.* **2017**, *63*, 687–699. [CrossRef]
34. Elzaawely, A.A.; Ahmed, M.E.; Maswada, H.F.; Al-Araby, A.A.; Xuan, T.D. Growth traits, physiological parameters and hormonal status of snap bean (*Phaseolus vulgaris* L.) sprayed with garlic cloves extract. *Arch. Agron. Soil Sci.* **2018**, *64*, 1068–1082. [CrossRef]
35. Petropoulos, S.A.; Taofiq, O.; Fernandes, Â.; Tzortzakis, N.; Ciric, A.; Sokovic, M.; Barros, L.; Ferreira, I.C. Bioactive properties of greenhouse-cultivated green beans (*Phaseolus vulgaris* L.) under biostimulants and water-stress effect. *J. Sci. Food Agric.* **2019**, *99*, 6049–6059. [CrossRef] [PubMed]
36. Pereira, C.; Dias, M.I.; Petropoulos, S.A.; Plexida, S.; Chrysargyris, A.; Tzortzakis, N.; Calhelha, R.C.; Ivanov, M.; Stojković, D.; Soković, M.; et al. The Effects of Biostimulants, Biofertilizers and Water-Stress on Nutritional Value and Chemical Composition of Two Spinach Genotypes (*Spinacia oleracea* L.). *Molecules* **2019**, *24*, 4494. [CrossRef]
37. AOAC International. *Official Methods of Analysis of AOAC International*; Horwitz, W., Latimer, G., Eds.; AOAC International: Gaithersburg, MD, USA, 2016.
38. Chrysargyris, A.; Nikolaidou, E.; Stamatakis, A.; Tzortzakis, N. Vegetative, physiological, nutritional and antioxidant behavior of spearmint (*Mentha spicata* L.) in response to different nitrogen supply in hydroponics. *J. Appl. Res. Med. Aromat. Plants* **2017**, *6*, 52–61. [CrossRef]
39. Pereira, C.; Barros, L.; Carvalho, A.M.; Ferreira, I.C.F.R. Use of UFLC-PDA for the analysis of organic acids in thirty-five species of food and medicinal plants. *Food Anal. Methods* **2013**, *6*, 1337–1344. [CrossRef]

40. Barros, L.; Pereira, E.; Calhelha, R.C.; Dueñas, M.; Carvalho, A.M.; Santos-Buelga, C.; Ferreira, I.C.F.R. Bioactivity and chemical characterization in hydrophilic and lipophilic compounds of *Chenopodium ambrosioides* L. *J. Funct. Foods* **2013**, *5*, 1732–1740. [CrossRef]
41. Barros, L.; Pereira, C.; Ferreira, I.C.F.R. Optimized analysis of organic acids in edible mushrooms from Portugal by Ultra-Fast Liquid Chromatography and Photodiode Array Detection. *Food Anal. Methods* **2013**, *6*, 309–316. [CrossRef]
42. Aimo, S.; Gosetti, F.; D'Agostino, G.; Gamalero, E.; Gianotti, V.; Bottaro, M.; Gennaro, M.C.; Berta, G. Use of arbuscular mycorrhizal fungi and beneficial soil bacteria to improve yield and quality of saffron (*Crocus sativus* L.). *Acta Hortic.* **2010**, 159–164. [CrossRef]
43. Weber, N.; Schmitzer, V.; Jakopic, J.; Stampar, F. First fruit in season: Seaweed extract and silicon advance organic strawberry (*Fragaria×ananassa* Duch.) fruit formation and yield. *Sci. Hortic. (Amsterdam)* **2018**, *242*, 103–109. [CrossRef]
44. Arthur, G.D.; Stirk, W.A.; Van Staden, J. Effect of a seaweed concentrate on the growth and yield of three varieties of *Capsicum annuum*. *S. Afr. J. Bot.* **2003**, *69*, 207–211. [CrossRef]
45. Alam, M.Z.; Braun, G.; Norrie, J.; Hodges, D.M. Effect of *Ascophyllum* extract application on plant growth, fruit yield and soil microbial communities of strawberry. *Can. J. Plant. Sci.* **2013**, *93*, 23–36. [CrossRef]
46. Alam, M.Z.; Braun, G.; Norrie, J.; Mark Hodges, D. *Ascophyllum* extract application can promote plant growth and root yield in carrot associated with increased root-zone soil microbial activity. *Can. J. Plant. Sci.* **2014**, *94*, 337–348. [CrossRef]
47. Battacharyya, D.; Babgohari, M.Z.; Rathor, P.; Prithiviraj, B. Seaweed extracts as biostimulants in horticulture. *Sci. Hortic. (Amsterdam)* **2015**, *196*, 39–48. [CrossRef]
48. Korir, H.; Mungai, N.W.; Thuita, M.; Hamba, Y.; Masso, C. Co-inoculation effect of rhizobia and plant growth promoting rhizobacteria on common bean growth in a low phosphorus soil. *Front. Plant. Sci.* **2017**, *08*, 1–10. [CrossRef]
49. Farouk, S.; Abdul Qados, A.M.S. Enhancing seed quality and productivity as well as physio-anatomical responses of pea plants by folic acid and/or hydrogen peroxide application. *Sci. Hortic. (Amsterdam)* **2018**, *240*, 29–37. [CrossRef]
50. Poorter, H.; Nagel, O. The role of biomass allocation in the growth response of plants to different levels of light, CO2, nutrients and water: A quantitative review. *Aust. J. Plant. Physiol.* **2000**, *27*, 595–607.
51. Colla, G.; Nardi, S.; Cardarelli, M.; Ertani, A.; Lucini, L.; Canaguier, R.; Rouphael, Y. Protein hydrolysates as biostimulants in horticulture. *Sci. Hortic. (Amsterdam)* **2015**, *196*, 28–38. [CrossRef]
52. Przygocka-Cyna, K.; Grzebisz, W. Effect of biofertilizers on nutrient uptake by vegetables grown in a short cropping sequence. *J. Elem.* **2018**, *23*, 807–823. [CrossRef]
53. Elsheikh, E.A.E.; El Tinay, A.H.; Fadul, I.A. Effect of nutritional status of faba bean on proximate composition, anti-nutritional factors and in vitro protein digestibility (IVPD). *Food Chem.* **1999**, *67*, 379–383. [CrossRef]
54. Kocira, A.; Świeca, M.; Kocira, S.; Złotek, U.; Jakubczyk, A. Enhancement of yield, nutritional and nutraceutical properties of two common bean cultivars following the application of seaweed extract (*Ecklonia maxima*). *Saudi, J. Biol. Sci.* **2018**, *25*, 563–571. [CrossRef] [PubMed]
55. Castellanos-Barriga, L.G.; Santacruz-Ruvalcaba, F.; Hernández-Carmona, G.; Ramírez-Briones, E.; Hernández-Herrera, R.M. Effect of seaweed liquid extracts from *Ulva lactuca* on seedling growth of mung bean (*Vigna radiata*). *J. Appl. Phycol.* **2017**, *29*, 2479–2488. [CrossRef]
56. Mancuso, S.; Azzarello, E.; Mugnai, S.; Briand, X. Marine bioactive substances (IPA extract) improve foliar ion uptake and water stress tolerance in potted *Vitis vinifera* plants. *Adv. Hortic. Sci.* **2006**, *20*, 156–161.
57. Gosling, P.; Mead, A.; Proctor, M.; Hammond, J.P.; Bending, G.D. Contrasting arbuscular mycorrhizal communities colonizing different host plants show a similar response to a soil phosphorus concentration gradient. *New Phytol.* **2013**, *198*, 546–556. [CrossRef]
58. Colla, G.; Cardarelli, M.; Bonini, P.; Rouphael, Y. Foliar applications of protein hydrolysate, plant and seaweed extracts increase yield but differentially modulate fruit quality of greenhouse tomato. *HortScience* **2017**, *52*, 1214–1220. [CrossRef]
59. Huang, D.; Ou, B.; Rior, R.L. The chemistry behind antioxidant capacity assays. *J. Agric. Food Chem.* **2005**, *53*, 1841–1856. [CrossRef]
60. Kellős, T.; Tímár, I.; Szilágyi, V.; Szalai, G.; Galiba, G.; Kocsy, G. Stress hormones and abiotic stresses have different effects on antioxidants in maize lines with different sensitivity. *Plant. Biol.* **2008**, *10*, 563–572. [CrossRef]

61. Kan, L.; Nie, S.; Hu, J.; Wang, S.; Bai, Z.; Wang, J.; Zhou, Y.; Jiang, J.; Zeng, Q.; Song, K. Comparative study on the chemical composition, anthocyanins, tocopherols and carotenoids of selected legumes. *Food Chem.* **2018**, *260*, 317–326. [CrossRef]
62. Elkelish, A.A.; Alnusaire, T.S.; Soliman, M.H.; Gowayed, S.; Senousy, H.H.; Fahad, S. Calcium availability regulates antioxidant system, physio-biochemical activities and alleviates salinity stress mediated oxidative damage in soybean seedlings. *J. Appl. Bot. Food Qual.* **2019**, *92*, 258–266.
63. El-Bassiouny, H.M.S.; El-Monem, A.A.A.; Abdallah, M.M.-S.; Soliman, K.M. Role of arbuscular mycorrhiza, α-tocopherol and nicotinamide on the nitrogen containing compounds and adaptation of sunflower plant to Water stress. *Biosci. Res.* **2018**, *15*, 2068–2088.
64. Laxa, M.; Liebthal, M.; Telman, W.; Chibani, K.; Dietz, K.J. The role of the plant antioxidant system in drought tolerance. *Antioxidants* **2019**, *8*. [CrossRef] [PubMed]
65. Casadesús, A.; Polo, J.; Munné-Bosch, S. Hormonal effects of an enzymatically hydrolyzed animal protein-based biostimulant (pepton) in water-stressed tomato plants. *Front. Plant. Sci.* **2019**, *10*, 1–11. [CrossRef] [PubMed]
66. Zushi, K.; Matsuzoe, N.; Kitano, M. Developmental and tissue-specific changes in oxidative parameters and antioxidant systems in tomato fruits grown under salt stress. *Sci. Hortic. (Amsterdam)* **2009**, *122*, 362–368. [CrossRef]
67. Palmieri, F.; Estoppey, A.; House, G.L.; Lohberger, A.; Bindschedler, S.; Chain, P.S.G.; Junier, P. Chapter Two - Oxalic acid, a molecule at the crossroads of bacterial-fungal interactions. *Adv. Appl. Microbiol.* **2019**, *106*, 49–77.
68. Weber, H.; Borisjuk, L.; Wobus, U. Sugar import and metabolism during seed development. *Trends Plant. Sci.* **1997**, *2*, 169–174. [CrossRef]
69. Rathinasabapathi, B. Metabolic engineering for stress tolerance: Installing osmoprotectant synthesis pathways. *Ann. Bot.* **2000**, *86*, 709–716. [CrossRef]
70. Begum, N.; Qin, C.; Ahanger, M.A.; Raza, S.; Khan, M.I.; Ashraf, M.; Ahmed, N.; Zhang, L. Role of Arbuscular Mycorrhizal Fungi in Plant Growth Regulation: Implications in Abiotic Stress Tolerance. *Front. Plant. Sci.* **2019**, *10*, 1–15. [CrossRef]
71. Rosa, M.; Prado, C.; Podazza, G.; Interdonato, R.; González, J.A.; Hilal, M.; Prado, F.E. Soluble sugars-metabolism, sensing and abiotic stress. A complex network in the life of plants. *Plant. Signal. Behav.* **2009**, *4*, 388–393. [CrossRef]
72. Petropoulos, S.; Ntatsi, G.; Levizou, E.; Barros, L.; Ferreira, I. Nutritional profile and chemical composition of *Cichorium spinosum* ecotypes. *LWT - Food Sci. Technol.* **2016**, *73*, 95–101. [CrossRef]
73. López-Bucio, J.; Pelagio-Flores, R.; Herrera-Estrella, A. Trichoderma as biostimulant: Exploiting the multilevel properties of a plant beneficial fungus. *Sci. Hortic. (Amsterdam)* **2015**, *196*, 109–123. [CrossRef]

6

Biostimulant Application Enhances Fruit Setting in Eggplant—An Insight into the Biology of Flowering

Alicja Pohl, Aneta Grabowska, Andrzej Kalisz and Agnieszka Sękara *

Department of Vegetable and Medicinal Plants, University of Agriculture in Krakow, 29-Listopada 54, 31-425 Kraków, Poland

* Correspondence: agnieszka.sekara@urk.edu.pl

Abstract: Eggplant (*Solanum melongena* L.) is a warm climate crop. Its cultivation extends to temperate regions where low temperatures can affect the course of the generative phase, which is primarily sensitive to abiotic stress. The novelty of the present investigation consisted of characterising the heterostyly, pollination, and fertilisation biology of eggplants in field cultivations, which provided a basis for explaining the effect of a protective biostimulant on these processes. We aimed to investigate the flowering biology of three eggplant hybrids treated with Göemar BM-86®, containing *Ascophylum nodosum* extract, to determine the crucial mechanisms behind the increased flowering and fruit set efficiency and the final effect of increased yield. The flower phenotype (long, medium or short styled), fruit setting, and the number of seeds per fruit were recorded during the two vegetation periods. The numbers of pollen tubes and fertilised ovules in ovaries were evaluated during the generative stage of development to characterise the course of pollination and fertilisation for all types of flowers depending on the cultivar and biostimulant treatment. The expression of heterostyly depended on the eggplant genotype, age of the plant, fruit load, and biostimulant treatment. Domination by long-styled flowers was observed, amounting to 41%, 42%, and 55% of all flowers of "Epic" F_1, "Flavine" F_1, and "Gascona" F_1, respectively. This flower phenotype contained the highest number of pollen tubes in the style and the highest number of fertilised ovules. The biostimulant had a positive effect on the flower and fruit set numbers, as well as on the pollination efficiency in all genotypes. *Ascophylum nodosum* extract could be used as an efficient stimulator of flowering and fruit setting for eggplant hybrids in field conditions in a temperate climatic zone.

Keywords: *Ascophyllum nodosum*; *Solanum melongena*; heterostyly; pollination efficiency

---

1. Introduction

Eggplant is a photoperiodically inert plant with bisexual and partially self-pollinating flowers, although cross-pollination increases the effectiveness of fruit setting [1].

The downward-facing flowers are born solitary or in clusters. The eggplant produces three types of flowers: With a long-style pistil, where the stigma is localised above the anthers; with a medium-style pistil, where the stigma is at the same level as the anthers; and with a short-style pistil, where the stigma is below the anthers (Figure 1). This flower character promotes outcrossing between morphs via delivery and uptake of pollen by pollinators [2,3]. The stamen pores of the long- and medium-styled pistils are localised above or close to the stigma, favouring self-pollination. On the contrary, the stigmas of short-styled pistils are inside the downward-facing anther cone, making self-pollination difficult [4–6].

a b c

Figure 1. Stylar heteromorphism in eggplant: The flower with long-styled pistil of "Flavine" F_1 (**a**), medium-styled pistil of "Gascona" F_1 (**b**), short-styled pistil of "Epic" F_1 (**c**).

Anthers are ready to release pollen and the stigma is receptive from the first opening of the flower (Figure 2). Stigma receptiveness gradually decreases with the plant's age, and by the fifth day of flowering, receptiveness is negligible, and the stigma turns brown [2].

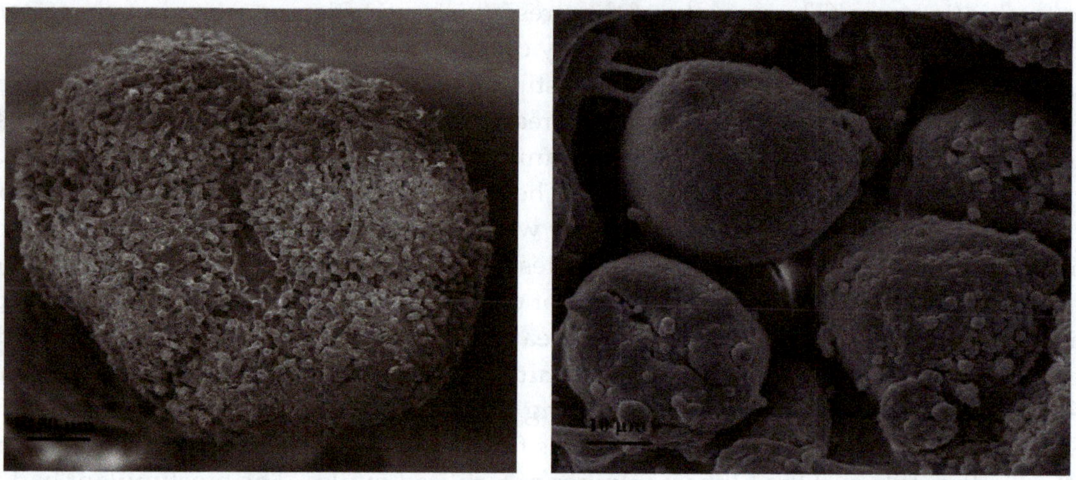

a b

Figure 2. Stigma in eggplant pistil with visible papillae in a receptive phase, (**a**), and pollen grains (**b**) in "Epic" F_1 by scanning electron microscopy.

All types of flowers are found in the same plant and even within the same cluster. The expression of heterostyly depends on the plant genotype, the age of the plant, fruit load, environmental conditions, and growing practices. Generally, domination by long-styled flowers has been reported, amounting to 50–100% of all flowers [7–9]. The higher fruit setting efficiency of this phenotype results from well-developed nodules with high pollen absorption capacity. However, the development of the ovules and their position in the placenta, as well as pollen grain shape, size, and amount in anthers, were nondifferentiated among long-, medium-, and short-styled pistils [6,10], although Wang et al. [11] demonstrated that lower fruit setting from short-styled flowers resulted from stigma-pollen incompatibility. The bumblebee (*Bombus terrestris*) is the most effective eggplant pollinator for plants under covers. Yield increase and better fruit quality are considered to be the major benefits of bumblebee application as compared to self-pollination or inflorescence vibrating [8,12]. Optimisation of eggplant yield in unfavourable conditions could also be achieved by introducing the cultivation of parthenocarpic cultivars [13]. Pollination leading to fruit and seed formation is associated with the production of endogenous growth regulators such as auxins. In this respect, the use of fruit-setting using auxin-based growth regulators has also been recommended to enhance fruit setting under suboptimal temperatures [5,14]. Investigations on the control of eggplant flowering through growth regulators have been successively performed since the end of the 20th century, but their results have been inconclusive [15,16]. Eggplant tolerance to biotic and abiotic stresses can be managed through grafting. The effects of rootstock/scion combinations on eggplant performance were investigated in terms of yield and fruit quality [17,18]. It can be assumed that this technique affects the flowering biology as well, but this issue needs future investigation. In Poland, eggplants are cultivated mainly under unheated foil covers from spring to autumn. To lower costs, cultivation is also performed in open fields where air temperatures may fall below the optimum, causing a reduction in flowering and fruit setting [19]. A promising way to control eggplant generative development could be biostimulant application. Biostimulants have been a focus of global interest of the scientific community since the end of the 20th century, giving promising results in different branches of agriculture as stimulators of crops growth, stress tolerance and yield [20,21]. Seaweed extracts (SWE) are among the main biostimulants, recognised as nontoxic, nonpolluting and nonhazardous to various organisms [22,23]. The majority of the SWE formulations are based on the extract of the brown algae *Ascophyllum nodosum* (L.) Le Jolis. Although seaweed extracts are heterogeneous in nature, the leading companies standardise their chemical composition to ensure consistent product quality [24,25]. Some authors have reported the stimulatory effect of seaweed extracts on eggplant yield [26,27], but there are no references on the flowering biology of this species as affected by SWE biostimulation. SWE action is extremely complex, but interdisciplinary investigation of biostimulant vs plant interactions may shed new light on the effective utilisation of these promising bioproducts in horticulture.

We hypothesise that seaweed extract affects the flowering and fruit setting of eggplant in a multidirectional manner. The reaction of plants to biostimulant treatment depends on the flowering biology of the cultivars, particularly the proportions of different flower phenotypes and their fertility. We aimed to investigate the flowering biology of three eggplant hybrids treated with seaweed extract Göemar BM-86® (Arysta LifeScience North America, LLC) to determine the crucial mechanisms behind the final effect of increased yield.

2. Materials and Methods

2.1. Experimental Arrangement

A two-factorial experiment was set up using randomised blocks in three replications, in the years 2013 and 2015, at the University of Agriculture in Krakow, Poland. The investigated eggplant hybrids, "Epic" F_1 (Seminis Vegetable Seeds), "Flavine" F_1 (Gautier Semences), "Gascona" F_1 (Gautier Semences), were selected on the basis of preliminary studies evaluating their performance in field cultivation under temperate climate conditions [26–28], determined by the earliness, vigour, and yield potential of those plants. Biostimulant Göemar BM-86® (Arysta LifeScience North America, LLC) was applied three times in two week intervals as a foliar application, in a dose of 1.5 $dm^3\ ha^{-1}$. Control plants were sprayed with distilled water. Goemar BM 86® is standardised *Ascophyllum nodosum* (L.) Le Jolis extract, which provides a constant and balanced formulation containing (in %): N, 5.0; Mg, 2.4, S, 3.2, B, 2.07; and Mo, 0.02 [29].

2.2. Cultivation Procedures

Eggplant seeds were sown on 1 March 2013 and 3 March 2015 in seed boxes. After three weeks, the seedlings with one fully developed leaf were transplanted into black 40-cell multipots (VEFI, Norway) with a single cell volume of 0.23 dm^3. Seedlings were grown in a greenhouse, in temperatures of 20/17 ± 2 °C day/night. The growing medium was peat substrate KlasmanTS2 (Klasmann-Deilmann GmbH, Geeste, Germany). The foliar fertiliser Kristalon Green (Yara, Szczecin, Poland) was applied twice in a dose of 10 g dm^{-3} water during seedling production. A gradual decrease in temperature and irrigation was used for the hardening of seedlings seven days before being transplanted to the experimental field (50°04′ N, 19°51′ E) on 7 May 2013 and 15 May 2015, with spacing of 0.75 × 0.60 m. Experimental plots covered 15 plants per treatment for observations of flowering and fruit setting and an additional 15 plants per treatment for flower collection for microscopic observations. Plots were surrounded by shelterbelts. The soil of the experimental field was Fluvic Cambisol (Humic) according to the FAO (Food and Agriculture Organization of the United Nations) classification with a C_{org} level of 2% and pH_{KCl} 6.11. Before the field experiment was established, the soil samples were analysed, and doses of fertilisers were applied to achieve a stable content of nutrients (in mg dm^3): N, 100; P, 90; K, 220; Ca, 1,100; Mg, 70. Cultivation procedures of weeding, irrigation, and plant protection were performed according to the standard recommendations for eggplant cultivated in field conditions in Poland, described by Sękara [30].

2.3. Weather Conditions

The climate of the experimental station is humid continental (Dfb) according to the Köppen's classification. Detailed data concerning the mean air temperature, photosynthetically active radiation (PAR), and the total rainfall during the vegetation seasons in 2013 and 2015 are presented in Table 1. Data were collected from automatic HOBO Pro RH/Temp loggers to assess temperature and a HOBO Weather Station (Onset Comp. Corp., Cape Cod, USA) to assess light characteristics and rainfall at the experimental site. The growing season in 2015 was generally warmer than that in 2013 regarding mean monthly temperatures, with the exception of June. In 2015, precipitation was distributed evenly, while in 2013, 45% of rainfall was recorded in June. A cool September in both years and low PAR caused a continuous decline in eggplant yield (Table 1).

Table 1. Mean monthly temperature, photosynthetically active radiation (PAR) and sum of rainfall in vegetation seasons 2013 and 2015.

Month	2013			2015		
	Temperature (°C)	**PAR (μmol m^{-2} s^{-1})**	**Sum of Rainfall (mm)**	**Temperature (°C)**	**PAR (μmol m^{-2} s^{-1})**	**Sum of Rainfall (mm)**
May	14.3	345	83	13.1	357	93
June	17.6	392	188	17.5	401	40
July	19.4	477	28	20.4	489	39
August	18.8	396	51	21.2	357	58
September	12.1	256	63	14.9	283	60

2.4. Procedures for Flowering and Fruit Setting Observations

The observations were conducted on 5 plants per replication (15 plants per treatment and cultivar) during the flowering period, from June to September. Single flowers were labeled according to the order of appearance on each plant. The numbers of flowers of particular phenotypes (with long-styled, medium-styled, or short-styled pistil) were recorded after the opening of petals. Then, the number of fruits set from flowers of a particular phenotype was also recorded about one week after fertilisation, when fruit sets reached 1–2 cm in diameter. Flowers which did not set fruits were naturally aborted. Fruits in a stage of harvest maturity were picked to reflect the standard cultivation conditions and to exclude excessive metabolite sink by ripening fruits. Data were calculated and presented as a sum of flowers and fruits per plant per month for the two experimental years separately.

2.5. Procedures for Microscopic Observations

At full flowering, 20 pollinated flowers of each phenotype per treatment and cultivar were collected in 2013 and 2015. Data are presented as a sum of observations for investigated seasons, $N = 40$. The styles were isolated and fixed in FAA (formalin-acetic-alcohol), according to Martin's method [31] adapted by Sękara [30]. The germination of pollen on stigmas, growth of pollen tubes, and fertilisation of ovules were examined under fluorescence microscopy with the use of SteREO LUMAR V12 microscope (Carl Zeiss AG, Jena, Germany) (Figure 3). The number of pollen tubes in half of the style and the number of fertilised ovules were evaluated. The numbers of pistils having a number of pollen tubes in the ranges 0–100; 100–200; 200–300; 300–400; 400–500; 500–600; 600–700; 700–800, 800–900; 900–1000 were determined. For fertilised ovules, the following ranges were included: 0–50; 50–100; 100–150; 150–200; 200–250; 250–300; 350–400.

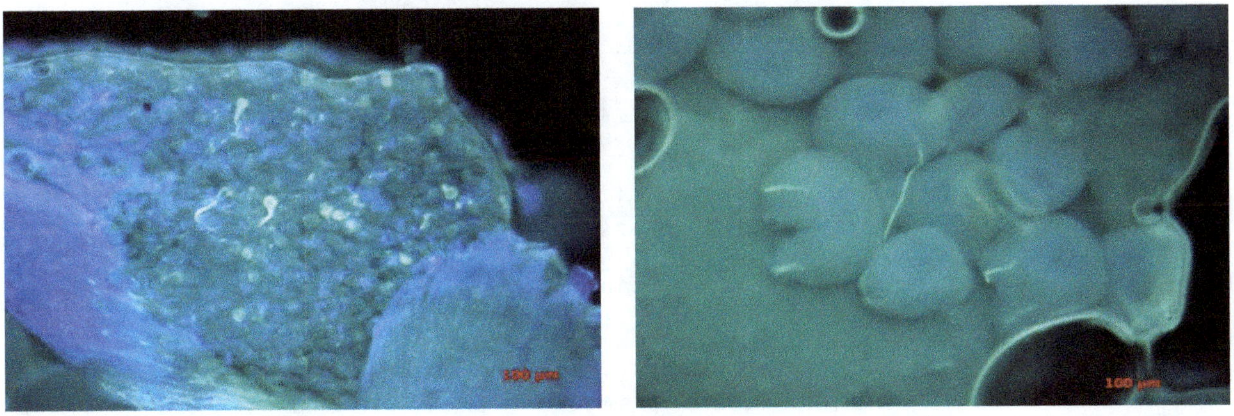

a b

Figure 3. Germination of the pollen on stigmas (**a**) and fertilisation of ovules (**b**) of eggplant observed under fluorescence microscopy after Martin's aniline blue fluorescence technique.

2.6. Statistical Analyses

Statistical analyses were performed using the Statistica 12.0 software package (StatSoft Inc., Tulsa, OK, USA). A three-way analysis of variance followed by Tukey's honest significance test was used to determine the main effects of the type of flower, biostimulant, and time of sampling, as well as interactions between main effects, at the $p \leq 0.05$ significance level. Data shown in the tables and figures are averages of three replicates.

3. Results

In the conditions of the present experiment, the eggplants started flowering at the beginning of June, while the period of the most intensive flowering fell in August. The investigated hybrids showed flower heterostyly—the presence of long-, medium-, and short-styled flowers was observed for all investigated plants. Moreover, heterostyly expression significantly depended on biostimulant treatment and the age of the plants (Table 2).

Table 2. Chosen aspects of flowering and fruit setting of eggplant as depended on fruit type and biostimulant treatment.

Parameter	Year	Type of Flower					
		With Long-Styled Pistil		With Medium-Styled Pistil		With Short-Styled Pistil	
		C *	B	C	B	C	B
"Epic" F_1							
Number of particular types of flowers per plant	2013 2015	12.2 d ** 12.1 c	13.4 d 13.5 c	5.6 b 6.4 b	8.2 c 7.6 b	3.4 a 3.5 a	3.5 ab 3.6 a
Number of fruits per plant set from particular types of flowers	2013 2015	2.4 b 3.2 b	4.2 d 5.6 c	0.8 a 1.0 a	2.0 b 3.2 b	1.0 a 1.0 a	1.4 ab 1.2 a
Effectiveness of fruit setting as depended on type of flower (%)	2013 2015	20.7 33.7	27.6 48.8	15.6 18.9	24.7 46.7	26.7 26.6	40.0 36.7
Number of seeds per fruit	2013 2015	332 d 358 e	365 e 372 f	245 b 255 c	289 c 312 d	102 a 95 a	125 a 138 b
"Flavine" F_1							
Number of particular types of flowers per plant	2013 2015	7.8 cd 8.4 b	10.0 d 11.2 c	6.6 bc 6.8 b	10.0 d 11.0 c	4.2 a 3.8 a	4.8 ab 4.8 a
Number of fruits per plant set from particular types of flowers	2013 2015	2.0 abc 1.8 a	2.8 b 3.6 b	1.4 ab 1.2 a	2.4 bc 3.2 b	1.2 a 1.2 a	1.6 ab 2.0 a
Effectiveness of fruit setting as depended on type of flower (%)	2013 2015	26.9 26.0	28.7 36.2	22.8 16.7	26.0 27.2	20.0 30.0	33.3 41.1
Number of seeds per fruit	2013 2015	312 c 322 c	328 c 333 d	258 b 289 b	325 c 316 c	142 a 134 a	155 a 143 a

Table 2. *Cont.*

Parameter	Year	Type of Flower					
		With Long-Styled Pistil		With Medium-Styled Pistil		With Short-Styled Pistil	
		C *	B	C	B	C	B
		"Gascona" F_1					
Number of particular types of flowers per plant	2013 2015	7.6 c 7.6 bc	8.0 c 7.8 c	5.6 b 6.0 b	8.2 c 7.8 c	3.4 a 4.2 a	3.6 a 3.8 a
Number of fruits per plant set from particular types of flowers	2013 2015	2.0 abc 2.4 ab	3.2 c 3.0 b	1.6 ab 1.6 a	2.4 bc 2.2 ab	1.0 a 1.4 a	0.8 a 1.4 a
Effectiveness of fruit setting as depended on type of flower (%)	2013 2015	26.2 38.7	41.5 37.1	22.6 11.7	29.1 30.6	23.3 36.7	20.0 33.3
Number of seeds per fruit	2013 2015	289 d 321 e	322 e 356 f	257 c 269 c	285 d 297 d	78 a 98 a	127 b 159 b

* C, control; B, biostimulant; ** Means within rows, followed by different letters, are significantly different at $p \leq 0.05$, $N = 3$. Comparisons were performed with the use of Tukey's honest significance test.

Among 23 flowers set by "Epic" F_1 plant during one vegetation period, 55% had long-style pistils, 30% had medium-style pistils, and 15% had short-style pistils. "Epic" F_1 plants produced 7 fruits during the vegetation season, on average; 57% of these were from long-styled flowers, 26% from medium-styled flowers, and 17% from short-styled flowers. Biostimulant treatment significantly increased the number of only medium-styled flowers in 2013 and the number of fruits set by long- and medium-styled flowers in both vegetation periods. The most effective in fruit setting were long-styled flowers. The biostimulant positively affected the percentage of fruits set by all flower phenotypes and the number of seeds, with the exception of flowers with short-styled pistils in 2013. The first fruits were collected at the end of July. The highest number of long- and medium-styled flowers was observed in August; the lowest was observed in September (Figure 4, Table 3).

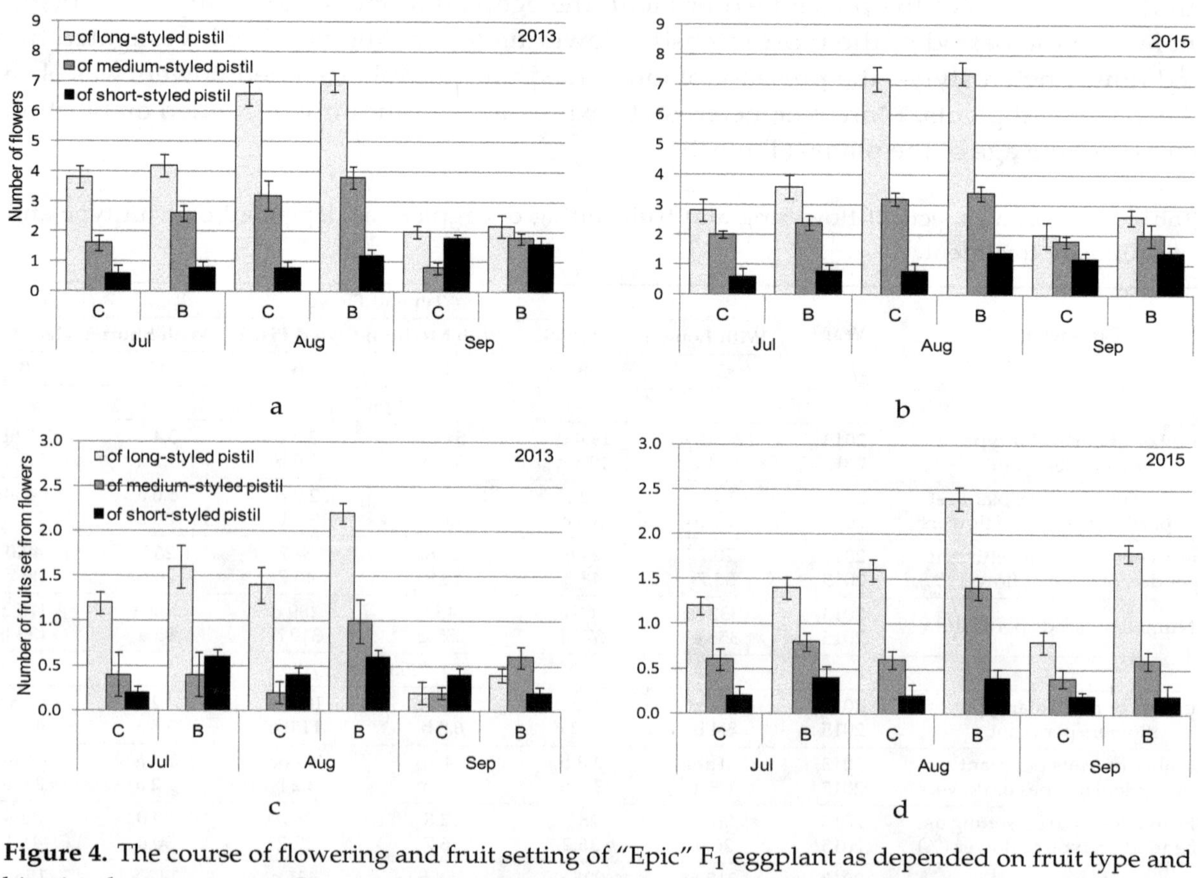

Figure 4. The course of flowering and fruit setting of "Epic" F_1 eggplant as depended on fruit type and biostimulant treatment. C, control; B, biostimulant. Bars represent mean number of flowers per plant in 2013 (**a**), 2015 (**b**) and fruits per plant in 2013 (**c**), and 2015 (**d**) (error bars indicate SE).

Table 3. Results of ANOVA for parameters of flowering and fruit setting of "Epic" F_1 eggplant presented in Figure 4.

ANOVA Source of Variation	"Epic" F_1			
	No of Flowers 2013	No of Fruits 2013	No of Flowers 2015	No of Fruits 2015
Type of flower (F)	***	***	***	***
Biostimulant (B)	**	*	***	***
Month (M)	***	***	*	**
F × B	ns	*	ns	*
F × M	***	***	***	ns
B × M	ns	ns	ns	*

Levels of significance for ANOVA: * $p \leq 0.05$; ** $p \leq 0.01$; *** $p \leq 0.001$; ns, not significant; $N = 3$. Comparisons were performed with the use of Tukey's honest significance test.

The number of short-styled flowers increased in line with aging of the plants. The number of fruits set from long- and medium-styled flowers increased from July to August, then decreased in September. We observed, on average, 106 pollen tubes in the short-styled pistils, 422 in medium-styled pistils, and 610 in long-styled pistils collected from control plants and 129, 490, and 778 pollen tubes, respectively, collected from biostimulant-treated plants (Figure 5). The ovaries of the short-styled flowers contained approximately 36 fertilised ovules, and more fertilised ovules were found in the remaining types of flowers: 199 and 225 in medium- and long-styled flowers, respectively, produced by control plants. The flowers of biostimulant-treated plants contained 39%, 32%, and 36% more fertilised ovules in the ovaries of short-, medium-, and long-styled flowers, respectively.

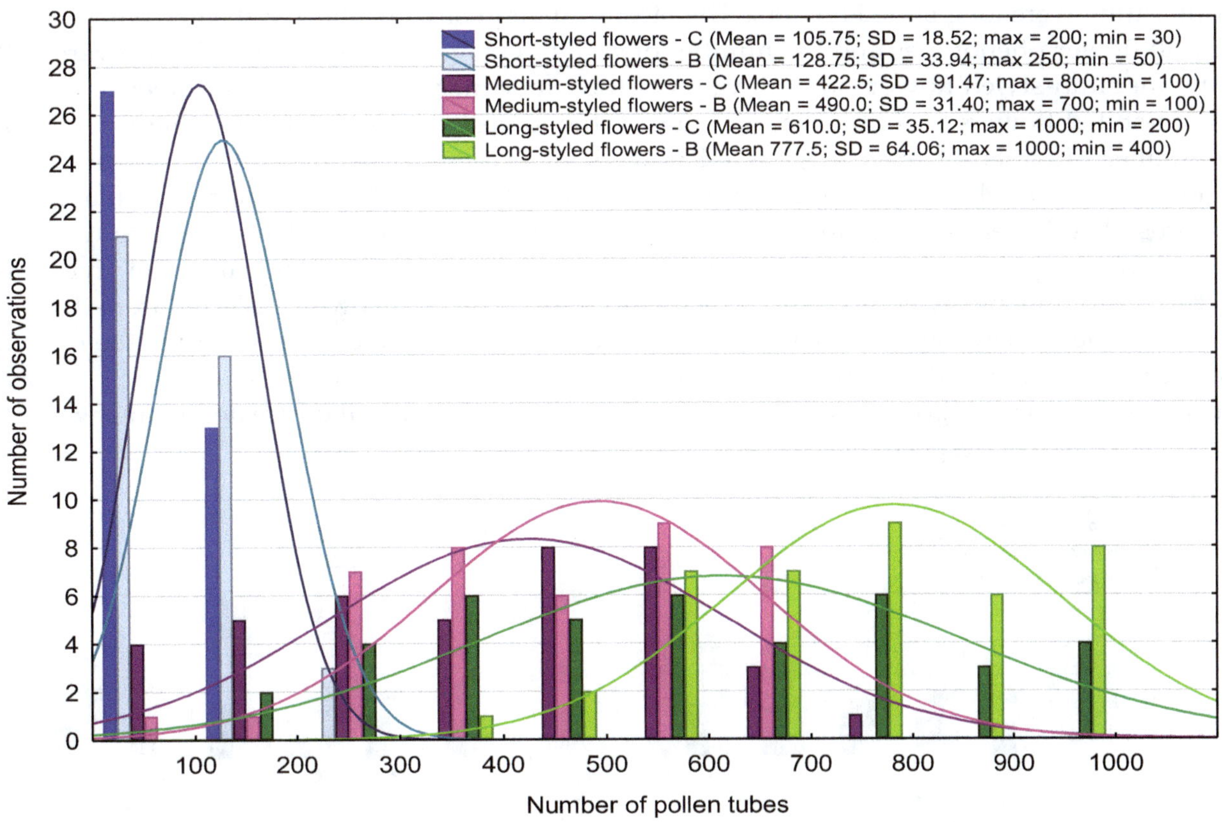

a

Figure 5. *Cont.*

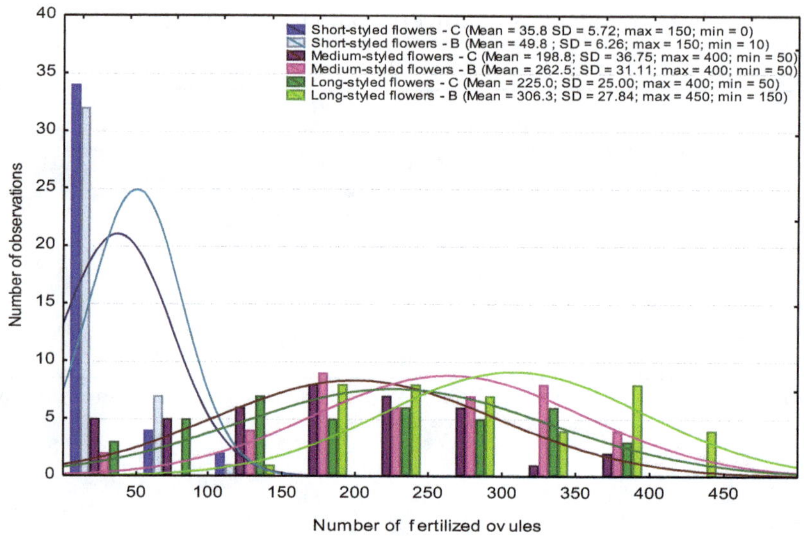

b

Figure 5. Number of pollen tubes (**a**) and fertilised ovules (**b**) in the styles of different flower types in "Epic" F_1.

The "Flavine" F_1 plants produced 22 flowers during the vegetation period; 41% had long-styled pistils, 38% had medium-styled pistils, and 19% had short-styled ones (Table 2). The number of fruits collected from a plant was 6 on average; 42% of these were from long-styled flowers, 34% from medium-styled flowers, and 24% from short-styled flowers. Biostimulant treatment significantly increased the number of long-styled flowers in 2015, the numbers of medium-styled flowers in 2013 and 2015, and the numbers of fruits set by long- and medium-styled flowers in 2015. The most effective in fruit setting were long-styled flowers. The biostimulant positively affected the percentage of fruits set by all flower phenotypes and the number of seeds in fruits born by long-styled flowers in 2015 and by medium-styled flowers in both years of the experiment. The highest number of long- and medium-styled flowers was observed in August; the lowest was in September (Figure 6, Table 4). The number of fruits set from long-styled flowers was the highest in August. We observed, on average, 149 pollen tubes in the middle of the style in the short-styled flowers of control plants, 410 in medium-styled flowers, and 595 in long-styled ones (Figure 7). In biostimulant-treated flowers, a 23% higher number of pollen tubes was observed in short-styled pistils, 16% higher in medium-styled pistils, and 20% higher in long-styled pistils. The ovaries of the short-styled flowers collected from the control plants contained, on average, 54 fertilised ovules; more fertilised ovules were found in the remaining types of flowers: 208 and 243 in medium- and long-styled flowers, respectively. The numbers of fertilised ovules in analogous types of flowers collected from biostimulant-treated plants were 102%, 24%, and 23% higher, respectively.

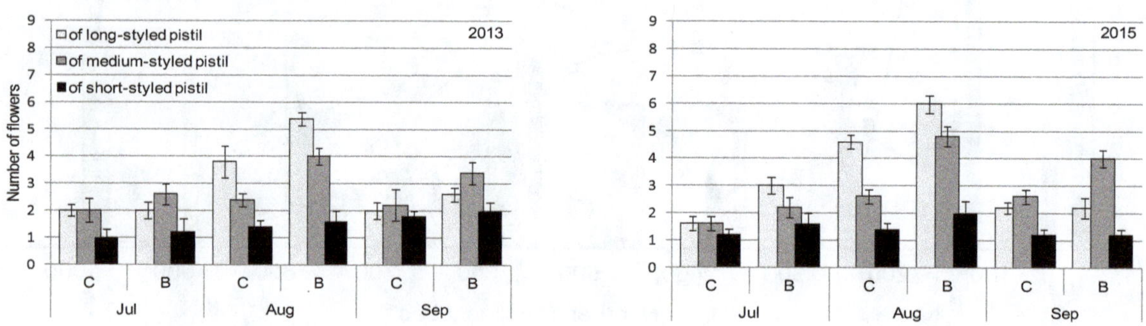

a b

Figure 6. *Cont.*

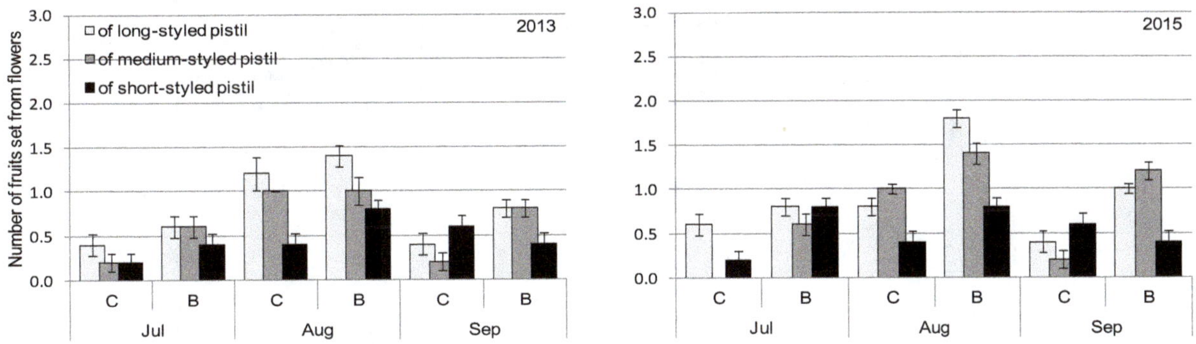

c d

Figure 6. The course of flowering and fruit setting of "Flavine" F_1 eggplant as depended on fruit type and biostimulant treatment. C, control; B, biostimulant. Bars represent mean number of flowers per plant in 2013 (**a**), 2015 (**b**) and fruits per plant in 2013 (**c**), and 2015 (**d**) (error bars indicate SE).

Table 4. Results of ANOVA for parameters of flowering and fruit setting of "Flavine" F_1 eggplant presented in Figure 6.

ANOVASource of Variation	"Flavine" F_1			
	No of Flowers 2013	No of Fruits 2013	No of Flowers 2015	No of Fruits 2015
Type of flower (F)	***	***	***	**
Biostimulant (B)	**	***	***	***
Month (M)	***	***	***	***
F × B	ns	*	*	ns
F × M	***	***	***	ns
B × M	ns	*	*	ns

Levels of significance for ANOVA: * $p \leq 0.05$; ** $p \leq 0.01$; *** $p \leq 0.001$; ns, not significant; $N = 3$. Comparisons were performed with the use of Tukey's honest significance test.

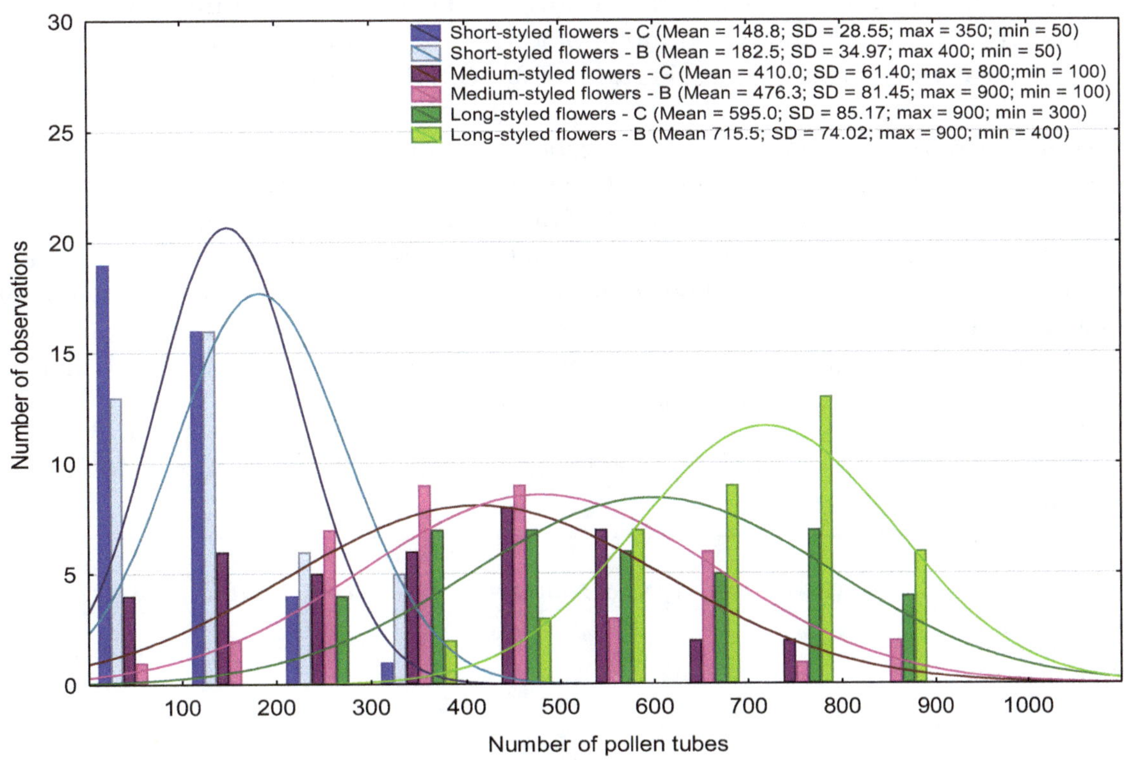

a

Figure 7. *Cont.*

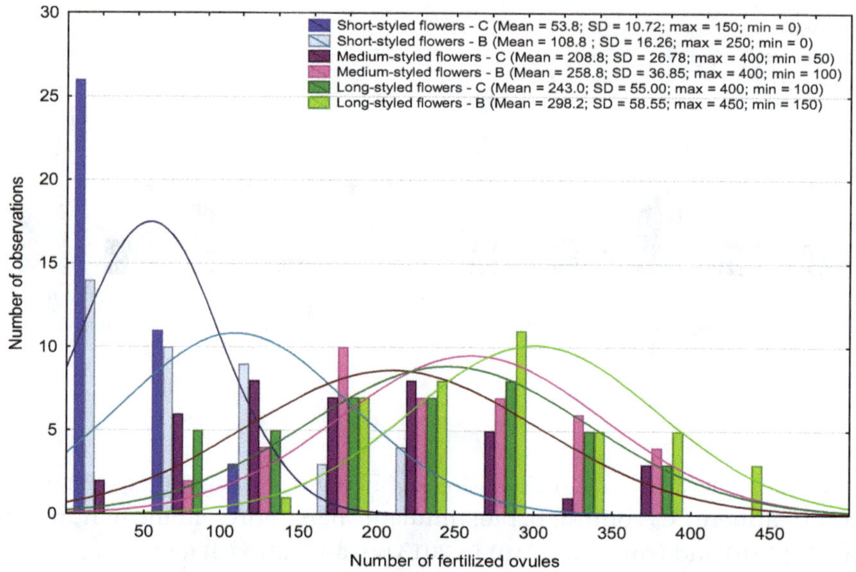

Figure 7. Number of pollen tubes (**a**) and fertilised ovules (**b**) in the styles of different flower types in "Flavine" F_1.

"Gascona" F_1 plants produced 18 flowers during the vegetation period, with 42% of these being long styled, 38% medium styled, and 20% short styled (Table 2). The number of fruits collected from a plant was 6, on average; 46% of these were from long-styled flowers, 34% from medium-styled flowers, and 20% from short-styled flowers. Biostimulant treatment significantly increased the number of medium-styled flowers in 2013 and 2015 but did not affect the number of fruits. The most effective in fruit setting were long-styled flowers. The biostimulant positively affected the percentage of fruits set by long-styled flowers in 2015 and medium-styled flowers in both years, but it negatively affected the effectiveness of fruit setting by short-styled flowers. Biostimulant treatment positively affected the seed number (Table 2). The highest number of long-, and medium-styled flowers was observed in August; the lowest was in September (Figure 8, Table 5). The number of fruits set from long- and medium-styled flowers increased from July to August, then decreased in September. Long-, medium-, and short-styled flowers were analysed regarding the number of pollen tubes in the styles. Differences in the course of pollination and fertilisation between investigated cultivars concerned the number of pollen tubes and fertilised ovules in the pistils. For control "Gascona" F_1 plants, we observed, on average, 119 pollen tubes in the middle of the style in the short-styled flowers, 418 in medium-styled flowers, and 595 in long-styled ones (Figure 9). The ovaries of the short-styled flowers contained approximately 0–50 fertilised ovules, while more fertilised ovules were found in the remaining types of flowers: 200–400 in medium- and long-styled flowers. The flowers of control plants contained lower numbers of both pollen tubes and fertilised ovules in all types of flowers.

Table 5. Results of ANOVA for parameters of flowering and fruit setting of "Gascona" F_1 eggplant presented in Figure 8.

ANOVASource of Variation	"Gascona" F_1			
	No of Flowers 2013	No of Fruits 2013	No of Flowers 2015	No of Fruits 2015
Type of flower (F)	***	***	***	*
Biostimulant (B)	*	***	ns	ns
Month (M)	***	ns	***	***
F × B	*	ns	*	ns
F × M	***	ns	***	*
B × M	ns	ns	ns	ns

Levels of significance for ANOVA: * $p \leq 0.05$; *** $p \leq 0.001$; ns, not significant; $N = 3$. Comparisons were performed with the use of Tukey's honest significance test.

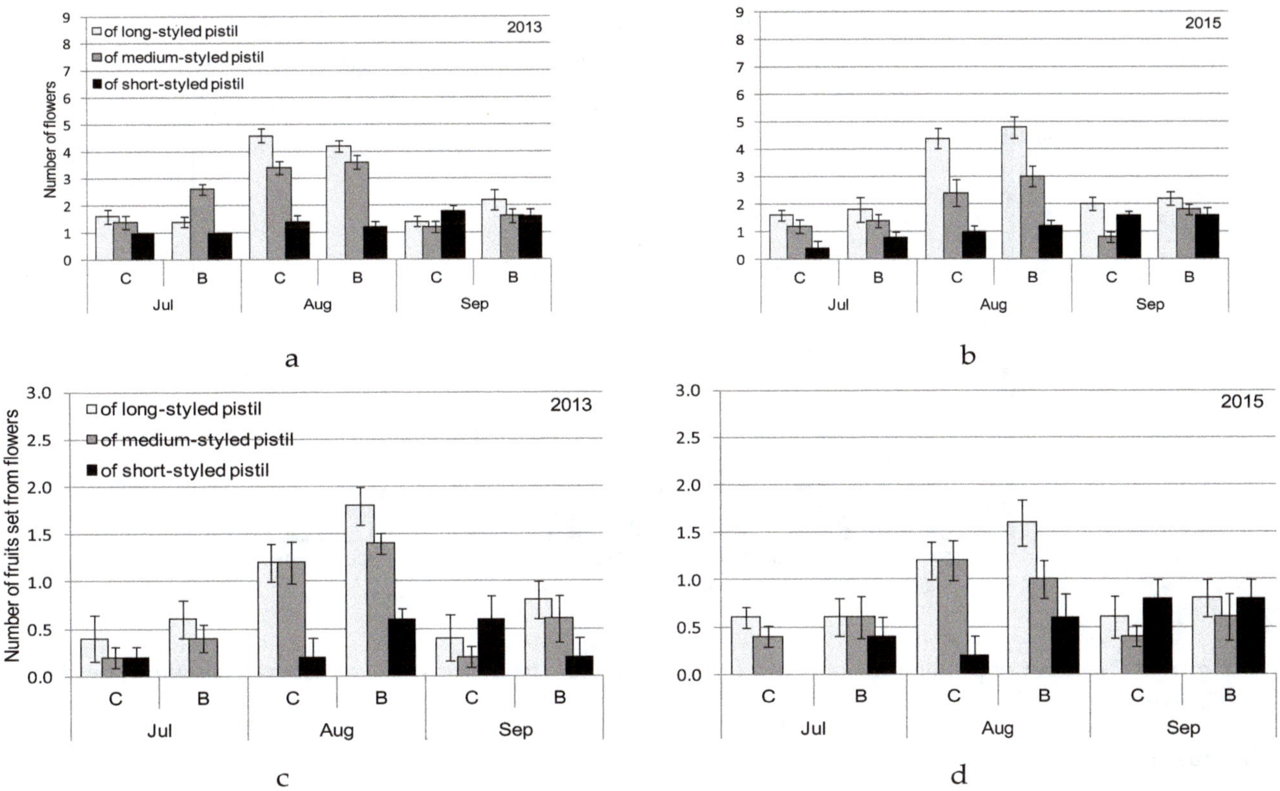

Figure 8. The course of flowering and fruit setting of "Gascona" F_1 eggplant as depended on fruit type and biostimulant treatment. C, control; B, biostimulant. Bars represent mean number of flowers per plant in 2013 (**a**), 2015 (**b**) and fruits per plant in 2013 (**c**), and 2015 (**d**) (error bars indicate SE).

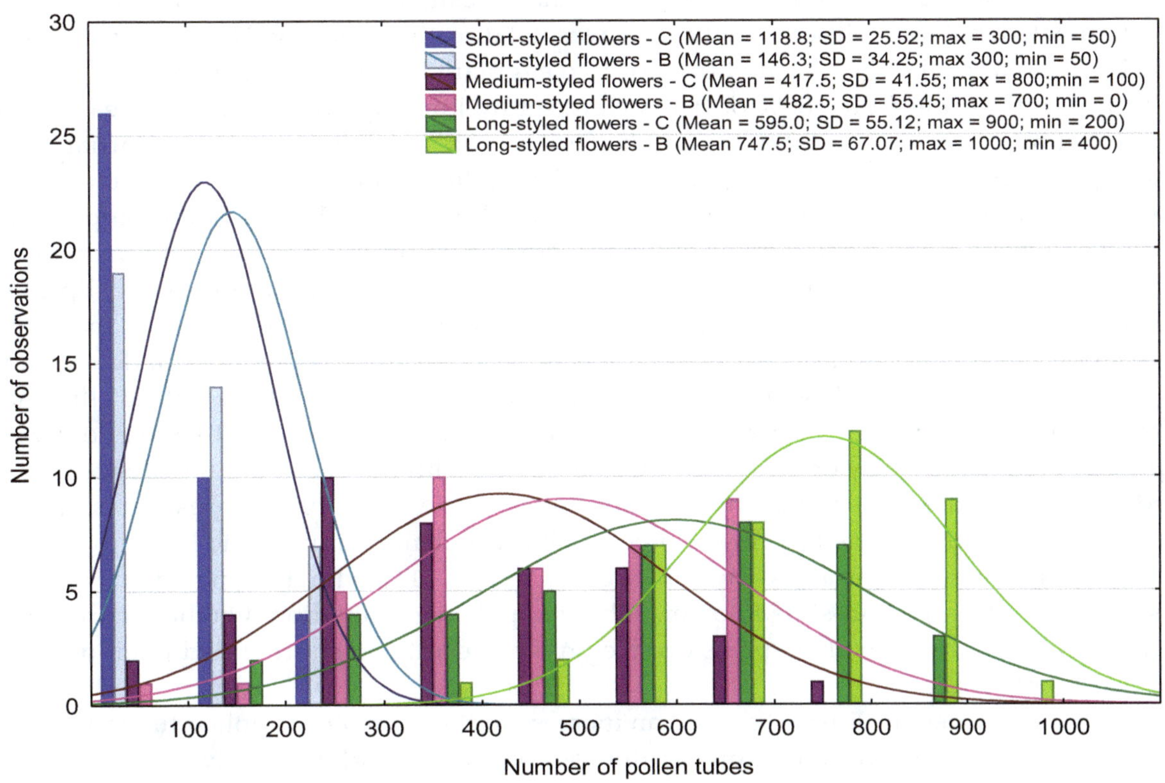

a

Figure 9. *Cont.*

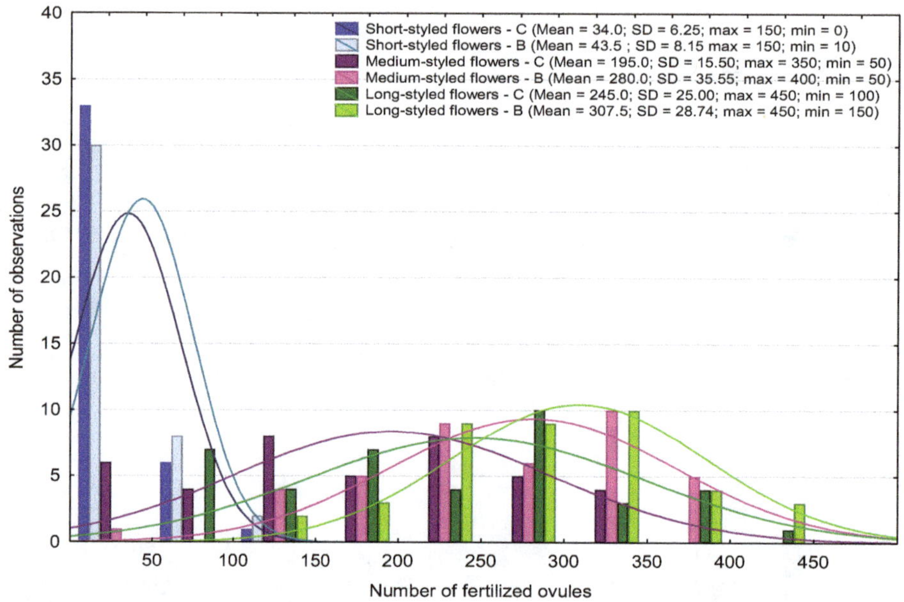

Figure 9. Number of pollen tubes (**a**) and fertilised ovules (**b**) in the styles of different flower types in "Gascona" F_1.

4. Discussion

4.1. Heterostyly Expression in Eggplant as Affected by Biostimulant Treatment and Cultivar

The recent research aimed to develop an overview of the heterostyly phenomenon in eggplant, and its implications on fruit setting biology. We demonstrated the presence of three phenotypes of flowers and the differentiated fertility of them, specific to the investigated hybrids. Generally, long-styled flowers dominated, but the fruit setting efficiency was not directly determined by the flower phenotype. A study by Srinivas et al. [32] indicated that for two eggplant hybrids of Indian breeding, with long, green and round, purple fruits, 80% of fruits were set by long-styled flowers, whereas 20% of the fruits were set by medium-styled flowers and no fruit by short-styled flowers. The partial sterility of short-styled flowers demonstrated in the cited research was due to small stigmas with under-developed papillae on which pollen grains failed to germinate. The short-styled flowers of the eggplant hybrids which are the subject of the present investigations were fertile, although the lowest number of pollen tubes and fertilised ovules was observed in this flower phenotype. Despite this fact, "Epic" F_1 and "Gascona" F_1 plants set about 20% of fruits from short-styled flowers. For "Flavine" F_1, the percentage of fruits set by this mentioned flower phenotype was 30%. Sękara and Bieniasz [6] determined that the ovules of short-styled pistils were typically developed, but that their fruit setting efficiency was low. On the contrary, results by Hazra et al. [33] indicated full sterility of short-styled flowers due to some problem related to ovary development. Observations with the use of a fluorescence microscope allowed us to verify the correct growth of pollen tubes in the styles of all types of pistils but their number was significantly affected by flower type and biostimulant treatment and by cultivar to a lesser extent. This observation is contrary to the results of Wang et al. [11], who determined that the structure of the stigmatic surface in short-styled flowers inhibited pollen germination. On the grounds of highly genotype-dependent heterostyly expression in eggplant, results on short-styled pistil performance may be divergent.

Application of Göemar BM-86® caused an increase in the numbers of pollen tubes and fertilised ovules. This phenomenon was common for all types of flowers and is directly attributable to pistil characteristics. Biostimulant-treated and control plants were not isolated, so they both could act as pollen donors. The effect of biostimulants on pollen production and fertility should also be an object of future research. Based on the available literature, we can only conclude that a wide pool of bioactive seaweed extract compounds provided balanced development and enhanced the flowering and fruiting of the investigated eggplant hybrids. The biostimulant-treated plants could be able to develop a better canopy for effective interception of light and—through a significant reduction in interplant competition for solar energy and nutrients—build suitable carbohydrate reserves earlier. Such mechanisms beyond increased flowering and fruit setting in seaweed extract treated plants were proposed by Arthur et al. [34] for bell peppers and by Helaly et al. [35] for tomatos.

The increasing number of short-styled flowers in line with plant aging, in the conditions of the present experiment, could be the result of increasing fruit load with the vegetation season's flow. Having well-developed anthers, short-styled flowers act as pollen donors to provide reproductive success. Araméndiz Tatis et al. [9] demonstrated that short-styled flowers of the "Lilac" eggplant landrace and "Long Purple" increased male fitness and thus produced an imbalance in functioning between male and hermaphrodite flowers. According to Khah et al. [36], fruit load negatively affected style length but not anther cone length in eggplant, even under favourable climatic conditions. The investigated hybrids could reduce energy outlines by creating flowers with reduced pistil and decreased fertility at the end of the vegetation period, but do so while producing pollen in the normally shaped anthers, promoting male behaviour. Short-styled flowers could be borne by fruit-loaded plants as a source of pollen for insects. The construction of the eggplant stamens is an expression of adaptation to pollination through vibrations. Such an adaptation limits the potential pollinators to species that are able to introduce vibrations into anthers, including bumblebees [13]. Bumblebees commonly visited eggplant flowers in the conditions of the presented experiment.

4.2. *Biostimulant-Affected Flower and Fruit Set Effectiveness*

Biostimulants have shown promising results in promoting flowering and reducing the fruit drop agents in many fruit trees, like apple, avocado, clementine, orange, olive, and pomegranate [37–40]. In this respect, seaweed extracts enriched in microelements are the most effective [41]. Vegetables with edible fruits are characterised by competition between the flowers and fruits at different stages of growth and in different positions in relation to inflowing assimilates [42]. Dropping of flowers, typical for eggplant, could be the mixed effect of lack of pollination or limited inflow of assimilates and the phenomenon of domination of fruit producing growth regulators. In the present research, hybrids treated with seaweed extract bore more flowers and fruits than did untreated ones. More intensive flower setting was elicited either by improved plant growth through seaweed extract application or by endogenous components, especially cytokinins, which enhance nutrient partitioning in vegetative plant organs and increase in the transport of assimilates to the growing fruits. A similar effect was observed for eggplant treated with seaweed extract by Abd El-Gawad and Osman [43]. Under the conditions of the presented experiment, biostimulant application also increased the number of pollen tubes and fertilised ovules in all types of flowers of the investigated cultivars. The overall positive influence of seaweed extracts on the plants resulted in better reproductive effectiveness and increased fruit yield and quality, described in detail by Pohl et al. [26,27]. Gómez-Cadenas et al. [44] investigated the effect of a biostimulant product containing macronutrients on citrus fruit set abscission. The beneficial effects of the biostimulant resulted from an increase in the photosynthetic efficiency which led to better transport of carbohydrates from leaves to fruit sets. Seaweed-treated apple trees also showed higher leaf chlorophyll contents and increased rates of photosynthesis and respiration due to treatment decreasing the oscillations in yield between "on" and "off" years and increasing the average fruit weight on plants affected by too high a crop load [45]. Pollination and fertilisation are very stress-sensitive stages of development [1]. Based on research on tomato, low temperatures, especially during the night, are not detrimental to ovule development but could affect stigma and style function [5]. Pollen viability is the highest at 20–22 °C [14], while the mean temperatures for the flowering period (July–September) were 16.8 and 18.8 °C in 2013 and 2015, respectively. Pollen development and viability depend on carbohydrate supply [5], so the increased photosynthetic performance of biostimulant-treated plants could improve sugar partitioning to developing pollen grains. The bioactive compounds of seaweed extracts enhance the tolerance of eggplant to abiotic stresses [46] and this tolerance can also cover the generative reproduction of this crop in temperate regions.

5. Conclusions

Eggplant is a warm climate crop and is cultivated for fruits, widely used in many world cuisines because of their unique taste and dietetic value. Nonoptimal growing conditions, especially in temperate climatic zones, affect plant flowering and fruit setting. Biostimulant application in the experiments presented herein affected the flowering biology of eggplant cultivars in different ways. Generally, the biostimulant positively affected the percentages of the most fertile medium- and long-styled flowers and the effectiveness of fruit setting by all flower phenotypes. Increased numbers of pollen tubes and fertilised ovules in all types of flowers of the investigated cultivars were noted. The overall positive influence of Göemar BM-86® on the plants resulted in increased reproductive effectiveness. Biostimulant application seems to be a promising solution to improve eggplant flowering and fruit setting in unfavourable growing conditions.

Author Contributions: Conceptualization, A.P. and A.S.; methodology, A.P. and A.S.; software, A.P. and A.S.; validation, A.P. and A.S. and A.K.; formal analysis, A.P.; investigation, A.P. and A.G.; resources, A.P. and A.S.; data curation, A.P. and A.S.; writing—original draft preparation, A.P. and A.S.; writing—review and editing, A.P. and A.S. and A.K.; visualization, A.P. and A.S.; supervision, A.S.

References

1. Pandit, M.K.; Thapa, H.; Akhtar, S.; Hazra, P. Evaluation of brinjal genotypes for growth and reproductive characters with seasonal variation. *J. Crop. Weed* **2010**, *6*, 31–34.
2. Jagatheeswari, D. Morphological studies on flowering plants (Solanaceae). *Int. Lett. Nat. Sci.* **2014**, *10*, 36–43. [CrossRef]
3. Keller, B.; Thomson, J.D.; Conti, E. Heterostyly promotes disassortative pollination and reduces sexual interference in Darwin's primroses: Evidence from experimental studies. *Funct. Ecol.* **2014**, *28*, 1413–1425. [CrossRef]
4. Glover, B.J.; Bunnewell, S.; Martin, C. Convergent evolution within the genus *Solanum*: The specialised anther cone develops through alternative pathways. *Gene* **2004**, *331*, 1–7. [CrossRef] [PubMed]
5. Karapanos, I.C.; Mahmood, S.; Thanopoulos, C. Fruit set in solanaceous vegetable crops as affected by floral and environmental factors. *Eur. J. Plant Sci. Biotechnol.* **2008**, *2*, 88–105.
6. Sękara, A.; Bieniasz, M. Pollination, fertilization and fruit formation in eggplant (*Solanum melongena* L.). *Acta Agrobot.* **2008**, *61*, 107–113.
7. Kowalska, G. The influence of heterostyly, pollination method and harmonization on eggplant's (*Solanum melongena* L.) flowering and fruiting. *Acta Agrobot.* **2003**, *56*, 61–76. [CrossRef]
8. Kowalska, G. Eggplant (*Solanum melongena* L.) flowering and fruiting dynamics depending on pistil type as well as way of pollination and flower harmonization. *Folia Hortic.* **2006**, *18*, 17–29.
9. Araméndiz Tatis, H.; Cardona Ayala, C.; Espitia Camacho, M. Floral morphology characterization of two cultivars of eggplant (*Solanum melongena* L.) (Solanaceae). *Rev. Fac. Nac. Agron. Medellin* **2009**, *62*, 5125–5134.
10. Rylski, I.; Nothmann, J.; Arcan, L. Differential fertility in short-styled eggplant flowers. *Sci. Hortic.* **1984**, *22*, 39–46. [CrossRef]
11. Wang, Y.; Liu, A.; Li, W.; Jiang, Y.; Song, S.; Li, Y.; Chen, R. Comparative proteomic analysis of eggplant (*Solanum melongena* L.) heterostylous pistil development. *PLoS ONE* **2017**, *12*, e0179018. [CrossRef] [PubMed]
12. Gemmill-Herren, B.; Ochieng, A.O. Role of native bees and natural habitats in eggplant (*Solanum melongena*) pollination in Kenya. *Agric. Ecosyst. Environ.* **2008**, *127*, 31–36. [CrossRef]
13. Kikuchi, K.; Honda, I.; Matsuo, S.; Fukuda, M.; Saito, T. Stability of fruit set of newly selected parthenocarpic eggplant lines. *Sci. Hortic.* **2008**, *115*, 111–116. [CrossRef]

14. Makrogianni, D.I.; Karapanos, I.C.; Passam, H.C. Seasonal fluctuations in pollen production and viability in eggplant and the quality of seed-containing and seedless (auxin-set) fruits. *J. Plant Growth Regul.* **2018**, *37*, 937–946. [CrossRef]
15. Sharma, M.D. Effect of plant growth regulators on growth and yield of brinjal at Khajura, Banke. *J. Inst. Agric. Anim. Sci.* **2006**, *27*, 153–156. [CrossRef]
16. Netam, J.L.; Sharma, R. Efficacy of plant growth regulators on growth characters and yield attributes in brinjal (*Solanum melongena* L.) cv. Brinjal. 3112. *J. Agric. Veterin. Sci.* **2014**, *7*, 27–30. [CrossRef]
17. Sabatino, L.; Iapichino, G.; D'Anna, F.; Palazzolo, E.; Mennella, G.; Rotino, G.L. Hybrids and allied species as potential rootstocks for eggplant: Effect of grafting on vigour, yield and overall fruit quality traits. *Sci. Hort.* **2018**, *228*, 81–90. [CrossRef]
18. Sabatino, L.; Iapichino, G.; Rotino, G.L.; Palazzolo, E.; Mennella, G.; D'Anna, F. *Solanum aethiopicum* gr. gilo and its interspecific hybrid with *S. melongena* as alternative rootstocks for eggplant: Effects on vigor, yield, and fruit physicochemical properties of cultivar 'Scarlatti'. *Agronomy* **2019**, *9*, 223. [CrossRef]
19. Caruso, G.; Pokluda, R.; Sękara, A.; Kalisz, A.; Jezdinský, A.; Kopta, T.; Grabowska, A. Agricultural practices, biology and quality of eggplant cultivated in Central Europe. A Review. *Hort. Sci.* **2017**, *44*, 21–42. [CrossRef]
20. Grabowska, A.; Kunicki, E.; Sękara, A.; Kalisz, A.; Jezdinský, A.; Gintrowicz, K. The effect of biostimulants on the quality parameters of tomato grown for the processing industry. *Agrochimica* **2015**, *59*, 203–217.
21. Kopta, T.; Pavlikova, M.; Sękara, A.; Pokluda, R.; Marsalek, B. Effect of bacterial-algal biostimulant on the yield and internal quality of lettuce (*Lactuca sativa* L.) produced for spring and summer crop. *Not. Bot. Horti. Agrobot. Cluj-Napoca* **2018**, *46*, 615–621. [CrossRef]
22. DuJardin, P. Plant biostimulants: Definition, concept, main categories and regulation. *Sci. Hortic.* **2015**, *196*, 3–14. [CrossRef]
23. Di Stasio, E.; Van Oosten, M.J.; Silletti, S.; Raimondi, G.; Dell'Aversana, E.; Carillo, P.; Maggio, A. *Ascophyllum nodosum*-based algal extracts act as enhancers of growth, fruit quality, and adaptation to stress in salinized tomato plants. *J. Appl. Phycol.* **2018**, *30*, 1–12. [CrossRef]
24. Sharma, H.S.; Fleming, C.; Selby, C.; Rao, J.R.; Martin, T. Plant biostimulants: A review on the processing of macroalgae and use of extracts for crop management to reduce abiotic and biotic stresses. *J. Appl. Phycol.* **2014**, *26*, 465–490. [CrossRef]
25. Yakhin, O.I.; Lubyanov, A.A.; Yakhin, I.A.; Brown, P.H. Biostimulants in plant science: A global perspective. *Front Plant Sci.* **2017**, *7*, 2049–2080. [CrossRef] [PubMed]
26. Pohl, A.; Grabowska, A.; Kalisz, A.; Sękara, A. Preliminary screening of biostimulative effects of Göemar BM-86 on eggplant cultivars grown under field conditions in Poland. *Acta Agrobot.* **2018**, *71*, 1752. [CrossRef]
27. Pohl, A.; Grabowska, A.; Kalisz, A.; Sękara, A. The eggplant yield and fruit composition as affected by genetic factor and biostimulant application. *Not. Bot. Horti. Agrobot. Cluj-Napoca* **2019**, *47*. [CrossRef]
28. Sękara, A.; Bączek-Kwinta, R.; Gawęda, M.; Kalisz, A.; Pokluda, R.; Jezdinský, A. Sequential abiotic stress applied to juvenile eggplant modifies the seedlings parameters, plant ontogeny and yield. *Hort. Sci.* **2016**, *43*, 149–157. [CrossRef]
29. UPL Limited, Arysta Lifescience. Available online: https://www.arystalifescience.com/ (accessed on 23 January 2019).
30. Sękara, A. *Biology of the Vegetative and Generative Development of Eggplant (Solanum melongena L.) in the Field Production. Chosen Aspects*; Zesz Nauk UR Wydawnictwo Uniwersytetu Rolniczego: Krakow, Poland, 2010; Volume 459, pp. 19–24.
31. Martin, F.W. Staining and observing pollen tubes in the style by means of fluorescence. *Stain. Technol.* **1959**, *34*, 125–128. [CrossRef]
32. Srinivas, G.; Jayappa, A.H.; Patel, A.I. Heterostyly: A threat to potential fruit yield in brinjal (*Solanum melongena* L.). *Adv. Life Sci.* **2016**, *5*, 1211–1215.
33. Hazra, P.; Mandal, J.; Mukhopadhyay, T.P. Pollination behaviour and natural hybridization in *Solanum melongena* L., and utilization of functional male sterile line in hybrid seed production. *Caps. Eggpl. Newsl.* **2003**, *22*, 143–146.
34. Arthur, G.D.; Stirk, W.A.; Van Staden, J. Effect of a seaweed concentrate on the growth and yield of three varieties of *Capsicum annuum*. *S. Afr. J. Bot.* **2003**, *69*, 207–211. [CrossRef]
35. Helaly, M.N.; Arafa, A.A.; Ibrahim, H.M.; Ghoniem, K.H. Improving growth and productivity of tomato by some biostimulants and micronutrients with or without mulching. *J. Phytol.* **2018**, *10*, 15–23.

36. Khah, E.M.; Antonopoulos, A.; Passam, H.C. Floral behaviour and fruit set in four cultivars of aubergine. *Acta Hortic.* **2002**, *579*, 259–264. [CrossRef]
37. Fornes, F.; Sánchez-Perales, M.; Guardiola, J.L. Effect of a seaweed extract on the productivity of 'de Nules' Clementine mandarin and Navelina orange. *Bot. Mar.* **2002**, *45*, 486–489. [CrossRef]
38. Basak, A. Effect of preharvest treatment with seaweed products, Kelpak® and Goëmar BM 86®, on fruit quality in apple. *Int. J. Fruit Sci.* **2008**, *8*, 1–14. [CrossRef]
39. Abubakar, A.R.; Naira, A.; Moieza, A. Effect of plant biostimulants on flowering, fruit drop, yield and return bloom of pomegranate cv. Kandhari Kabuli. *Asian J. Hortic.* **2012**, *7*, 473–477.
40. Morales-Payan, J.P.; Candelas, C.D. Increasing organic avocado fruit yield using a *Ascophyllum nodosum* biostimulant and fertilization. *Acta Hortic.* **2014**, *1042*, 121–124. [CrossRef]
41. Chouliaras, V.; Tasioula, M.; Chatzissavvidis, C.; Therios, I.; Tsabolatidou, E. The effects of a seaweed extract in addition to nitrogen and boron fertilization on productivity, fruit maturation, leaf nutritional status and oil quality of the olive (*Olea europaea* L.) cultivar Koroneiki. *J. Sci. Food Agric.* **2009**, *89*, 984–988. [CrossRef]
42. Marcelis, L.F.M.; Elings, A.; Bakker, M.J.; Brajeul, E.; Dieleman, J.A.; de Visser, P.H.B.; Heuvelink, E. Modelling dry matter production and partitioning in sweet pepper. *Acta Hortic.* **2006**, *718*, 121–128. [CrossRef]
43. Abd El-Gawad, H.G.; Osman, H.S. Effect of exogenous application of boricacid and seaweed extract on growth, biochemical content and yield of eggplant. *J. Hortic. Sci. Ornam. Plants* **2014**, *6*, 133–143.
44. Gómez-Cadenas, A.; Pérez-Santamarina, R.; Ghorbel, R. Effect of a biostimulant product containing macronutrients and a carboxylic acid (AMEC®) on citrus fruitlet abscission. *Acta Hort.* **2012**, 189–193.
45. Spinelli, F.; Fiori, G.; Noferini, M.; Sprocatti, M.; Costa, G. Perspectives on the use of a seaweed extract to moderate the negative effects of alternate bearing in apple trees. *J. Hort. Sci. Biotechnol.* **2009**, *84*, 131–137. [CrossRef]
46. Van Oosten, M.J.; Pepe, O.; De Pascale, S.; Silletti, S.; Maggio, A. The role of biostimulants and bioeffectors as alleviators of abiotic stress in crop plants. *Chem. Biol. Technol. Agric.* **2017**, *4*, 5. [CrossRef]

7

Bacillus subtilis CBR05 for Tomato (*Solanum lycopersicum*) Fruits in South Korea as a Novel Plant Probiotic Bacterium (PPB): Implications from Total Phenolics, Flavonoids and Carotenoids Content for Fruit Quality

Murugesan Chandrasekaran [1], Se Chul Chun [2], Jae Wook Oh [3,*], Manivannan Paramasivan [4], Ramesh Kumar Saini [5] and Jesudoss Joseph Sahayarayan [6]

1 Department of Food Science and Biotechnology, Sejong University, Gwangjin-gu, Seoul 05006, Korea; chandrubdubio@gmail.com
2 Department of Environmental Health Science, Konkuk University, Gwangjin-gu, Seoul 05029, Korea; scchun@konkuk.ac.kr
3 Department of Stem Cell and Regenerative Biotechnology, Konkuk University, Gwangjin-gu, Seoul 05029, Korea
4 Department of Microbiology, Bharathidasan University, Tiruchirappalli 620024, India; manivannan3065@gmail.com
5 Department of Crop Science, Konkuk University, Gwangjin-gu, Seoul 05029, Korea; saini_1997@yahoo.com
6 Department of Bioinformatics, Alagappa University, Karaikudi 630 003, India; jjsrbioinformatics2016@gmail.com
* Correspondence: ohjw@konkuk.ac.kr

Abstract: Plant growth-promoting rhizobacteria (PGPR) are naturally occurring soil biota which benefit plants by improving plant productivity and immunity. The aim of the present work was to evaluate the effect of the inoculation of PGPR strain, *Bacillus subtilis* CBR05 on the quality of tomato fruits produced under greenhouse conditions. Results were compared with mock-inoculated control and market sample. We found a significant increase in total phenol and flavonoid contents of tomato fruits in PGPR strain *B. subtilis* CBR05 inoculated plants compared to those of mock-inoculated control and market sample. Moreover, *B. subtilis* CBR05 inoculation stimulated antioxidant activities and levels of carotenoid (β carotene and lycopene) content in plants. In addition, the inoculation of the strain *B. subtilis* CBR05 produced the highest content of lycopene (21.08 μg/g FW) in tomato fruits as compared to mock-inoculated plants. Our results show that the PGPR strain *B. subtilis* CBR05 is a versatile soil bacterium that enhances tomato production by elevating antioxidant activities and carotenoid (β carotene and lycopene) levels in fruit.

Keywords: *Bacillus subtilis*; tomato; antioxidant activity; carotenoids; probiotics; PGPR

---

1. Introduction

Tomato (*Solanum lycopersicum*) is regarded as the second most vegetable crop next to potato in the agricultural implications of human consumption. According to agricultural statistics, tomatoes along with sweet corn and snap beans constitute 93% of crop production and processing strategies (Agricultural Statistics, United States Department of Agriculture (USDA), 2016). The positive benefits of tomato consumption have been rigorously proved against a variety of diseases like chronic degenerative diseases, owing to the escalated content of significant phytochemicals with potent health benefits, like the carotenoids (β-carotene and lycopene), the glycoalkaloids (dehydrotomatine and α-tomatine), ascorbic acid, tocopherols, and many phenolic and flavonoid compounds [1–3]. Tomato also contributes as a major dietary ingredient for Vitamin A and C which implies increased per individual consumption in the United States and many Western countries [2,4]. Fruit ripening in tomato comprises a cascade of events on biochemical, physiological, and structural perspectives involving the influence of secondary metabolites that confer flavor, aroma, texture, and appearance of the tomato [5,6]. Accumulation of large quantities of pigments, especially lycopene and β carotene, inside the plastoglobules of chromoplast provides a visual indication that the fruit is mature and suitable for consumption [4].

Agronomic practices are recognized as a vital factor in determining the nutritional quality of tomato crops [7,8]. The nutrient contents in tomato fruits depend on the environment in which they grow [9,10]. Nowadays, the use of crop modeling to identify effective farmer strategies to counteract adverse future climatic conditions has become a standard in climate change impact assessments [11–13]. Over the past few years, a variety of methods have been proposed to comprehensively assess fruit quality and its relationship with water, including principal component analysis (PCA), analytic hierarchy process (AHP), gray relational analysis (GRA), and technique for order preference by similarity to ideal solution (TOPSIS) [14–17]. The challenge of producing fresh fruits and vegetables is increasing both yield and quality to satisfy consumers as the environment changes in ways that are deleterious to crop species [18]. The quality of agricultural products is affected by many pre- and postharvest factors [9]. The utilization of biofertilizers that mitigate these adverse environmental effects has become a feasible and beneficial production practice. Plant growth-promoting rhizobacteria (PGPR) may be considered as preharvest biotic factors that mitigate adverse environmental effects and promote improved crop yield and quality [19,20].

Among various PGPR approaches, *Bacillus* species are considered as likely candidates due to their broad-spectrum antagonistic activity against phytopathogens, production of long-lived and stress-tolerant spores, secondary metabolites, lytic enzymes, resistance to adverse environments, and plant growth promotion [19,21–23]. *Bacillus subtilis* plays a significant role in improving plant growth and tolerance to both biotic and abiotic stresses. PGPR strains also act as bio-stimulants of phytohormones and peptide synthesis [20,24], but studies of the PGPR strain, *B. subtilis* CBR05 on tomato have not yet appeared. Preharvest factors that directly affect crop yield and quality can be summarized into biological factors comprising pathological, entomological, and animal issues, which was found to be nullified upon increased usage of PGPR. The dire need for assessment of tomato and tomato-based products is given significant attention concerning nutrition and quality relying on the nature of the variety, maturity at harvest, effective transport, and storage [25]. Characterization of the carotenoids, mainly β-carotene and lycopene during storage and various ripening stages, shows drastic developments in sustainable yield and quality parameters of tomato [26]. This information adds to our understanding of temporal differentiation of nutritionally significant phytochemicals during ripening of tomato fruits. The objective of this study was to evaluate the effects of *B. subtilis CBR05* on the quality of tomato fruits under greenhouse conditions.

2. Material and Methods

2.1. Chemicals and Reagents

Authentic standards of carotenoid, all-*E*-lutein, was purchased from Cayman Chemical Company, Michigan, USA. All-*E*-β-carotene were purchased from Sigma-Aldrich, St. Louis, MO, USA. Only the HPLC grade of organic solvents was employed in carotenoid extraction (Daejung, Siheung-*si*, Korea).

2.2. Bacterial Strain and Culture Conditions

B. subtilis CBR05 isolated from our lab maintained on tryptic soy agar (TSA) plates. For long-time storage, bacterial cultures were maintained in tryptic soy broth at −80 °C. For experimental purposes, the cultures were transferred to TSA (MBCell, Seoul, Korea) and incubated at 30 °C for 24 h. The inoculum mixture of the strain was prepared by culturing in nutrient broth and incubating at 28 °C with constant shaking at 130 rpm. The bacterial cells were centrifuged at 10,000 g for 10 min at 4 °C. The cell precipitate was resuspended in 10 mM $MgCl_2$ and the cell concentration of bacterial suspensions of *B. subtilis* CBR05 was adjusted to 10^8 colony-forming unit (CFU)/mL (OD_{600} = 1.0) for further studies.

2.3. Plant and Growth Conditions

Tomato seeds (Korean cultivar, Kwangbok) utilized for this study were obtained from a Korean seed resource center, Seoul, South Korea. They were surface sterilized in sodium hypochlorite, rinsed several times with distilled water, and planted onto pots containing sterilized growth media (Peat moss with perlite in a ratio of 3:1). Two sets of plants (three plants per set) were maintained, one without PGPR (mock-inoculated control) inoculation and the other with PGPR strain, *B. subtilis* CBR05 inoculum. All treatments were placed randomly in the greenhouse and replicated 3 times. Plant growth-promoting *B. subtilis* CBR05 were applied under sterile conditions to the base of the plants close to the roots to ensure better colonization. Plants were maintained under greenhouse conditions at a temperature of 25 °C with watering carried out every alternate day, to make a better availability for nutrition and plant growth promotion.

2.4. Antioxidant Assays

2.4.1. DPPH Assay

1,1-diphenyl-2-picrylhydrazyl (DPPH) assay was performed to assess the fractions exhibiting scavenging property of free radicals in vitro [27]. Then, 0.2 mM solution of DPPH in ethanol was added to the fraction of aliquot at concentrations (100 μg/mL). The mixture was allowed to stand for 30 min and the absorbance was measured at 517 nm using a UV-Visible spectrophotometer. The percent scavenging activity was determined and Trolox was used as the standard.

2.4.2. ABTS (2,2′-azino-bis (3-ethylbenzothiazoline-6-sulfonic acid) Assay

A Trolox equivalent antioxidant capacity (TEAC)/ABTS assay was conducted based on the method of Ramos et al. [28]. The ABTS solution (7 mM) was oxidized with potassium peroxodisulfate (2.45 mM) for 16–18h at room temperature. The ABTS solution was diluted with solvents. An aliquot (100 μg/mL) was mixed with diluted ABTS solution and the absorbance was read at 734 nm. Trolox and ascorbic acid were used as reference standards.

2.5. Determination of Total Phenolic Contents

The Folin–Ciocalteu method was used to estimate various concentrations concerned with the total phenolic content. The extracts were dissolved in absolute methanol and later 200 μL of the extract was mixed with 800 μL of 1 N Folin–Ciocalteu reagent (1: 10). After 5 min at room temperature, 3 mL of sodium carbonate (15%) was added to the extracts. Following incubation for 30 min at room temperature, the absorbance was read at 765 nm using a UV-spectrophotometer. A standard curve for gallic acid equivalents (GAE) (milligrams per gram of extract (mg GAE/ge)) was utilized to evaluate the concentration of total phenolic compounds. Analyses were performed in triplicate per each extract.

2.6. Determination of Total Flavonoid Contents

Screening the content of flavonoids was done by a modified protocol as reported previously [29]. The extracts were dissolved in absolute methanol. In a 15 mL conical tube, 1 mL of a sample was mixed with 0.3 mL of 5% sodium nitrite, followed by incubation for 5 min. After incubation, 0.3 mL of aluminum chloride (10%) and 2 mL of sodium hydroxide (1 mol/L) were added to the reaction mixture, and the absorbance was read at 496 nm with a UV-spectrophotometer, using catechin as the standard. Quercetin equivalents (QE) present per g of extract (mg QE/ge) was used to quantify the expression levels.

2.7. *Extraction and Quantification of Carotenoids*

Carotenoids were extracted in triplicates and quantified according to previously established protocol with minor modifications [30,31]. All the preparations were performed in low light conditions to avoid the degradation of carotenoids. Three independent biological samples were extracted separately. Briefly, one whole tomato fruit was finely chopped and mixed thoroughly. Five grams of chopped fruits (exact to 0.001 g) from each treatment were separately transferred into test tubes containing 20 mL of acetone and 0.1% butylated hydroxytoluene (BHT: *w/v*). The samples were homogenized with a mechanical homogenizer and centrifuged at 5000× *g* (5 min at 4 °C temperature). The supernatant was recovered and pelleted samples were repeatedly extracted until the pallets became colorless. Supernatants from all extractions were pooled and vacuum-dried in a rotary evaporator (Temperature < 35 °C) (Büchi RE 111, Switzerland).

2.8. *HPLC Analysis*

The extract was recovered with 10 mL of methylene chloride (CH_2Cl_2) containing 0.1% BHT and transferred to an amber color HPLC vial for HPLC analysis. The chromatographic separation was achieved using an Agilent Model 1100 HPLC instrument (Agilent Technologies Canada Inc., Mississauga, ON, Canada) equipped with a degasser, autosampler, dual pump, and diode array detector (DAD). Samples were scanned (200–800 nm) with 0.05 min (1 s) response time, 8.0 mm slit width, and a detection wavelength of 450 (for most of the carotenoids) and 470 nm (for lycopene). The bandwidth was ±16 nm for all detection wavelengths. Similarly, 600 nm was used as a reference wavelength with ±50 nm bandwidth in all detections. The column used was a YMC, C30 carotenoid column, 250 × 4.6 mm, 5 µm (YMC, Wilmington, NC, USA), and the chromatographic data were recorded with ChemStation LC 3D software. The column thermostat was maintained at 25 °C temperature. Then, 20 µL of standards and samples were injected with an autosampler. The solvent system consisted of Methanol: methyl tertiary butyl ether (MTBE): water (81:15:4) (Mobile phase A) and MTBE: Methanol (91:9) (Mobile phase B). The gradient elution was 0%–100% B in 90 min, and 5-min post-run at a flow rate of 1 mL/min.

2.9. *Statistics*

All of the experiments were conducted in triplicate and results were tabulated as the Mean ± standard deviation (SD). Statistical significance of the data was determined using one-way analysis of variance (one-way ANOVA) followed by Fisher LSD (Least Significant Difference) test. Data analyses were performed using Sigmastat v8.02 (Systat Software Inc., San Jose, CA, USA). A p value of ≤ 0.05 was considered significant.

3. Results

In the present investigation, the effects of the PGPR strain, *B. subtilis* CBR05 inoculation on the maintenance of carotenoids, total phenolics, flavonoid contents, and antioxidant properties were evaluated. The results revealed that the PGPR strain, *B. subtilis* CBR05 has the capacity to improve the plant growth and change some of the tomato fruit quality characteristics under greenhouse conditions. Tomato plants treated with PGPR strain, *B. subtilis* CBR05 showed significantly higher biomass compared with mock-inoculated controls. Significant increases in root length and dry weight, over mock-inoculated controls, were achieved in green-house conditions (data not shown). In addition, antioxidant activities of three tomato fruits (mock-inoculated, *B. subtilis* CBR05 inoculated, and market fruit) were determined using both DPPH and ABTS radical scavenging method. The extract of fruits from plants inoculated with *B. subtilis* CBR05 strain was the most active against DPPH and ABTS radical and that of fruits from the market, the differences being significant when compared to mock-inoculated control (Figure 1).

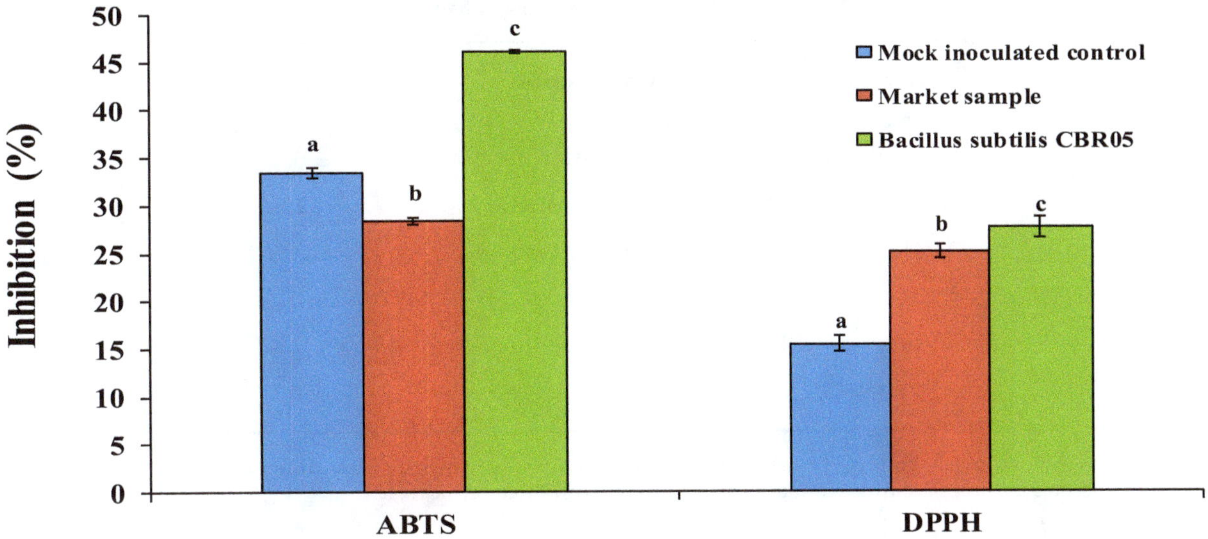

Figure 1. Antioxidant assays ((2,2′-azino-bis (3-ethylbenzothiazoline-6-sulfonic acid) (ABTS) and 1,1-diphenyl-2-picrylhydrazyl (DPPH)). Values are mean ± standard deviation from three replicates. Bars followed by the same letter(s) are not significantly different ($p \leq 0.05$).

The total phenolic content of the fruit extracts is shown in Figure 2. Among the fruits, the PGPR strain, *B. subtilis* CBR05 inoculated plants had the highest phenolic content followed by market fruits. Moreover, the flavonoid contents were also elevated in PGPR treated tomato fruits as compared to those of both mock-inoculated control and market fruits (Figure 2). We did not find any significant differences between mock-inoculated control and market fruit. The results of this study show that *B. subtilis* CBR05 is an effective probiotic agent for the promotion of tomato fruit quality.

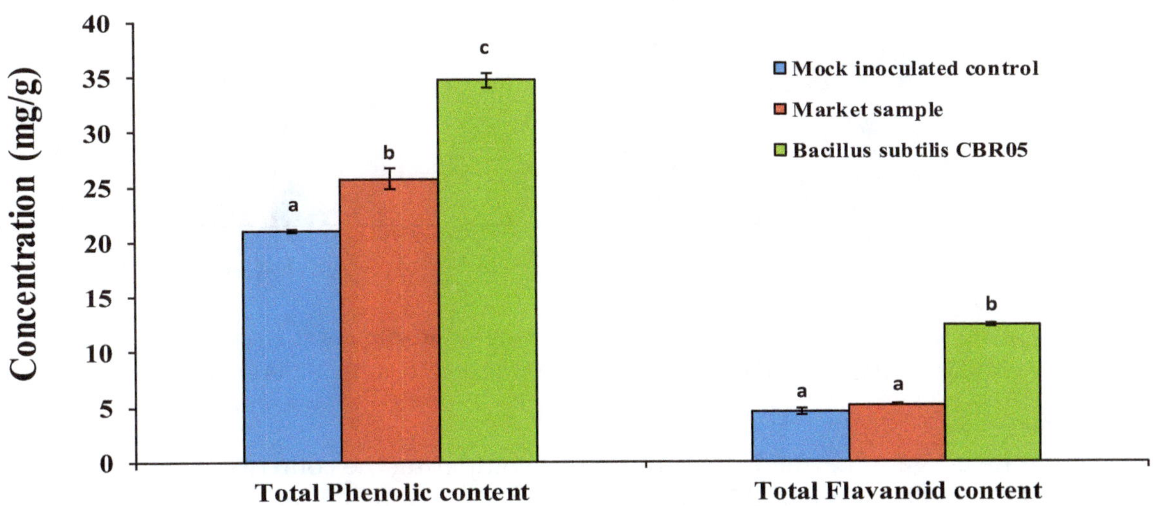

Figure 2. Total phenol and flavonoids contents. Values are mean ± standard deviation from three replicates. Bars followed by the same letter(s) are not significantly different ($p \leq 0.05$).

The carotenoids, such as lycopene and β-carotene, were extracted and quantified (Table 1). The contents and composition of carotenoids were found to differ among control, *B. subtilis* CBR05 inoculated, and market fruits (Figure 3). Carotenoid levels (lycopene) in tomato fruits in plants treated with PGPR strain, *B. subtilis* CBR05 were higher than both the control and market fruits (Table 1; Figure 3).

Table 1. Contents of carotenoids in tomato fruits.

Treatment	Carotenoids	Content (μg/g FW)
Mock inoculated control	All-*E*-β-Carotene	5.63 ± 0.28 [a]
	All-*E*-Lycopene	7.48 ± 1.96 [a]
Market tomato	All-*E*-β-Carotene	4.65 ± 0.96 [a]
	All-*E*-Lycopene	10.51 ± 3.34 [a]
Bacillus subtilis CBR05	All-*E*-β-Carotene	3.53 ± 0.12 [a]
	All-*E*-Lycopene	21.08 ± 0.32 [b]

In each experiment, means followed by different superscript letters are significantly different among the treatment groups ($p \leq 0.05$). Values are mean ± standard deviation from three replicates.

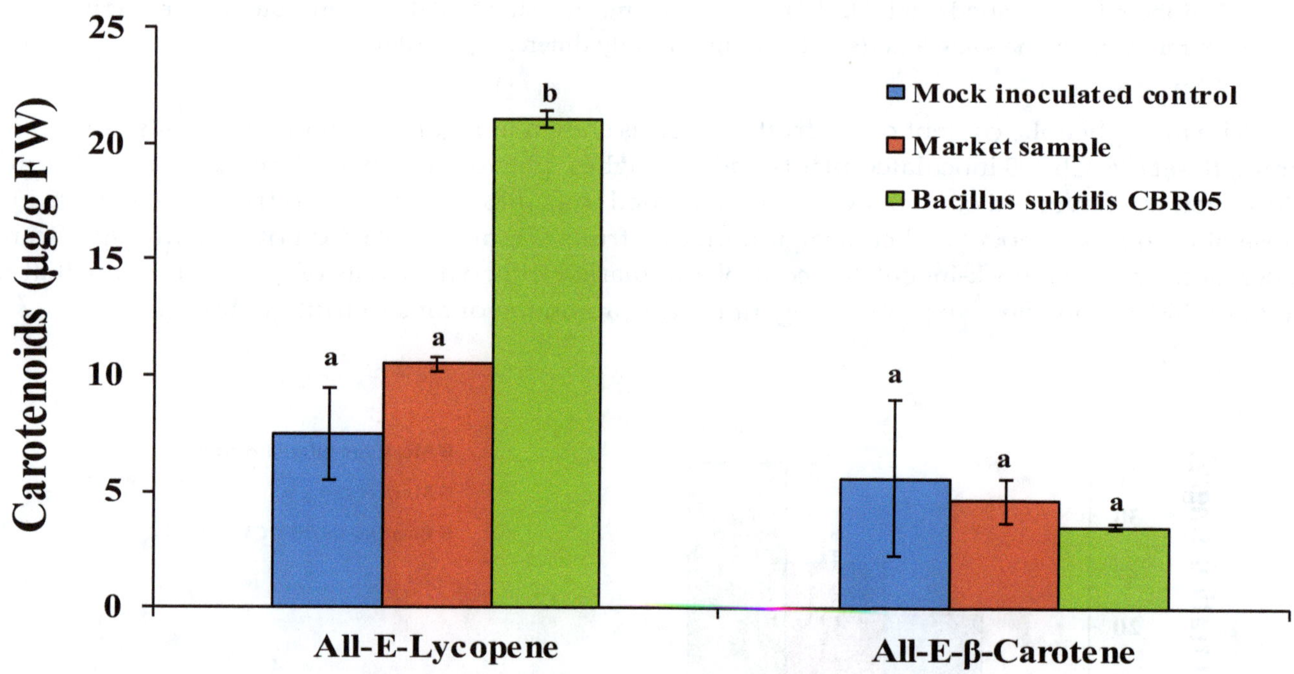

Figure 3. Carotenoids' contents (lycopene and β-carotene). Values are mean ± standard deviation from three replicates. Bars followed by the same letter(s) are not significantly different ($p \leq 0.05$).

The present investigation extends their results by comparing the contents of carotenoids from the fruits using HPLC–DAD (Figure 4). Using this methodology, all-E- β -carotene and all-E- lycopene were identified as the major carotenoids in tomato fruits, based on retention time with standards and by comparing the peak spectra recorded with a DAD during the analysis. The chromatograms (470 nm) and the peak spectra of major identified peaks were shown in Figure 4. The other minor carotenoids were not quantified due to the unavailability of standard compounds. We did not find any significant differences in β-carotene. However, in the present study, we have recorded a significantly higher amount of lycopene (All-E-lycopene) in fruits during ripening in PGPR strain, *B. subtilis* CBR05 inoculated plants than those of both mock-inoculated control and market fruits. These results validate the productive roles of the *B. subtilis* CBR05 in enhancing the nutritional potential of tomato fruits.

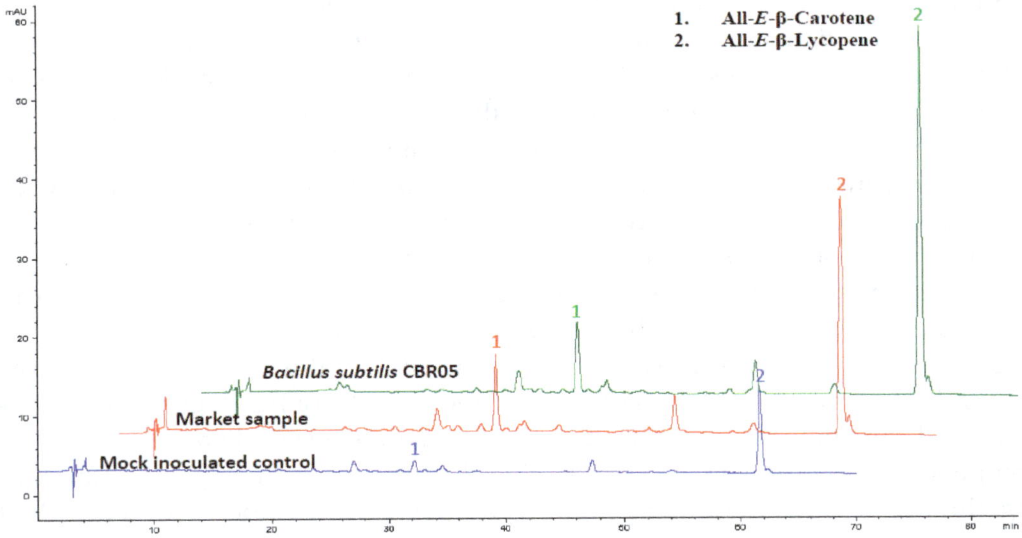

Figure 4. HPLC chromatogram of carotenoids.

4. Discussion

The use of PGPR is increasing in agriculture and may offer an attractive alternative to synthetic chemicals and fertilizers. Plant growth-promoting microorganisms are efficient microbial competitors that can promote plant growth by producing phytohormones and/or by increasing available nutrients through production of secondary metabolites or act as biocontrol agents to protect plants from infection by phytopathogens [19,21–24]. There have been many reports on PGPR and their effective roles [19–24]. Insufficient experimental work has been reported to speculate on the mechanisms of PGPR effects on fruit quality. In the present study, PGPR strain, *B. subtilis* CBR05 isolated from rice were used as inoculants for tomato plants grown under greenhouse conditions. Our results showed that PGPR inoculations significantly increased the total biomass and root length compared to those in the control. *B. subtilis* CBR05 appears to impart plant growth promotion effects that are distinct from other commercial biocontrol agents. Tomato fruits are a good source of antioxidant compounds that can reduce harmful oxidation reactions in the human body, thus preventing various diseases associated with free radical oxidation, such as cardiovascular and neurological disorders and cancer [1–4].

Antioxidant activity has been widely used to test the ability of plant extracts to act as free radical scavengers [32]. In the present investigation, *B. subtilis* CBR05 had a net positive effect on the antioxidant activity measured by the DPPH and ABTS scavenging capacity, which seems to indicate that the bacteria acted as a regulator of the synthesis of antioxidant compounds in the plant (Figure 1). Strong scavenging of ion radical was exhibited by the inoculated tomato fruits, thus showing that *B. subtilis* CBR05 inoculation increased the radical scavenging capacity of tomato fruits. In a previous study, *B. licheniformis* inoculated plants had increased antioxidant profiles in tomato plants under greenhouse conditions [32]. Similar results concerned with enhanced fruit quality and marketable grade have been reported for other crops under the influence of PGPR [33] but this report is the first of its kind to specifically address PGPR strain, *B. subtilis* CBR05 in improvement of fruit quality. Moreover, PGPR enhances fruit characteristics based on the mediation of increased availability of nutrients to plants like phosphorous and iron, enhancing the nutritional status of the plants in the rhizosphere [33–35].

Phenolic contents perform an essential role in plant resistance and defense against phytopathogens, which are closely linked with reactive oxygen species (ROS). This study reveals that among the selected tomato fruits, PGPR strain, *B. subtilis* CBR05 inoculated tomato fruits had the highest amount of phenolics (Figure 2). Some phenolic compounds may prevent oxidative damage in vivo and thus protect against the development of the disease such as cardiac disease and cancer [1,2]. This might be considered as useful for health purposes. In addition, inoculation of *B. subtilis* significantly increased flavonoid content compared to those of both control and market fruits. Similarly, increases in total flavonoids content by *B. licheniformis* have been reported for tomato fruits [32]. In our previous studies, we also found that *B. subtilis* CBR05 inoculation enhanced the accumulation of peroxidase and polyphenol oxidase enzymes, which are involved in the metabolism of phenols and flavonoids [22,23]. Hence, this shows that *B. subtilis* colonization induces resistance against biotic and abiotic stress agents. The results of antioxidant assays also revealed that tomatoes are a rich source of antioxidants, thus their habitual consumption can potentially help to combat the oxidative stress.

Regulation of carotenoid biosynthesis and high-accumulation lycopene during tomato fruit development is widely studied [36–38]. In the present study, we found a significantly higher amount of lycopene (All-E-lycopene) in *B. subtilis* CBR05 inoculated tomato fruits (Figures 3 and 4). Lycopene possesses the highest antioxidant potential among the carotenoids and several other antioxidants found in fruits and vegetables [39]. Thus, the addition of PGPR enhances lycopene content in tomato fruits and can potentially contribute to antioxidant levels of diets. This potent antioxidant activity of lycopene protects from a variety of ROS and reactive nitrogen species (RNS), thus helping in preventing chronic diseases in humans [31,36]. Similar to the lycopene contents, the DPPH and ABTS antioxidant activity of *B. subtilis* CBR05 inoculated tomato fruits was much higher than both the control and market fruits. Earlier, we also reported that defense-related enzymes in tomato after treatment with *B. subtilis* CBR05 efficiently combated *X. campestris* pv, *vesicotoria* [40] and induction of defense-related enzymes like superoxide dismutase, catalase, peroxidase, and polyphenol oxidase assessment revealed the up-regulation of glucanase and phenyl ammonia lyase indicating induced systemic resistance (ISR) in tomato. The earlier results established that the antioxidant capabilities of tomatoes are naturally present. Further, it was specifically proved that *B. subtilis* CBR05 mechanism of disease resistance against *X. campestris* pv, *vesicotoria* was confirmed for the involvement of the de novo pathway involved in Vitamin B6 biosynthesis [41]. In addition, known bacterial elicitors of ISR are microbial associated molecular patterns triggering immunity. When this fails, microbial effector-triggered immunity is induced and leads to programmed cell death. It increases the plant's systemic resistance to subsequent pathogen challenge by PGPR. Moreover, plant probiotic bacterium (PPB) could be used to reduce the use of chemicals (fertilizers, pesticides) in agriculture. This could lead to improved quality at reduced costs and could provide the basis for more sustainable agriculture [42]. Organic agriculture has been widely promoted and adopted to establish better sustainability in food production and crop protection. *Micromonospora* has been regarded as a PPB due to rhizobia helper bacteria properties in *Medicago sativa* L. [43]. Similarly, *Phyllobacterium* and *B. licheniformis* also show promising benefits for increasing vitamin C content across various functional foods and have been considered as PPB, devoid of economic loss [32,44].

In the present investigation, we extracted and quantified phenols, flavonoids, and carotenoids, in the mock-inoculated control, *B. subtilis* CBR05 inoculated plants and market fruits. Among them, PGPR strain, *B. subtilis* CBR05 inoculated tomato fruits were found to have the richest source of lycopene, total phenolics, and flavonoids contents. Additionally, the PGPR strain, *B. subtilis* CBR05 inoculated tomato fruits showed potent antioxidant activities. The significantly higher lycopene content and radical-quenching activity of PGPR strain, *B. subtilis* CBR05 inoculation confer tomato fruits with more nutritional value, and their consumption can minimize oxidative stress-mediated chronic diseases. Thus, biofertilizers based on PGPR may be a viable alternative to improve the nutraceutical quality of greenhouse-produced tomato fruits.

Author Contributions: Conceptualization, M.C., and S.C.C.; methodology, M.C., R.K.S., and S.C.C.; formal analysis, M.C., R.K.S., M.P., and S.C.C.; writing—Original draft preparation, M.C., M.P., J.J.S., and J.W.O.; writing and editing, M.C., J.J.S., M.P., and J.W.O.

References

1. Choi, S.H.; Kim, D.S.; Kozukue, N.; Kim, H.J.; Nishitani, Y.; Mizuno, M.; Levin, C.E.; Friedman, M. Protein, free amino acid, phenolic, β-carotene, and lycopene content, and antioxidative and cancer cell inhibitory effects of 12 greenhouse-grown commercial cherry tomato varieties. *J. Food Compos. Anal.* **2014**, *34*, 115–127. [CrossRef]

2. Franceschi, S.; Bidoli, E.; La Vecchia, C.; Talamini, R.; D'Avanzo, B.; Negri, E. Tomatoes and risk of digestive-tract cancers. *Int. J. Cancer* **1994**, *59*, 181–184. [CrossRef] [PubMed]
3. Saini, R.K.; Zamany, A.J.; Keum, Y.S. Ripening improves the content of carotenoid, alpha-tocopherol, and polyunsaturated fatty acids in tomato (*Solanum lycopersicum* L.) fruits. *3 Biotech* **2017**, *7*, 43. [CrossRef] [PubMed]
4. Klee, H.J.; Giovannoni, J.J. Genetics and control of tomato fruit ripening and quality attributes. *Annu. Rev. Genet.* **2011**, *45*, 41–59. [CrossRef]
5. Alexander, L.; Grierson, D. Ethylene biosynthesis and action in tomato: A model for climacteric fruit ripening. *J. Exp. Bot.* **2002**, *53*, 2039–2055. [CrossRef]
6. Konozy, E.H.E.; Rogniaux, H.; Causse, M.; Faurobert, M. Proteomic analysis of tomato (*Solanum lycopersicum*) secretome. *J. Plant Res.* **2012**, *126*, 251–266. [CrossRef]
7. Barrett, D.M.; Weakley, C.; Diaz, J.V.; Watnik, M. Qualitative and nutritional differences in processing tomatoes grown under commercial organic and conventional production system. *J. Food Sci.* **2007**, *72*, 441–451. [CrossRef]
8. Erba, D.; Casiraghi, M.C.; Ribas-Agustí, A.; Cáceres, R.; Marfà, O.; Castellari, M. Nutritional value of tomatoes (*Solanum lycopersicum* L.) grown in greenhouse by different agronomic techniques. *J. Food Compos. Anal.* **2013**, *31*, 245–251. [CrossRef]
9. Dumas, Y.; Dadomo, M.; Di Lucca, G.; Grolier, P. Effects of environmental factors and agricultural techniques on antioxidant content of tomatoes. *J. Sci. Food Agric.* **2003**, *83*, 369–382. [CrossRef]
10. Wang, X.; Xing, Y. Evaluation of the effects of irrigation and fertilization on tomato fruit yield and quality: A principal component analysis. *Sci. Rep.* **2017**, *7*, 350. [CrossRef] [PubMed]
11. Beveridge, L.; Whitfield, S.; Challinor, A. Crop modelling: Towards locally relevantand climate-informed adaptation. *Clim. Chang.* **2018**, *147*, 475–489. [CrossRef]
12. Challinor, A.J.; Müller, C.; Asseng, S.; Deva, C.; Nicklin, K.J.; Wallach, D.; Vanuytrecht, E.; Whitfield, S.; Ramirez-Villegas, J.; Koehler, A.K. Improving the use of crop models for risk assessment and climate change adaptation. *Agric. Syst.* **2018**, *159*, 296–306. [CrossRef] [PubMed]
13. Giuliani, M.M.; Gatta, G.; Cappelli, G.; Gagliardi, A.; Donatelli, M.; Fanchini, D.; De Nart, D.; Mongiano, G.; Bregaglio, S. Identifying the most promising agronomic adaptation strategies for the tomato growing systems in Southern Italy via simulation modeling. *Eur. J. Agron.* **2019**, *111*, 125937. [CrossRef]
14. Wang, F.; Kang, S.; Du, T.; Li, F.; Qiu, R. Determination of comprehensive quality index for tomato and its response to different irrigation treatments. *Agric. Water Manag.* **2011**, *98*, 1228–1238. [CrossRef]
15. Shi, S.; Chen, D.; Fu, J.; Liu, L.; Wei, Y.; Shu, B.; Li, W.; Liang, Q.; Liu, C. Comprehensive evaluation of fruit quality traits in longan 'Fengliduo' × 'Dawuyuan' sexual progenies. *Sci. Hortic.* **2015**, *192*, 54–59. [CrossRef]
16. Wang, C.; Gu, F.; Chen, J.; Yang, H.; Jiang, J.; Du, T.; Zhang, J. Assessing the response of yield and comprehensive fruit quality of tomato grown in greenhouse to deficit irrigation and nitrogen application strategies. *Agric. Water Manag.* **2015**, *161*, 9–19. [CrossRef]
17. Luo, H.; Li, F. Tomato yield, quality and water use efficiency under different drip fertigation strategies. *Sci. Hortic.* **2018**, *235*, 181–188. [CrossRef]
18. Mader, P.; Fliessbach, A.; Dubois, D.; Gunst, L.; Fried, P.; Niggli, U. Soil fertility and biodiversity in organic farming. *Science* **2002**, *296*, 1694–1697. [CrossRef] [PubMed]
19. Vessey, J.K. Plant growth promoting rhizobacteria as biofertilizers. *Plant Soil* **2003**, *255*, 571–586. [CrossRef]
20. Mena-Violante, H.G.; Olalde-Portugal, V. Alteration of tomato fruit quality by root inoculation with plant growth-promoting rhizobacteria (PGPR): *Bacillus subtilis* BEB-13bs. *Sci. Hortic.* **2007**, *113*, 103–106. [CrossRef]
21. Zehnder, G.W.; Murphy, I.F.; Sikora, E.J.; Kloepper, J.W. Application to rhizobacteria for induced resistance. *Eur. J. Plant Pathol.* **2001**, *107*, 39–50. [CrossRef]
22. Chandrasekaran, M.; Chun, S.C. Induction of defence-related enzymes in tomato (*Solanum lycopersicum*) plants treated with *Bacillus subtilis* CBR05 against *Xanthomonas campestris* pv. *vesicatoria*. *Biocontrol Sci. Technol.* **2016**, *26*, 1366–1378. [CrossRef]
23. Chandrasekaran, M.; Chun, S.C. Expression of PR-protein genes and induction of defense-related enzymes by *Bacillus subtilis* CBR05 in tomato (*Solanum lycopersicum*) plants challenged with *Erwinia carotovora* subsp. *carotovora*. *Biosci. Biotechnol. Biochem.* **2016**, *80*, 2277–2283. [CrossRef] [PubMed]
24. Glick, B.R.; Penrose, D.M.; Li, J. A model for the lowering of plant ethylene concentrations by plant growth promoting bacteria. *J. Theor. Biol.* **1998**, *190*, 63–68. [CrossRef] [PubMed]

25. Verheul, M.; Maessen, H.; Grimstad, O.S. Optimizing a year-round cultivation system of tomato under artificial light. *Acta Hortic.* **2012**, *956*, 389–394. [CrossRef]
26. Moco, S.; Capanoglu, E.; Tikunov, Y.; Bino, R.J.; Boyacioglu, D.; Hall, R.D.; Vervoort, J.; De Vos, R.C. Tissue specialization at the metabolite level is perceived during the development of tomato fruit. *J. Exp. Bot.* **2007**, *58*, 4131–4146. [CrossRef]
27. Blois, M.S. Antioxidant determinations by the use of a stable free radical. *Nature* **1958**, *181*, 1199–1200. [CrossRef]
28. Ramos, A.; Rivero, R.; Victoria, M.C.; Visozo, A.; Piloto, J.; Garcia, A. Assessment of mutagenicity in *Parthenium hysterophorus* L. *J. Ethnopharmacol.* **2001**, *77*, 25–30. [CrossRef]
29. Chen, L.; Xin, X.; Yuan, Q.; Su, D.; Liu, W. Phytochemical properties and antioxidant capacities of various colored berries. *J. Sci. Food Agric.* **2014**, *94*, 180–188. [CrossRef]
30. Rodriguez-Amaya, D.B. *A Guide to Carotenoid Analysis in Foods*; ILsi Press: Washington, DC, USA, 2001.
31. Saini, R.K.; Nile, S.H.; Park, S.W. Carotenoids from fruits and vegetables: Chemistry, analysis, occurrence, bioavailability and biological activities. *Food Res. Int.* **2015**, *76*, 735–750. [CrossRef]
32. Ochoa-Velasco, C.E.; Valadez-Blanco, R.; Salas-Coronado, R.; Sustaita-Rivera, F.; Hernández-Carlos, B.; García-Ortega, S.; Santos-Sánchez, N.F. Effect of nitrogen fertilization and *Bacillus licheniformis* biofertilizer addition on the antioxidants compounds and antioxidant activity of greenhouse cultivated tomato fruits (*Solanum lycopersicum* L. var. Sheva). *Sci. Hortic.* **2016**, *201*, 338–345. [CrossRef]
33. Baset Mia, M.A.; Shamsuddin, Z.H. Nitrogen fixation and transportation by rhizobacteria: A scenario of rice and banana. *Int. J. Bot.* **2010**, *6*, 235–242.
34. Bar-Ness, E.; Chen, Y.; Hadar, Y.; Marchner, H.; Romheld, V. Siderophores of Pseudomonas putida as an iron source for dicot and monocot plants. *Plant Soil* **1991**, *130*, 231–241. [CrossRef]
35. Lucas, G.J.A.; Probanza, A.; Ramos, B.; Palomino, M.R.; Gutierrez Mañero, F.J. Effect of inoculation of *Bacillus licheniformis* on tomato and pepper. *Agronomie* **2004**, *24*, 169–176.
36. Martí, R.; Roselló, S.; Cebolla-Cornejo, J. Tomato as a source of carotenoids and polyphenols targeted to cancer prevention. *Cancer* **2016**, *8*, 58. [CrossRef]
37. Ronen, G.; Cohen, M.; Zamir, D.; Hirschberg, J. Regulation of carotenoid biosynthesis during tomato fruit development: Expression of the gene for lycopene epsilon-cyclase is downregulated during ripening and is elevated in the mutant Delta. *Plant J. Cell Mol. Biol.* **1999**, *17*, 341–351. [CrossRef]
38. Su, L.; Diretto, G.; Purgatto, E.; Danoun, S.; Zouine, M.; Li, Z.; Roustan, J.P.; Bouzayen, M.; Giuliano, G.; Chervin, C. Carotenoid accumulation during tomato fruit ripening is modulated by the auxin–ethylene balance. *BMC Plant Biol.* **2015**, *15*, 114. [CrossRef]
39. Zhang, Y.; Liu, Y.; Lv, Q. DFT study on the quenching mechanism of singlet oxygen by lycopene. *RSC Adv.* **2016**, *6*, 98498–98505. [CrossRef]
40. Chandrasekaran, M.; Belachew, S.T.; Yoon, E.; Chun, S.C. Expression of β-1,3-glucanase (GLU) and phenylalanine ammonia-lyase (PAL) genes and their enzymes in tomato plants induced after treatment with *Bacillus subtilis* CBR05 against *Xanthomonas campestris* pv. *vesicatoria*. *J. Gen. Plant Pathol.* **2016**, *83*, 7–13. [CrossRef]
41. Chandrasekaran, M.; Manivannan, P.; Chun, S.C. *Bacillus subtilis* CBR05 induces Vitamin B6 biosynthesis in tomato through the *de novo* pathway in contributing disease resistance against *Xanthomonas campestris* pv. *vesicatoria*. *Sci. Rep.* **2019**, *9*, 6495. [CrossRef]
42. Jiménez-Gómez, A.; Flores-Félix, J.D.; García-Fraile, P.; Mateos, P.F.; Menéndez, E.; Velázquez, E.; Rivas, R. Probiotic activities of rhizobium laguerreae on growth and quality of spinach. *Sci. Rep.* **2018**, *8*, 295. [CrossRef]
43. Martínez-Hidalgo, P.; García, J.M.; Pozo, M.J. Induced systemic resistance against *Botrytis cinerea* by *Micromonospora* strainsisolated from root nodules. *Front. Microbiol.* **2015**, *6*, 922. [CrossRef] [PubMed]
44. Flores-Felix, J.D.; Silva, L.R.; Rivera, L.P.; Marcos-Garcia, M.; Garcia-Fraile, P.; Martinez-Molina, E.; Mateos, P.F.; Velazquez, E.; Andrade, P.; Rivas, R. Plants probiotics as a tool to produce highly functional fruits: The case of *Phyllobacterium* and vitamin C in strawberries. *PLoS ONE* **2015**, *10*, e0122281. [CrossRef]

8

A Composite Bioinoculant based on the Combined Application of Beneficial Bacteria and Fungi

Henrietta Allaga [1], Bettina Bóka [1], Péter Poór [2], Viktor Dávid Nagy [1], Attila Szűcs [1], Istvánné Stankovics [3], Miklós Takó [1], László Manczinger [1], Csaba Vágvölgyi [1], László Kredics [1,*] and Péter Körmöczi [4]

1 Department of Microbiology; Faculty of Science and Informatics; University of Szeged; Közép fasor 52, 6726 Szeged, Hungary; henrietta.allaga@gmail.com (H.A.); boka.tina@gmail.com (B.B.); viktor.david.nagy@gmail.com (V.D.N.); a.szunyi@windowslive.com (A.S.); tako78@bio.u-szeged.hu (M.T.); manczing@bio.u-szeged.hu (L.M.); csaba@bio.u-szeged.hu (C.V.)

2 Department of Plant Biology, Faculty of Science and Informatics, University of Szeged, Közép fasor 52, 6726 Szeged, Hungary; poorpeti@bio.u-szeged.hu

3 Felső-Bácskai Agrolabor Ltd., Rákóczi Ferenc u. 19., 6430 Bácsalmás, Hungary; agrolabor@agrolabor.hu

4 Department of Field Crops Research, National Agricultural Research and Innovation Centre, Alsó kikötő sor 9, 6726 Szeged, Hungary; kormoczipeti@gmail.com

* Correspondence: kredics@bio.u-szeged.hu

Abstract: A composite soil bioinoculant containing beneficial bacteria and fungi was developed for biocontrol of plant pathogens, phosphorous mobilization, stem degradation, humification, and nitrogen fixation. A *Trichoderma asperellum* isolate with outstanding in vitro antagonistic abilities toward a series of plant pathogenic fungi was included as a potential biocontrol component. The selected strain was also shown to promote growth and increase photosynthetic activity of tomato plants. For phosphorous mobilization and stem degradation, a *Trichoderma atrobrunneum* strain was selected, which produced cellulose-degrading enzymes even in the absence of stem residues, while this ability increased 10–15-fold in the presence of ground maize stem. The strain was also shown to produce large amounts of enzymes liberating organically bound phosphorous, as well as cellulase and xylanase activities in solid-state fermentation on various plant residues. A *Streptomyces albus* strain with excellent peroxidase-producing abilities was selected as a potential humus-producing component, while an *Azotobacter vinelandii* strain with the potential to provide excess nitrogen for crops was included for nitrogen fixation. The assembled soil bioinoculant had positive effect on the uptake of certain important macro- and microelements (potassium, sodium, and manganese) from the soil by field-grown tomato plants. The applied screening strategy proved to be applicable for the assembly of a composite soil bioinoculant with notable application potentials.

Keywords: biocontrol; plant growth promotion; soil inoculant; *Trichoderma*; *Azotobacter*; *Streptomyces*

---

1. Introduction

Chemical pesticides and fertilizers are applied world-wide in agricultural production. Pesticides are used for the prevention and control of plant pests and diseases in order to reduce or eliminate yield losses and maintain product quality. However, there are serious concerns regarding the risks resulting from occupational exposure to them, as well as from environmental pollution leading to the presence of their residues in the food-chain and drinking water [1]. Chemical fertilizers are used to supply plants with necessary elements (primarily phosphorous and nitrogen), thereby improving crop productivity; however, their application is resulting in pollution with phosphates and nitrates. The agricultural run-off of phosphates deriving from fertilizers contributes to the eutrophication of fresh water bodies and also presents a serious threat to the biodiversity in terrestrial ecosystems [2], while the increased run-off of nitrogen fertilizers results in nitrate pollution of surface and groundwater [3]. Therefore, the need for alternative, environment-friendly, microbial soil treatment strategies with favorable effects on crop plants is emerging all over the world. Microbial abilities of biocontrol, plant growth promotion, stem degradation, phosphorous solubilization, humification, and nitrogen-fixation can be exploited for the development of microbial soil inoculants to be applied in sustainable agricultural production.

One of the main challenges in the agricultural use of beneficial microorganisms as plant growth promoters and/or biocontrol agents (BCAs) is their frequently inconsistent field performance [4,5], which may be due to a series of biotic and abiotic factors. Among the abiotic factors, physicochemical properties of the rhizosphere such as pH, temperature, water activity, and the chemical composition of the soil are varying in space and time, which substantially influences the performance of biocontrol and plant growth promoting microorganisms. Particular agents may exert different activities under different soil environmental conditions. Inconsistent field performance has long been identified as the major impediment to the wide-scale commercialization of beneficial microorganisms for agricultural applications [6]. A possible strategy to counteract inconsistencies due to varying environmental conditions is the development of consortial soil inoculants consisting of multiple beneficial organisms. The combination of efficient plant growth promoting microorganisms and BCAs may result in an increased consistency of field performance during different periods of the growing season, thereby enabling a more predictable increase in crop yields [7].

The aim of this study was to assemble a consortial soil bioinoculant based on the combined application of beneficial bacteria and fungi with the potential of increasing pathogen control, plant growth and crop yield, stem residue degradation, phosphorous mobilization, humification, and nitrogen fixation in treated agricultural soils.

2. Materials and Methods

2.1. Examined Strains

The microbial strains involved in this study derived from the Szeged Microbiology Collection, Szeged, Hungary (SZMC). The *Trichoderma* strains included in the study (*Trichoderma asperellum* SZMC 20866, and SZMC 20786; *Trichoderma harzianum* species complex (THSC) members SZMC 20761, SZMC 20762, and SZMC 20869; *Trichoderma atroviride* SZMC 20780 and SZMC 20781; *Trichoderma virens* SZMC 20779; *Trichoderma gamsii* SZMC 20783; and *Trichoderma hamatum* SZMC 20784) were isolated from Hungarian agricultural soil samples and initially identified by sequence analysis of the internal transcribed spacer (ITS) region [8]. However, as ITS sequence analysis is not able to discriminate between species belonging to THSC, the species level identity of strains SZMC 20761, SZMC 20762, and SZMC 20869 was determined during this study by the sequence analysis of a fragment of the *tef1* alpha gene [9] as *Trichoderma guizhouense, T. guizhouense* and *Trichoderma atrobrunneum* (GenBank accession numbers MN750371, MN750372, MN750373), respectively. Fungal isolates were maintained on yeast extract-glucose medium (5 g L^{-1} glucose, 5 g L^{-1} KH_2PO_4, 1 g L^{-1} yeast extract, 20 g L^{-1} agar).

Streptomyces sp. isolates (SZMC 0282, SZMC 0232, 00001, 00002, 00004, 00005, 00006, 00007, 00008, 00009, 00010, 00012, 00013, 00014, 00015, 00017, 00019, 00020, 00021, 00022, 00023, 00024, 00025, 00026, 00027, 00028, 00029, 00030, 00031, 00032, 00033, 00034, 00035, 00036, 00037, 00038, 00039, 00040, 00041, 00042, 00043, 00044, 00045, 00046, and 00047) and *Azotobacter vinelandii* SZMC 22195 were derived from soil samples. *Streptomyces microflavus* DSM 40561 was derived from the DSMZ (Deutsche Sammlung von Mikroorganismen und Zellkulturen) strain collection and was included as control in the peroxidase-producing assays. Bacterial strains were maintained on glucose-yeast extract-malt extract *Streptomyces* medium (GYM-STR: 4 g L^{-1} glucose, 4 g L^{-1} yeast extract, 10 g L^{-1} malt extract, 2 g L^{-1} $CaCO_3$, and 20 g L^{-1} agar).

2.2. *Determination of Biocontrol Index (BCI) Values*

Dual confrontation tests were performed in vitro in Petri dishes (90 mm in diameter) according to the method described by Szekeres et al. [10]. During the experiments, 11 different strains of plant pathogenic fungi were confronted with the *Trichoderma* strains, including 3 *Armillaria* species (*Armillaria mellea* SZMC 23638, *Armillaria ostoyae* SZMC 23080, and *Armillaria gallica* SZMC23076), 4 *Fusarium solani* species complex (FSSC) isolates (SZMC 11057F, SZMC 6241J, SZMC 11067F, and SZMC 11070F), as well as *Phoma cucurbitacearum* (SZMC 16088), *Alternaria alternata* (SZMC 16085), *Botrytis cinerea* (SZMC 6244J), and *Rhizoctonia solani* (SZMC 6252J). The tests were carried out in three replicates on malt extract agar (MEA) medium (10 g L^{-1} glucose, 2.5 g L^{-1} yeast extract, 20 mL L^{-1} 20% malt extract, 20 g L^{-1} agar). After the incubation period, digital photos were taken with a Nikon Coolpix P7700 camera (Nikon, Tokyo, Japan) from each Petri plate, and the area visibly covered by the *Trichoderma* strain as well as the area covered by *Trichoderma* and the pathogen together were calculated for each plate with the aid of the Image J software (http://imagej.nih.gov/ij). BCI values were calculated with Excel 2010 (Microsoft, Redmond, WA, USA) according to the formula: BCI = (area of *Trichoderma* colony/total area occupied by the colonies of both *Trichoderma* and the plant pathogenic fungus) × 100 [10].

2.3. *Liquid and Solid-State Fermentations*

For the investigation of cellulase and phosphatase activities, culturing was performed in 100 mL flasks containing 20 mL liquid minimal (10 g L^{-1} glucose, 5 g L^{-1} $(NH_4)_2SO_4$, 5 g L^{-1} KH_2PO_4, 0.1 g L^{-1} $MgSO_4 \times 7H_2O$) or maize stem medium (2 g L^{-1} dried maize stem ground with a coffee grinder (Bosch, Gerlingen, Germany), 1 g L^{-1} $NaNO_3$). The minimal medium was inoculated with *Trichoderma* conidia to a concentration of 2×10^5 mL^{-1}, while the maize stem medium was inoculated with the total amount of 5-days-old fungal mycelium pre-grown in 20 mL liquid minimal medium, filtered with the aid of filter paper and vacuum pump and washed with sterile distilled water. After 5 days of incubation at 25 °C in an IKA KS 4000 IC Control shaker (ProfiLab24, Berlin, Germany) at 180 rpm, samples were filtered, and the culture filtrates were used for extracellular enzyme activity measurements.

Extracellular enzyme activities of *T. atrobrunneum* SZMC 20869 from THSC and the industrially important *Trichoderma reesei* strain QM9414 (SZMC 22616) were also compared in solid-state fermentation (SSF) experiments using maize, wheat, sunflower, and canola stem residues as substrates. One gram amounts of ground plant residues were placed into 50 mL Erlenmeyers flask and moisturized with 5 mL distilled water. After sterilization, the substrates were inoculated with 2×10^5 *Trichoderma* conidia. On the 8th day of fermentation, extractions were performed by adding 20 mL distilled water to the cultures and incubating for 3 h at 4 °C. The fluid phases were filtered through sterile gauze sheets into 15 mL centrifuge tubes and centrifuged at 4600 rpm for 10 min, 2 times by transferring the fluid phase to a new centrifuge tube. The 8× dilution of the fluid samples in distilled water were used for extracellular enzyme activity measurements.

2.4. *Enzyme Activity Measurements*

Cellobiohydrolase, β-glucosidase, β-xylosidase, and acidic phosphatase enzyme activities were measured with the chromogenic substrates p-nitrophenyl-β-D-cellobioside, p-nitrophenyl-β-D-glucopyranoside, p-nitrophenyl-β-D-xylopyranoside, and p-nitrophenyl-phosphate (Sigma-Aldrich, Budapest, Hungary), respectively. Enzyme reactions were carried out in three replicates in the wells of 96-well microtiter plates (Sarstedt, Nümbrecht, Germany) by mixing 100 μL culture filtrate or SSF extract with 100 μL substrate solution (1 mg mL^{-1} in distilled water). After 1 h of incubation at room temperature, enzyme reactions were stopped with 10% (w/v) Na_2CO_3 and the optical densities measured at 405 nm with a Spectrostar Nano microplate reader (BMG Labtech, Ortenberg, Germany).

Peroxidase assays of bacteria were carried out in liquid *Streptomyces* induction medium (STR-IND) 6 g L^{-1} yeast extract, 8 g L^{-1} xylan, 0.1 g L^{-1} $(NH_4)_2SO_4$, 0.3 g L^{-1} NaCl, 0.1 g L^{-1} $MgSO_4$, 0.02 g

L^{-1} $CaCO_3$, 0.6 mL L^{-1} TE (0.1 g L^{-1} $FeSO_4$, 0.002 g L^{-1} $MnSO_4 \times 7H_2O$, 0.09 g L^{-1} $ZnSO_4 \times 7H_2O$) inoculated with the examined *Streptomyces* strains. After 7 days of incubation (28 °C, 150 rpm), the samples were centrifuged at 7000 rpm for 10 min. The reaction mixture contained 0.2 mL phosphate buffer (100 mM, pH 7.2), 0.2 mL 2,4-dichlorophenoxy acetic acid (25 mM), 0.2 mL 4-aminoantipyrine (4AAP, 16 mM), 0.2 mL ferment broth, and 0.2 mL hydrogen peroxide (50 mM). The reaction mixtures were put into a 53 °C thermostat for 1 min, which was followed by the measurement of the optical density at 510 nm with a Spectrostar Nano microplate reader (BMG Labtech, Ortenberg, Germany). All measurements were carried out in three replicates.

In order to investigate the peroxidase enzyme production, dye decolorization assays were also performed. For this purpose, we used STR-IND medium supplemented with 20 g L^{-1} agar and Remazol Brilliant Blue (RBB), Methyl Orange (MO), or Neutral Red (NR) dyes at a concentration of 1 g L^{-1}. The plates were inoculated with *Streptomyces* isolates with the aid of inoculation loop onto the middle of the Petri plates. Color changes were observed around the colonies after 1 week of incubation.

2.5. *Growth Assay in Nitrogen-Free Medium*

The growth kinetics of *A. vinelandii* strain SZMC 22195 were tested in nitrogen source-free liquid medium (5 g L^{-1} glucose, 5 g L^{-1} mannitol, 0.1 g L^{-1} $CaCl_2 \times 2H_2O$, 0.1 g L^{-1} $MgSO_4 \times 7H_2O$, 0.005 g L^{-1} $Na_2MoO_4 \times 2H_2O$, 0.9 g L^{-1} K_2HPO_4, 0.1 g L^{-1} KH_2PO_4, 0.01 g L^{-1} $FeSO_4 \times 7H_2O$, pH 7.3). During an incubation period of 1 week (30 °C, 120 rpm), the optical densities of the liquid cultures were measured on days 1, 2, 4, and 7 at 620 nm with a Spectrostar Nano microplate reader.

2.6. *Plant Material for Growth Chamber Experiments*

Seeds of tomato (*Solanum lycopersicum* Mill. L. cvar. Ailsa craig) were germinated at 26 °C for 3 days in the dark, and the seedlings were subsequently transferred to 6 × 6 cm pots filled with vermiculite (Terracult GmbH, Siegburg, Germany) for 6 weeks. Plants were irrigated every third day with nutrient solution containing 2 mM $Ca(NO_3)_2$, 1 mM $MgSO_4$, 0.5 mM KCl, 0.5 mM KH_2PO_4, and 0.5 mM Na_2HPO_4, pH 6.0. The concentrations of micronutrients were 0.001 mM $MnSO_4$, 0.005 mM ZnSO4, 0.0001 mM (NH4)6Mo7O24, 0.01 mM H_3BO_4, and 0.02 mM Fe(III)-EDTA. The plants were grown in a controlled environment under 300 µmol m^{-2} s^{-1} light intensity (emitted F36W/GRO lamps, Feilo Sylvania, Erlangen, Germany), with 12/12-h light/dark period, day/night temperatures of 24/22 °C, and relative humidity of 55–60%. Plants were treated with 20 µL of *Trichoderma* suspension (1×10^6 conidia mL^{-1}) after the 3-days-long germination. Samples for measurements were prepared in each replicate from the second, fully expanded young leaves of tomato plants. After harvest, the plant height and root length as well as the biomass production were recorded in 5 replicates.

2.7. *Measurement of Stomatal Conductance, CO_2 Assimilation, and Total Soluble Sugar Content*

Stomatal conductance and CO_2 assimilation were measured in 3 replicate samples by a portable photosynthesis system (LI-6400, LI-COR Inc., Lincoln, NE, USA), as described by Poór et al. [11]. Data were recorded after 15 min light adaptation on 300 µmol m^{-2} s^{-1} light intensity and under constant conditions (25 °C, 65 ± 10% relative humidity, and controlled CO_2 supply of 400 µmol mol^{-1}).

Total sugar contents were determined according to Dubois et al. [12]. One gram of leaf samples was homogenized in 10 mL distilled water and incubated in a 90 °C water bath for 45 min. Samples were centrifuged (12,000× *g* for 15 min, at 4 °C), and 40 µL of the supernatant was mixed with 400 µL of 1.8% phenol and 2 mL of concentrated sulfuric acid. The absorbance was measured from 5 replicate samples by a spectrophotometer at 490 nm.

2.8. *Chlorophyll a Fluorescence Measurements*

Chlorophyll a fluorescence was detected in 3 replicate samples with the portable photosynthesis system (LI-6400, LI-COR Inc., Lincoln, NE, USA) described above [11]. Leaves were dark-adapted for 15 min before the measurement of the minimal fluorescence (F0) using weak measuring light. The maximal fluorescence (Fm) was measured by applying a pulse (800 ms) of saturating light (12,000 μmol m^{-2} s^{-1}). The leaves were then illuminated continuously with actinic light (300 μmol m^{-2} s^{-1}). After 20 min, the light-adapted steady-state fluorescence (Fs) was recorded and the maximum fluorescence level (Fm′) in the light-adapted state was determined with saturating pulses. The actinic light was next turned off and the minimum fluorescence level in the light-adapted state (F0′) was determined by illuminating the leaf with 3-s far-red light (5 μmol m^{-2} s^{-1}). The following chlorophyll fluorescence parameters were calculated: the maximal quantum yield of PSII photochemistry, Fv/Fm = (Fm − F0/Fm); the actual quantum yield of PSII electron transport in the light adapted-state, ΦPSII = (Fm′ − Fs)/Fm′ [13]; the photochemical quenching coefficient, qP = (Fm′ − Fs)/(Fm′ − F0′) [14]; and the non-photochemical quenching NPQ = Fm/Fm′ − 1 [15].

2.9. *Pigment Analysis*

For pigment analysis, a two-step extraction was applied. Fifty milligrams of leaf samples were homogenized in ice-cold 100% (v/v) acetone (1 mL) and extracted for 24 h. Samples were centrifuged (12,000× *g* for 15 min at 4 °C). The pellet was extracted again with 80% (v/v) acetone (1 mL) for 24 h. After spinning down (12,000× *g*, 15 min, 4 °C), the supernatants were collected. The pigment composition was measured in 5 replicate samples as according to Lichtenthaler and Wellburn [16].

2.10. *Field Experiment*

A field study was performed in tomato culture (*Solanum lycopersicum* cvar. ACE-55) on sandy loam soil according to the yarn number of Arany (K_A), which is a humus-rich soil with good nutrient availability and water management. The GPS coordinates of the examined area are 46°05′01.05″ N, 19°26′28.83″ E. Three-week-old tomato seedlings were planted on 11 May 2019. A total of 220 seedlings were planted in the field with 40 cm row distance and 40 cm plant distance, with 20 seedlings in a row. At the beginning of the experiment, a bioinoculant preparation consisting of a mixture of two *Trichoderma* strains and two bacteria was prepared. The concentration of the bioinoculant was adjusted to 10^6 conidia mL^{-1} for both selected *Trichoderma* components, and 10^8 cells mL^{-1} for both selected bacteria. The application of the soil inoculant was performed after 100× dilution at a concentration of 100 mL L^{-1}, while the control area was not treated. Three control and eight treated rows were examined. No organic manure or chemical fertilizer was applied to the area during the soil preparation. Changes in the contents of soil macro- and microelements were measured three times during cultivation (I: 22 June, II: 6 July, and III: 3 August 2019) from rhizosphere samples taken from a depth of 15 cm. The total numbers of the control and treated plants were 60 and 160, respectively. During the experiment, the same plant protection measures were applied both in the control and treated area: Cuproxat FW (5 mL L^{-1}), Topaz (0.5 mL L^{-1}), Mospilan (200 mg L^{-1}), Wuxal (2 mL L^{-1}), Humusz (1 mL L^{-1}), as well as calcium (5 mL L^{-1}) and magnesium (5 mL L^{-1}) in the form of foliar fertilizer were used 3 times during the cultivation (20 May, 8 June, and 13 July 2019). Control and treated plants were examined separately.

2.11. *Soil Examination Methods*

Soil sampling and analysis were carried out according to the test methods prescribed by the Hungarian Standard [17,18]. The soil tests were carried out by the Felső-Bácskai Agrolabor Ltd., Bácsalmás, Hungary. Carbonate content was determined with a Scheibler's calcimetre (Bovimex, Székesfehérvár, Hungary). The total salt content was measured by the electric conductivity using a HI98311 conductivity meter (HANNA Instruments, Szeged, Hungary). These methods were based on the Hungarian Standard (MSZ) MSZ-08-0206-2:1978. The soil texture was determined by the yarn number of Arany (MSZ-08-0205:1978). The macro- and microelement content and the humus content were measured by a Lambda 25 UV/VIS Spectrophotometer (PerkinElmer, Waltham, MA, USA) according to the Hungarian standards MSZ 20135:1999 and MSZ-21470-52:1983, respectively. N-content was measured by the Kjeldahl method, K_2O with flame emission spectrophotometry (FES), while Mg and Ca content with flame atomic absorption spectroscopy (FAAS) after acidic (H_2SO_4-$HClO_4$)

digestion [19]. Samples were prepared with microwave digestion for microelement analysis (Cu, Mn, Zn, and Fe content) measured with FAAS [17,19,20].

2.12. Statistical Analysis

Data presented as average values resulted from at least 3 independent experiments. Statistical analyses were carried out for the measurement data with Sigma plot v11.0 software (Systat Software Inc., Erkrath, Germany) using Student's t test, and the differences were considered significant if $p < 0.05$ (*), $p < 0.01$ (**), or $p < 0.001$ (***). The statistical analyses of the soil results and crop yield data were performed with the GraphPad Prism v8.3.0 software (GraphPad Software Inc., San Diego, CA, USA) applying two-way ANOVA, and the results were considered significant if $p < 0.05$ (*), $p = 0.02$ (**), $p = 0.001$ (***), and $p < 0.0001$ (****).

3. Results

3.1. Selection of the Components for the Soil Inoculant

The potential biocontrol component of the soil inoculant was selected based on the results of dual confrontation tests between *Trichoderma* strains and plant pathogenic fungi. As shown in Table 1 and Figure 1, the two examined *T. asperellum* strains were the most effective against many of the tested pathogens. The BCI values of *T. asperellum* SZMC 20786 were the highest against FSSC SZMC 11067F and SZMC 11070F and *Alternaria alternata* 16085, while against *Rizoctonia solani* and *Armillaria gallica* they even reached 100, which means that the *Trichoderma* could completely overgrow and inhibit these plant pathogens. Except for *T. atrobrunneum* SZMC 20869 and *T. virens* SZMC 20779, all other *Trichoderma* strains could completely inhibit the *R. solani* strain SZMC 6252J. Based on the results, strain *T. asperellum* SZMC 20786 was selected as the potential biocontrol component of the soil inoculant.

Eight *Trichoderma* strains were included in the screening for cellulose-degrading and phosphatase-producing abilities (Figure 2). High levels of β-glucosidase and cellobiohydrolase enzyme activities could be measured in the case of three *Trichoderma* strains, which included two isolates of THSC (*T. guizhouense* SZMC 20761, *T. atrobrunneum* SZMC 20869) and one isolate of *T. hamatum* (SZMC 20784). The cellulolytic activities of SZMC 20869 were inducible with maize stem powder (Figure 2A,B), while the other two strains possessed high enzyme activity values only in liquid minimal medium. Only low phosphatase enzyme activities could be detected for the examined *Trichoderma* strains, except for the above mentioned *T. atrobrunneum* SZMC 20869 strain (Figure 2C), for which increased enzyme activity levels could be observed in liquid medium supplemented with maize stem powder. Based on the results, strain *T. atrobrunneum* SZMC 20869 from THSC was selected as the potential stem-degrading and phosphate-mobilizing component of the soil inoculant.

In the dye decolorization assays performed with *Streptomyces* isolates, color changes could be observed only when RBB was applied. Only the isolates *S. albus* SZMC 0232 and SZMC 0282 and *S. microflavus* DSM 40561 gave positive reactions with the RBB dye. The results of peroxidase assays are shown in Figure 3, indicating that only the isolates *S. albus* SZMC 0232 and SZMC 0282 showed an increased peroxidase production. The activities of these two strains were about twice as high as those of the control strain *S. microflavus* DSM 40561. Based on the results, strain *S. albus* SZMC 0282 was selected as the potential humus-producing component of the soil inoculant.

Table 1. Biocontrol index values of the examined *Trichoderma* strains against plant pathogenic fungi.

Examined *Trichoderma* Strains	Plant Pathogens										
	FSSC SZMC 11057F	FSSC SZMC 6241J	FSSC SZMC 11067F	FSSC SZMC 11070F	*Armillaria mellea* MUCL 31056	*Armillaria ostoyae* SZMC 23080	*Armillaria gallica* SZMC 23076	*Alternaria alternata* SZMC 16085	*Phoma cucurbitacearum* SZMC 16088	*Botrytis cinerea* SZMC 6244J	*Rhizoctonia solani* SZMC 6252J
T. asperellum SZMC 20866	79.77 ± 1.22	83.71 ± 1.35	71.38 ± 3.04	75.61 ± 5.35	33.91 ± 6.67	100.00 ± 0.00	100.00 ± 0.00	64.32 ± 0.53	81.36 ± 1.26	100.00 ± 0.00	100.00 ± 0.00
T. asperellum SZMC 20786	77.76 ± 0.96	83.50 ± 6.97	88.84 ± 5.48	85.76 ± 2.12	47.68 ± 5.22	92.33 ± 3.51	100.00 ± 0.00	65.77 ± 1.42	80.63 ± 0.50	100.00 ± 0.00	100.00 ± 0.00
T. atroviride SZMC 20780	62.41 ± 2.90	63.39 ± 5.10	37.63 ± 2.15	51.19 ± 0.96	93.76 ±2.89	100.00 ± 0.00	100.00 ± 0.00	51.43 ± 0.42	59.21 ± 1.37	100.00 ± 0.00	100.00 ± 0.00
T. atroviride SZMC 20781	63.91 ± 0.76	71.32 ± 4.40	39.93 ± 2.14	48.39 ± 2.53	97.43 ± 3.07	98.02 ± 2.00	100.00 ± 0.00	54.16 ± 0.27	56.82 ± 1.20	100.00 ± 0.00	100.00 ± 0.00
T. gamsii SZMC 20783	60.82 ± 1.42	67.42 ± 0.79	39.97 ± 1.74	59.36 ± 1.34	82.21 ± 9.56	100.00 ± 0.00	100.00 ± 0.00	49.22 ± 2.06	60.14 ± 0.75	50.70 ± 7.26	100.00 ± 0.00
T. hamatum SZMC 20784	68.22 ± 0.73	64.10 ± 0.67	47.88 ± 1.75	62.52 ± 0.63	96.71 ± 3.41	97.00 ± 0.00	92.75 ± 2.95	61.28 ± 1.38	67.96 ± 1.19	46.82 ± 1.06	100.00 ± 0.00
T. guizhouense SZMC 20761	63.02 ± 5.40	58.81 ± 1.72	63.39 ± 1.69	64.92 ± 0.87	96.70 ± 3.23	99.93 ± 0.11	100.00 ± 0.00	60.77 ± 1.06	61.19 ± 1.35	44.20 ± 4.21	100.00 ± 0.00
T. atrobrunneum SZMC 20869	42.19 ± 2.53	47.46 ± 2.08	37.83 ± 0.17	38.40 ± 0.95	54.38 ± 1.98	100.00 ± 0.00	88.43 ± 4.27	18.02 ± 1.02	14.92 ± 1.89	42.26 ± 4.55	74.87 ± 7.34
T. guizhouense SZMC 20762	63.69 ± 1.23	58.47 ± 5.30	47.91 ± 0.37	59.77 ± 1.79	97.42 ± 2.4	100.00 ± 0.00	100.00 ± 0.00	59.33 ± 0.53	57.84 ± 1.95	49.43 ± 3.03	100.00 ± 0.00
T. virens SZMC 20779	40.66 ± 1.51	37.20 ± 1.38	53.43 ± 0.62	49.10 ± 2.91	53.48 ± 2.76	100.00 ± 0.00	86.19 ± 5.77	25.07 ± 2.33	35.59 ± 1.91	45.32 ± 8.37	62.37 ± 5.36

FSSC: *Fusarium solani* species complex; MUCL: Belgian Coordinated Collections of Microorganisms/MUCL Agro-food & Environmental Fungal Collection; SZMC: Szeged Microbiology Collection.

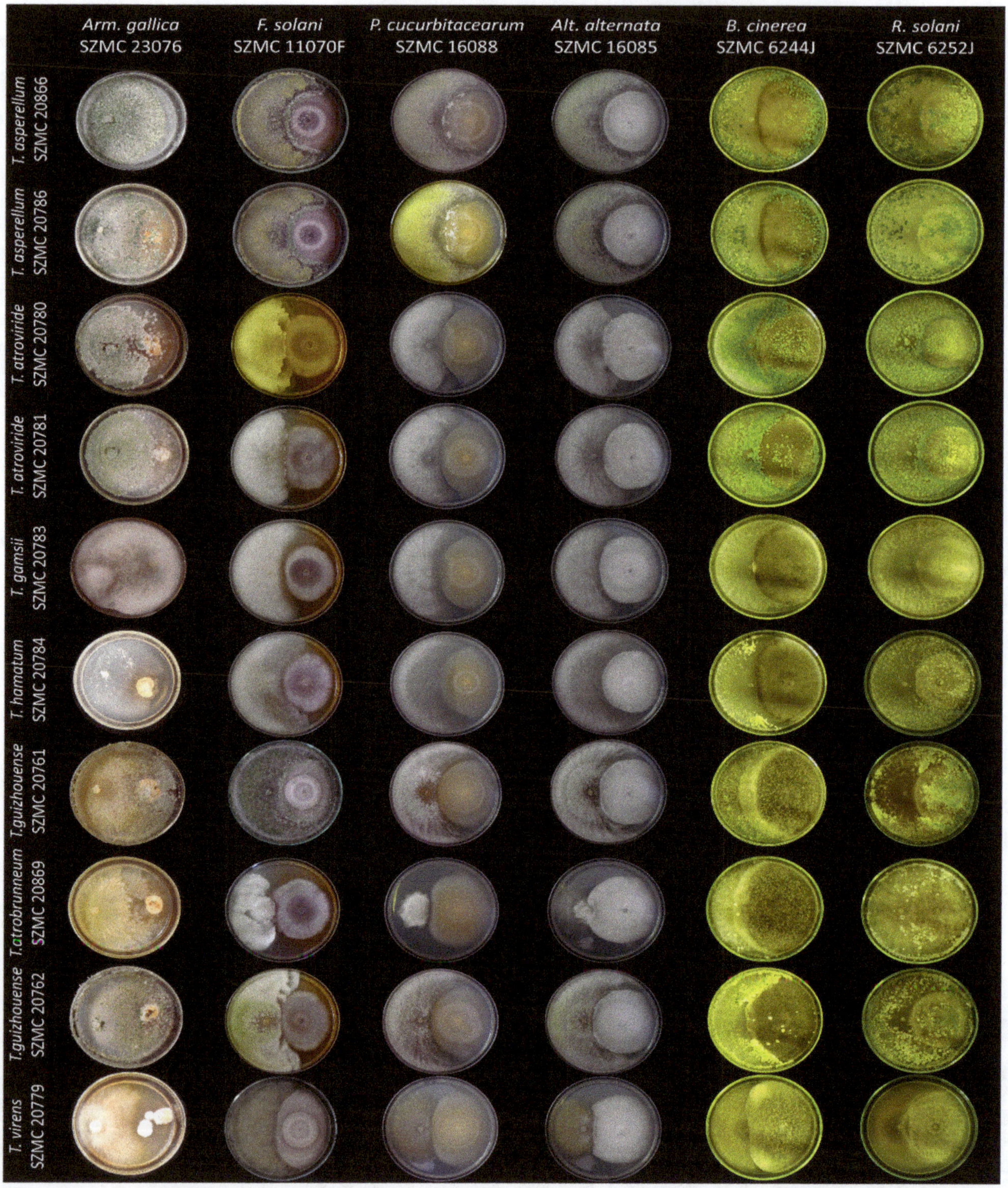

Figure 1. In vitro antagonism of *Trichoderma* strains against different plant pathogenic fungi examined in dual confrontation tests.

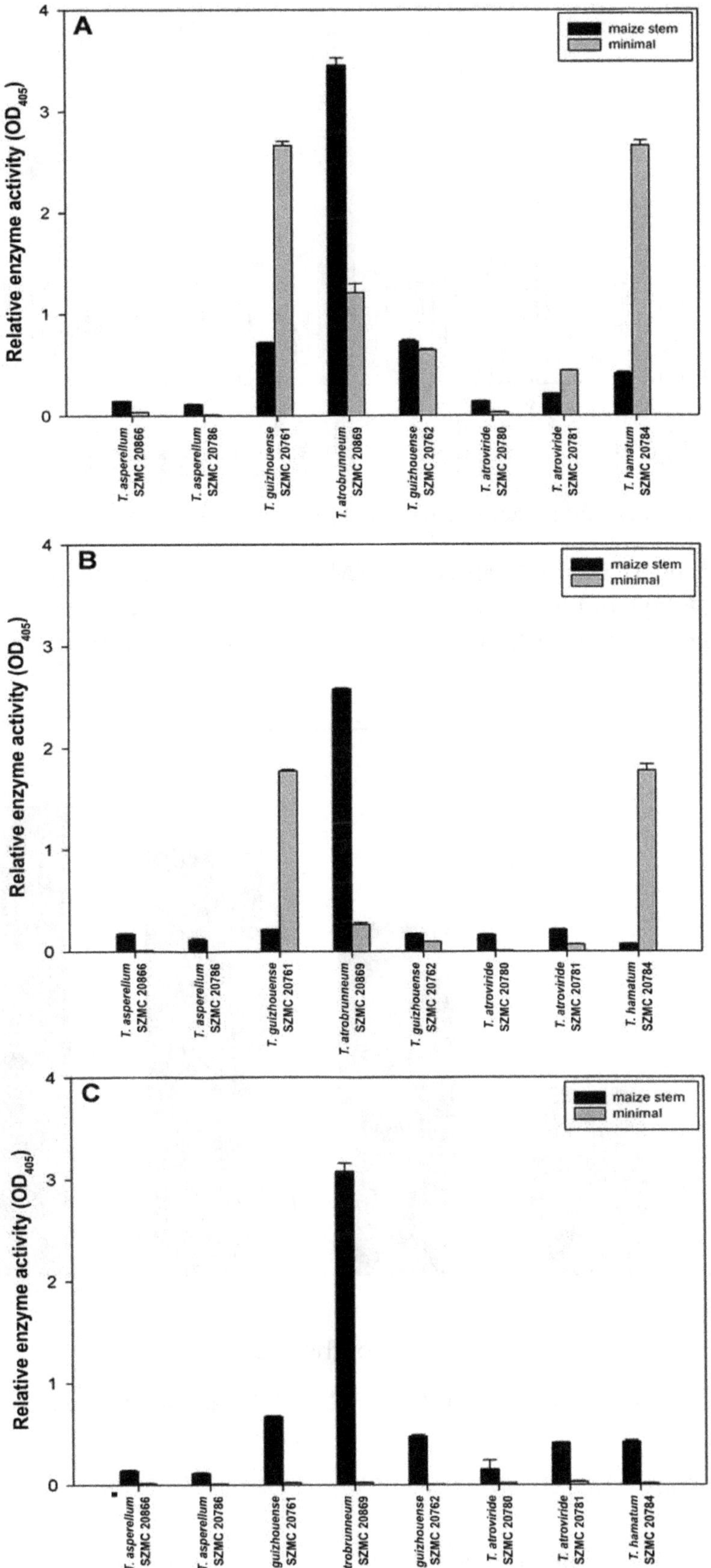

Figure 2. Extracellular enzyme activities of *Trichoderma* strains in liquid minimal and maize stem medium (mean ± SE, $n = 3$). (**A**): β-glucosidase, (**B**): cellobiohydrolase, (**C**): phosphatase.

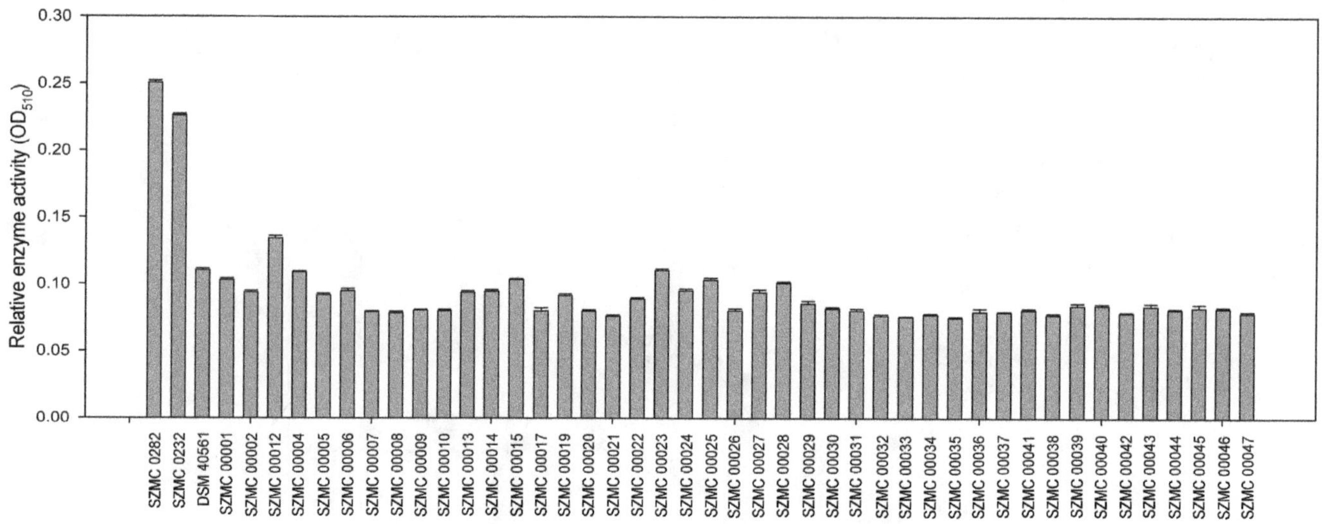

Figure 3. Peroxidase activities of *Streptomyces* isolates (mean ± SE, $n = 3$).

As the growth kinetics of strain *A. vinelandii* SZMC 22195 revealed that it reached the concentration of 10^8 cells ml^{-1} after 4 days of incubation in nitrogen-free liquid medium and the cell concentration increased further until day 7 (Figure 4), this strain was selected as the nitrogen-fixing component of the soil inoculant.

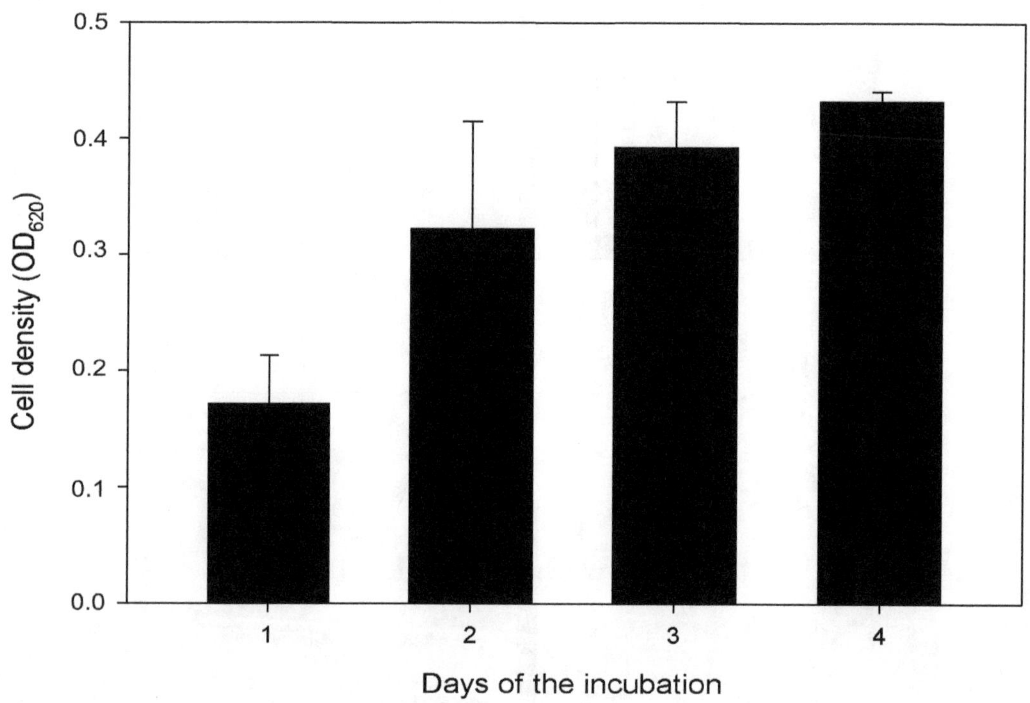

Figure 4. Growth kinetics of *Azotobacter vinelandii* SZMC 22195 in nitrogen-free liquid medium (mean ± SE, $n = 3$).

3.2. Influence of *T. asperellum* Strain SZMC 20786 on the Shoot and Root Growth and Photosynthetic Activity of Tomato Plants

In the case of the *T. asperellum* strain SZMC 20786 selected for the composite bioinoculant, direct plant growth promotion and effects on photosynthetic activity were examined on tomato plants. Significant increases in the fresh weight of the roots and shoots could be recorded in comparison to the control plants (Figure 5). The results deriving from the measurements of stomatal conductance and CO_2 assimilation are shown in Figure 6. Treatment with strain SZMC 20786 resulted in a non-significant increase in stomatal conductance (Figure 6A), coupled with a significant increase in both the CO_2 assimilation (Figure 6B) and the total sugar content (Figure 6C), indicating an increased photosynthetic activity in the plants treated with *T. asperellum*.

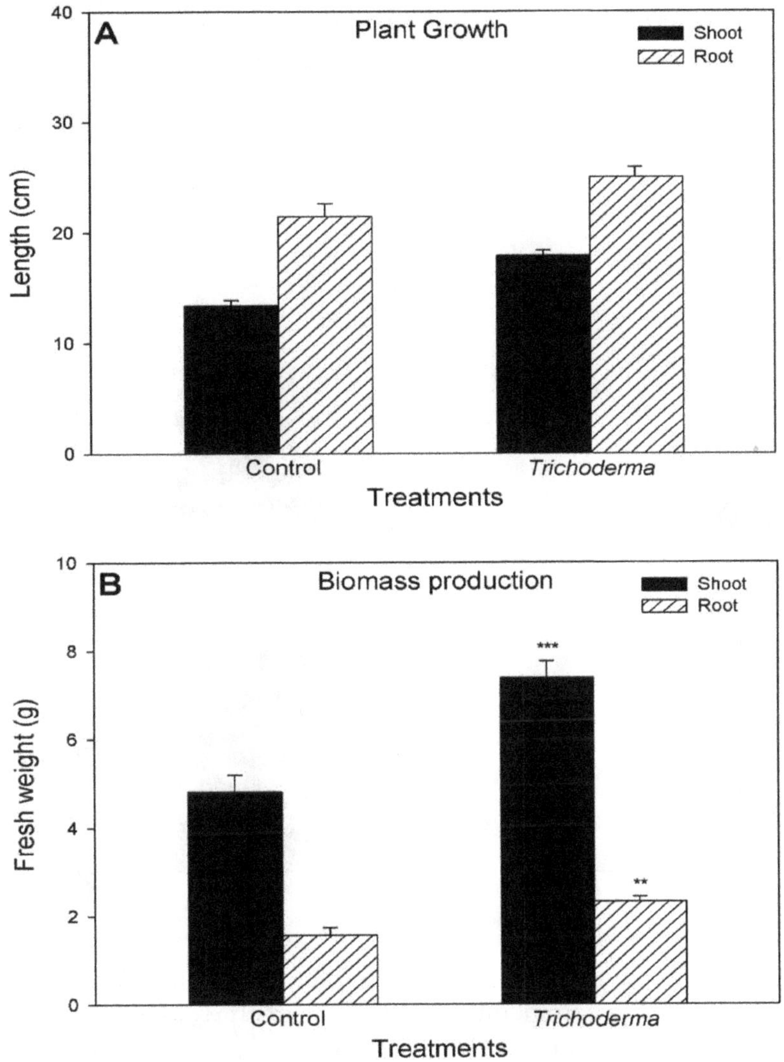

Figure 5. Effect of *T. asperellum* SZMC 20786 treatment on the growth parameters of tomato plants. (**A**) Root and shoot growth (mean ± SE, $n = 10$), (**B**) root and shoot biomass (mean ± SE, $n = 10$). Treated samples marked with asterisks are significantly different from the untreated control at $p \leq 0.01$ (**), or $p \leq 0.001$ (***).

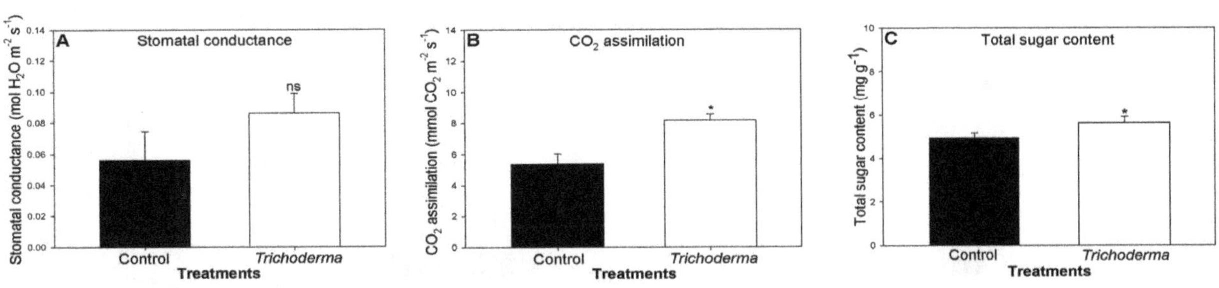

Figure 6. Changes in stomatal conductance (**A**), CO_2 assimilation (**B**), and total sugar content (**C**) in the leaves of tomato plants 6 weeks after treatment with *T. asperellum* SZMC 20786 (mean ± SE, $n = 5$). Treated samples marked with asterisks are significantly different from the untreated control at $p \leq 0.05$ (*), ns: not significant.

This is also supported by the changes of the chlorophyll a fluorescence induction parameters: the maximal quantum yield of PSII photochemistry (Fv/Fm), the actual quantum yield of PSII electron transport in the light-adapted state (ΦPSII), the photochemical quenching coefficient (qP), and the non-photochemical quenching (NPQ) were increased in plants treated with *T. asperellum* strain SZMC 20786; however, a significant change could be measured only for qP (Figure 7). The *T. asperellum* treatment also resulted in non-significant increases in the levels of chlorophyll a + b and carotenoids (Figure 8).

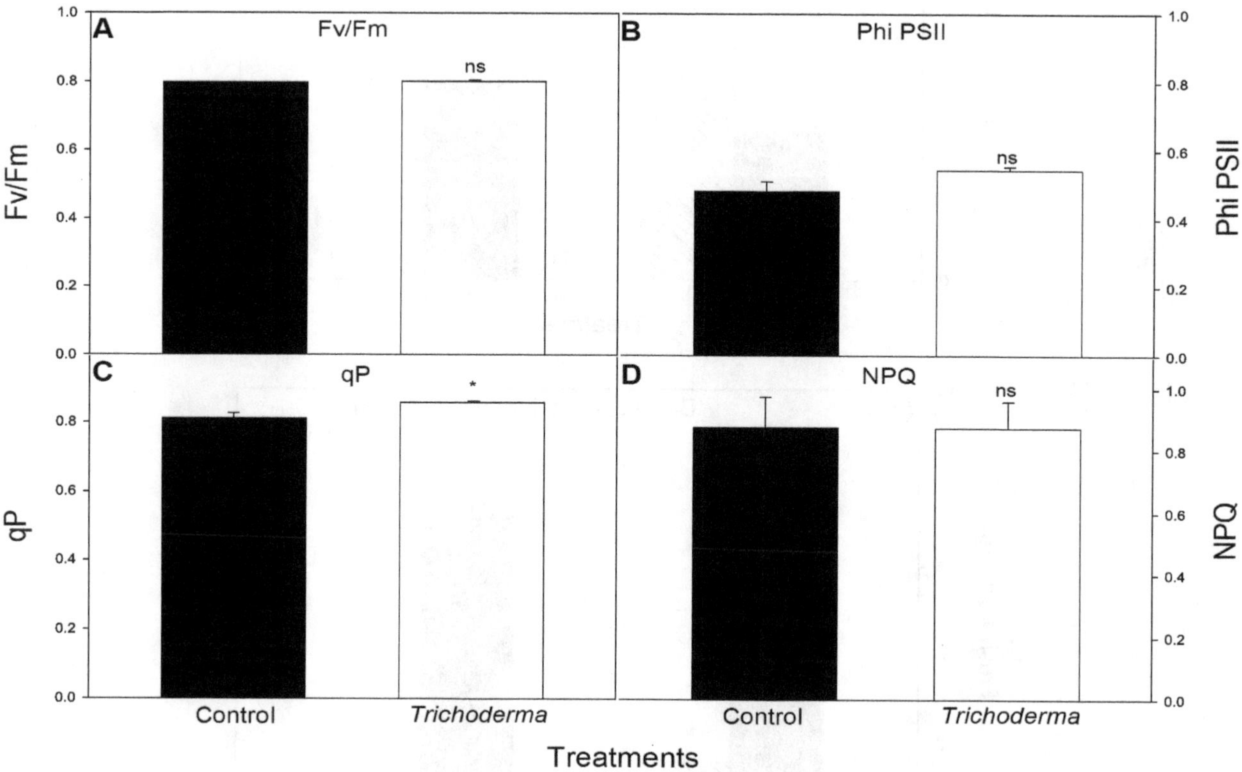

Figure 7. Changes of chlorophyll *a* fluorescence induction parameters ((**A**): Fv/Fm; (**B**): ΦPSII; (**C**): qP; (**D**): NPQ) in the leaves of tomato plants 6 weeks after treatment with *T. asperellum* SZMC 20786 (mean ± SE, $n = 5$). Treated samples marked with asterisks are significantly different from the untreated control at $p \leq 0.05$ (*), ns: not significant.

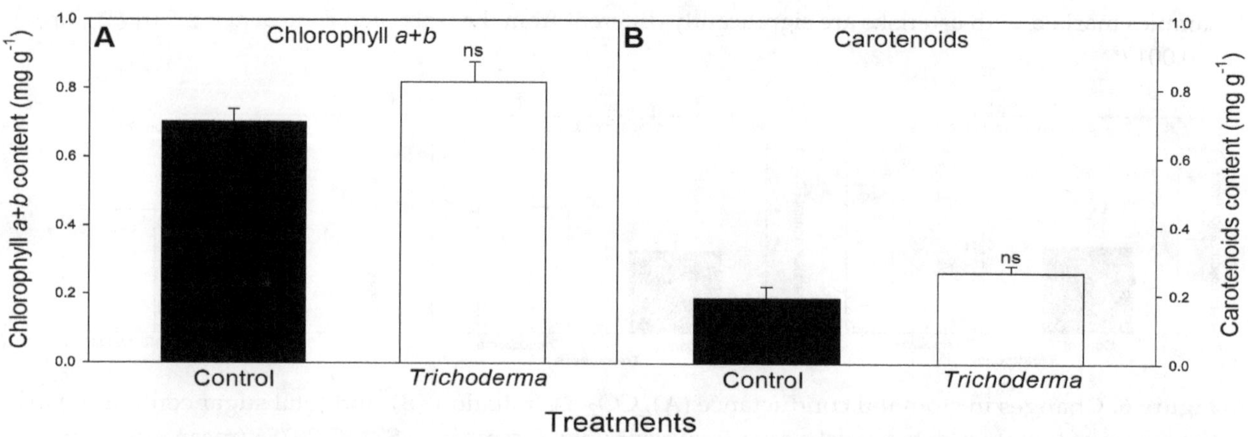

Figure 8. Changes in the content of chlorophyll a + b (**A**) and carotenoids (**B**) in the leaves of tomato plants 6 weeks after treatment with *T. asperellum* SZMC 20786 (mean ± SE, $n = 5$). ns: not significant.

3.3. Solid State Fermentation of Plant Stem Residues with *T. atrobrunneum* in Comparison with *T. reesei*

The extracellular enzyme activities of *T. atrobrunneum* SZMC 20869 (THSC) selected for the composite bioinoculant were compared with those of the industrially important, hypercellulolytic *T. reesei* strain QM9414 (SZMC 22616) in SSF experiments performed on the residues of 4 different crop plants (wheat, maize, sunflower, and canola) as substrates (Figure 9). Strain SZMC 20869 was able to produce β-glucosidase, cellobiohydrolase, β-xylosidase, and phosphatase activities on all four examined stem residues. Although the industrial strain of *T. reesei* produced larger amounts of β-glucosidase, cellobiohydrolase, and β-xylosidase—the three examined plants' cell-wall-degrading enzymes (PCWDEs)—on maize stem and canola stem residues, and was also a better producer of the two cellulolytic enzyme activities on sunflower stem residues than *T. atrobrunneum* SZMC 20869; the selected *T. atrobrunneum* strain was more efficient in production of all examined PCWDEs on wheat straw as the substrate. Regarding phosphatase activities, strain SZMC 20869 had been proved to be better than *T. reesei* on sunflower stem residues and equal to it on maize stem residues.

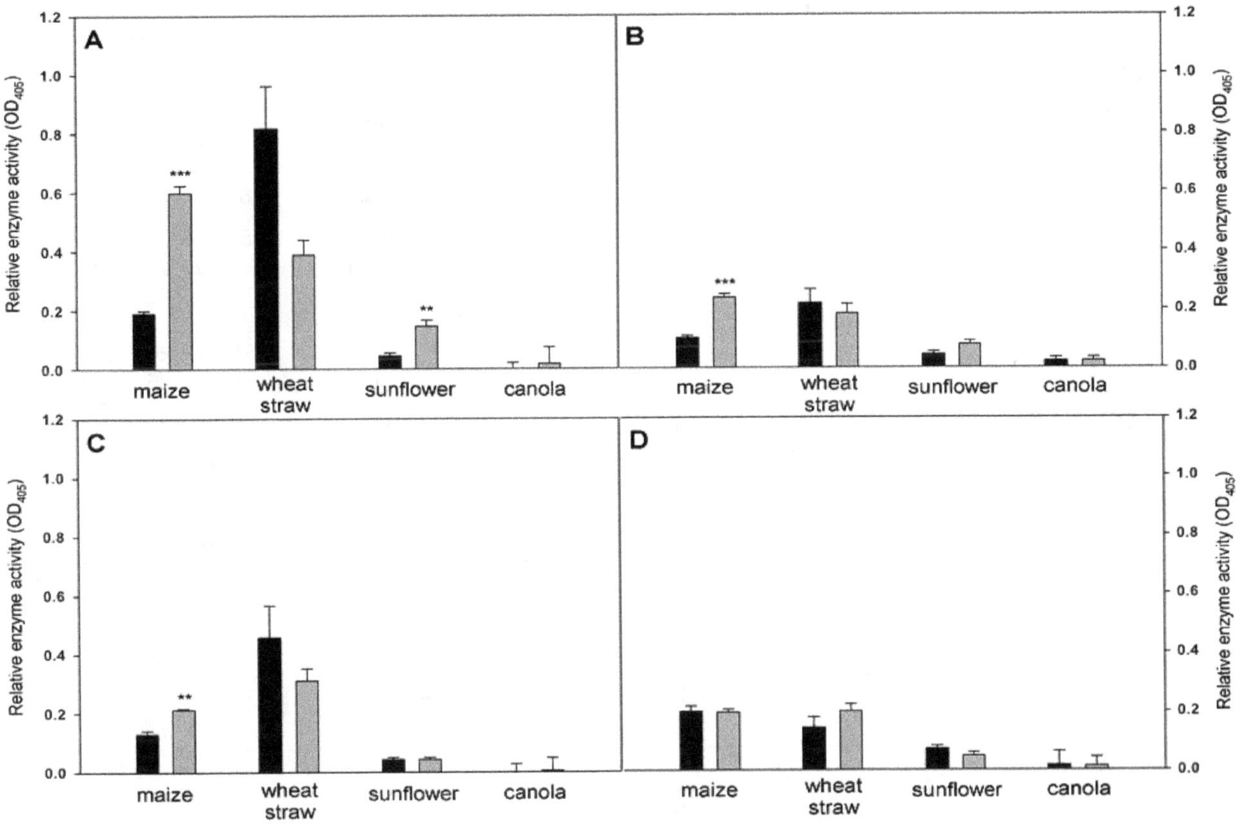

Figure 9. Extracellular enzyme activities of *T. atrobrunneum* SZMC 20869 and *Trichoderma reesei* SZMC 22616 after 8 days of solid-state fermentation on stem residues of wheat, maize, sunflower, and canola plants as substrates. (**A**) β-glucosidase, (**B**) cellobiohydrolase, (**C**) β-xylosidase, and (**D**) phosphatase (mean ± SE, $n = 3$). ■: *T. atrobrunneum*; ■: *T. reesei*. Columns of *T. reesei* marked with asterisks are significantly different at $p \leq 0.01$ (**), or $p \leq 0.001$ (***).

3.4. Field Experiment with the Combination of the Selected Bioinoculant Strains in Tomato Culture

Soil initial chemical and physical characteristics of the experimental area were as follows: pH: 7.6 (KCl), K_A: 35, carbonate: 10.6 m/m%, humus content: 1.84 m/m%, P_2O_5: 2423 mg/kg, K_2O: 669 mg/kg, NOX-N: 11 mg/kg; SO_4-S: 39 mg/kg, Cu: 4.4 mg/kg, Zn: 7.6 mg/kg, Mn: 13 mg/kg, Mg: 233 mg/kg, and Na: 99 mg/kg.

Changes in the soil macroelement, microelement, and humus content were monitored three times during the cultivation (Table 2). The initial potassium, sodium, and manganese contents of the total area determined from soil samples taken on 11th May were 669 mg kg^{-1}, 99 mg kg^{-1}, and 13 mg kg^{-1}, respectively. According to the statistical analysis, the K_2O, Na, and Mn content of the soil changed significantly in the treated area compared to the control area. This could be the result of a more efficient uptake of K_2O, Na, and Mn in the case of the treated plants; however, it is also possible that the elements P, Na, and Mn could have been sequestered in the microbial biomass. The average tomato fruit crop yields per row were 22,470 g and 22,810 g for the control and treated rows, respectively; however, the increase in the crop size in the case of the treated rows did not prove to be significant. The average total green mass of control and treated rows were 11,613 g and 11,032 g respectively, with no significant difference.

Table 2. Changes in the macro- and microelement content of the soil during the cultivation period of tomato.

Parameter	Measure	Sampling Time	Mean ± SD	
			Control Rows	Treated Rows
Humus	m/m %	I.	1.89 ± 0.04	1.98 ± 0.11
W		II.	1.57 ± 0.09	1.90 ± 0.33
		III.	1.90 ± 0.10	1.93 ± 0.19
P_2O_5	mg kg^{-1}	I.	2478.33 ± 182.57	1895.88 ± 125.12
		II.	1677.67 ± 69.04	1546.88 ± 135.31
		III.	1694.00 ± 86.41	1145.25 ± 368.56
K_2O	mg kg^{-1}	I.	702.67 ± 23.70	558.00 ± 76.90 *
		II.	701.67 ± 26.66	451.75 ± 64.14 ***
		III.	821.33 ± 54.63	480.13 ± 85.53 ****
SO_4-S	mg kg^{-1}	I.	16.33 ± 0.47	17.00 ± 1.22
		II.	15.00 ± 0.82	16.88 ± 2.32
		III.	7.33 ± 0.94	7.25 ± 2.11
NO_x-N	mg kg^{-1}	I.	3.93 ± 0.54	4.68 ± 0.96
		II.	4.67 ± 1.96	4.64 ± 1.25
		III.	5.17 ± 0.66	5.24 ± 1.96
Na	mg kg^{-1}	I.	49.33 ± 0.94	45.00 ± 4.00
		II.	45.00 ± 12.75	25.88 ± 12.69 *
		III.	46.33 ± 4.11	33.88 ± 5.37
Mg	mg kg^{-1}	I.	209.67 ± 2.62	220.75 ± 2.59
		II.	194.33 ± 8.22	194.75 ± 12.16
		III.	229.00 ± 11.22	213.38 ± 25.70
Mn	mg kg^{-1}	I.	14.33 ± 2.05	10.75 ± 0.66
		II.	19.33 ± 1.25	17.63 ± 1.58
		III.	19.67 ± 1.25	15.88 ± 3.18 *
Zn	mg kg^{-1}	I.	6.20 ± 0.78	5.75 ± 0.72
		II.	5.70 ± 0.37	6.44 ± 0.74
		III.	4.73 ± 0.40	5.11 ± 1.16
Cu	mg kg^{-1}	I.	6.23 ± 1.11	4.36 ± 0.44
		II.	8.20 ± 0.33	7.76 ± 0.44
		III.	9.20 ± 1.55	9.15 ± 3.00

Soils samplings were performed on I: 22 June, II: 6 July, and III: 3 August 2019. Data marked with asterisks are significantly different from the untreated control at $p < 0.05$ (*), $p = 0.001$ (***), $p < 0.001$ (****).

4. Discussion

The soil inoculant developed during this study contains two *Trichoderma* strains (*T. asperellum* and *T. atrobrunneum*) and two bacteria (*A. vinelandii* and *S. albus*) with potentially synergistic beneficial effects.

Members of the genus *Trichoderma* are geographically widespread filamentous ascomycetes from Hypocreales [21], which have long been known as agriculturally important, beneficial fungi with antagonistic abilities toward plant pathogenic fungi. *Trichoderma* antagonism is based on a series of different mechanisms including the competition for space and nutrients, antibiosis, mycoparasitism [22], plant growth promotion [23], enhancement of plant resistance to diseases [24,25], and relieving abiotic stress in plants [26]. These properties make many representatives of the genus *Trichoderma* (e.g., the THSC, *T. asperellum*, *T. atroviride*, or *T. virens*) to potential ingredients of soil inoculant and biocontrol preparations. However, when the practical application of a *Trichoderma* strain is planned, an exact, sequence-based, species-level identification is important to prevent the spread of species known as the causal agents of the green mold disease in mushroom cultivation [27–29] or of opportunistic infections in immunocompromised humans [30].

Strain *T. asperellum* SZMC 20786 was selected as a component of the composite soil bioinoculant due to its good in vitro antagonistic performance against different plant pathogenic fungi and its abilities to promote the growth of tomato plants and increase their photosynthetic activities. According to the literature, one of the mostly studied strains of "*T. asperellum*" for plant growth promotion is strain T203 [31], which, however, was recently reidentified as *T. asperelloides* [32]; thus, the number of studies about the plant growth promoting activities of *T. asperellum* sensu stricto is restricted. Qi and Zhao [33] demonstrated the plant growth promoting activities of *T. asperellum* srain Q1 on cucumber plants, and the positive effects were detected even when the plants were subjected to salt stress. In our study, *T. asperellum* strain SZMC 20786 showed positive effects on the CO_2 assimilation, total sugar content, and the photochemical quenching coefficient of tomato leaves. Similar results were obtained by Doni et al. [34], who found plant growth promotion as well as increased stomatal conductance and CO_2 assimilation in rice plants treated with *Trichoderma* sp. isolates. Other studies reported about the positive effects on the photosynthetic pigments exerted by *T. harzianum* strains on tomato [35] and wheat plants [36], as well as by *T. hamatum* on mungbean [37].

Another beneficial trait of many *Trichoderma* strains is their efficient ability to produce PCWDEs including cellulases and xylanases, which can be exploited both in the biotechnological industry and in the agriculture for the degradation of cellulose and xylan-containing materials, e.g., stem residues [38,39]. In accordance with these, another *Trichoderma* component, a *T. atrobrunneum* isolate possessing good cellulase, xylanase, and phosphatase enzyme production capabilities has been included in the assembled soil inoculant.

The species *A. vinelandii* involves Gram-negative, aerobic, free-living soil-inhabiting gamma-proteobacteria from the *Pseudomonadaceae* family. This species is capable of direct nitrogen fixation from the atmosphere by three distinct nitrogenase systems under fully aerobic conditions, thereby providing plant roots with bioavailable nitrogen source [40]. This aerobic bacterium possesses various protection mechanisms for nitrogenase against oxygen, which include alginate formation [41]. Furthermore, phytohormone and siderophore synthesis as well as phosphate solubilization are also among the abilities of *Azotobacter* species. These properties were suggested to be directly involved in their plant growth promotion effect [42]. Considering the above facts, an *A. vinelandii* strain has also been included in the soil inoculant.

As peroxidases were shown to play an important role in the humification properties of *Streptomyces* species [43], the potential humus-producing component of the bioinoculant was selected from *Streptomyces* isolates with a peroxidase assay, which revealed a *S. albus* strain as the best peroxidase-producing isolate. The species *S. albus* is among the geographically most widely distributed members of the genus *Streptomyces*; it could be isolated from various habitats including sea sediments, sponges, and insects [44]. This species was shown to be able to biosynthesize heterologously diverse and important natural products and was suggested to encode important natural product gene clusters [44]. Based on the genome sequence of *S. albus*, the secretion of a series of degradative enzymes (including amylases, chitinases, glucanases, proteinases/peptidases, and a cellulase) with supposed roles in breaking down heterogeneous alternative food sources in soil could be predicted [45]. The feather-degrading abilities of *S. albus* could be exploited during the development of an eco-friendly biofertilizer feather compost [46].

Several publications are available in the literature about the combined application of beneficial microorganisms for plant growth promotion and biological control. There are studies about the co-application of multiple bacteria including the combination of *Bacillus* and *Pseudomonas* for growth promotion and biological control of soil-borne diseases in pepper and tomato [47], as well as to increase rice yields [48], or the application of a mixture of fluorescent pseudomonads to suppress take-all disease of wheat [49]. There are also examples of co-application of fungi and bacteria, for instance the combination of *T. koningii* with fluorescent pseudomonads for the control of take-all disease of wheat [50], or the combination of *T. harzianum* with an *Alcaligenes* strain for the reduction of the incidence of rot disease caused by *Phytophthora capsici* in black pepper [51]. Other microbial combinations have been applied to promote the growth of tomato [52], to control tobacco diseases [53], or to improve the salinity tolerance of *Vicia faba* [54]. The combination of microorganisms could also increase the dry matter yield and nutrient uptake by wheat grown in a sandy soil [55].

5. Conclusions

The selection of the components for the composite soil inoculant assembled in this study was driven by the idea to combine various crop protective and plant growth promoting traits (biocontrol against plant pathogenic fungi, phosphorous mobilization, stem degradation, humification, and nitrogen fixation) of different microorganisms into a preparation, which may also have the potential to exert an increased consistency of field performance under various environmental conditions. The screening strategy performed during this study proved to be applicable for the assembly of a promising composite soil bioinoculant with notable application potentials.

Author Contributions: H.A., P.K., C.V., L.M. and L.K. contributed to the design and implementation of the research and evaluation of the results, and participated in the preparation of the manuscript; P.K. and L.K. performed the in vitro antagonism and liquid fermentation experiments; B.B. and L.M. contributed to the selection of the bacterial components of the bioinoculant; P.K. and P.P. designed and performed the experiments on tomato plants; V.D.N. and M.T. contributed to the solid-state fermentation experiments; H.A. performed and evaluated the field experiment; I.S. performed the measurements of the soil samples taken; and A.S. performed data curation and statistical analyses. The Tables and Figures were prepared by H.A., P.K., B.B., P.P., V.D.N. and C.V. All authors have read and agreed to the published version of the manuscript.

Acknowledgments: The authors thank Sarolta Szabó and Gábor Tarnai (BioeGO Ltd.) for their support during the performance of the field experiment.

References

1. Damalas, C.A.; Eleftherohorinos, I.G. Pesticide exposure, safety issues, and risk assessment indicators. *Int. J. Environ. Res. Public Health* **2011**, *8*, 1402–1419. [CrossRef] [PubMed]
2. Ceulemans, T.; Stevens, C.J.; Duchateau, L.; Jacquemyn, H.; Gowing, D.J.; Merckx, R.; Wallace, H.; van Rooijen, N.; Goethem, T.; Bobbink, R.; et al. Soil phosphorus constrains biodiversity across European grasslands. *Glob. Chang. Biol.* **2014**, *20*, 3814–3822. [CrossRef] [PubMed]
3. Sekhon, G.S. Fertilizer-N use efficiency and nitrate pollution of groundwater in developing countries. *J. Contam. Hydrol.* **1995**, *20*, 167–184.
4. Meyer, S.L.F.; Roberts, D.P. Combinations of biocontrol agents for management of plant-parasitic nematodes and soilborne plant-pathogenic fungi. *J. Nematol.* **2002**, *34*, 1. [PubMed]

5. Syed Ab Rahman, S.F.; Singh, E.; Pieterse, C.M.J.; Schenk, P.M. Emerging microbial biocontrol strategies for plant pathogens. *Plant Sci.* **2018**, *267*, 102–111. [CrossRef] [PubMed]
6. Weller, D.M.; Thomashow, L.S.; Cook, R.J. Biological control of soil-borne pathogens of wheat: Benefits, risks and current challenges. In *Biological Control: Benefits and Risks*; Hokkanen, H.M.T., Lynch, J.M., Eds.; Cambridge University Press: Cambridge, UK, 1995; pp. 149–160.
7. Körmöczi, P.; Danilovic, G.; Manczinger, L.; Jovanovic, L.; Pankovic, D.; Vágvölgyi, C.; Kredics, L. Species composition of *Trichoderma* isolates from the rhizosphere of vegetables grown in Hungarian soils. *Fresen. Environ. Bull.* **2013**, *22*, 1736–1741.
8. Javaid, A. Effects of biofertilizers combined with different soil amendments on potted rice plants. *Chilean J. Agric. Res.* **2011**, *71*, 157–163. [CrossRef]
9. Chen, L.; Bóka, B.; Kedves, O.; Nagy, V.D.; Szűcs, A.; Champramary, S.; Roszik, R.; Patocskai, Z.; Münsterkötter, M.; Huynh, T.; et al. Towards the biological control of devastating forest pathogens from the genus *Armillaria*. *Forests* **2019**, *10*, 1013. [CrossRef]
10. Szekeres, A.; Leitgeb, B.; Kredics, L.; Manczinger, L.; Vágvölgyi, C. A novel, image analysis-based method for the evaluation of in vitro antagonism. *J. Microbiol. Meth.* **2006**, *65*, 619–622. [CrossRef]
11. Poór, P.; Gémes, K.; Horváth, F.; Szepesi, Á.; Simon, M.L.; Tari, I. Salicylic acid treatment via the rooting medium interferes with stomatal response, CO_2 fixation rate and carbohydrate metabolism in tomato, and decreases harmful effects of subsequent salt stress. *Plant Biol.* **2011**, *13*, 105–114.
12. Dubois, M.; Gibbs, K.A.; Hamilton, J.K.; Roberts, D.A.; Smith, F. Colorimetric methods for the determination of sugars and related substances. *Anal. Chem.* **1956**, *28*, 350–352. [CrossRef]
13. Genty, B.; Briantais, J.M.; Baker, N.R. The relationship between the quantum yield of photosynthetic electron transport and quenching of chlorophyll fluorescence. *Biochim. Biophys. Acta* **1989**, *990*, 87–92. [CrossRef]
14. Bilger, W.; Schreiber, U. Energy-dependent quenching of dark level chlorophyll fluorescence in intact leaves. *Photosynth. Res.* **1986**, *10*, 303–308. [CrossRef] [PubMed]
15. Bilger, W.; Björkman, O. Role of the xanthophyll cycle in photoprotection elucidated by measurements of light-induced absorbance changes, fluorescence and photosynthesis in *Hedera canariensis*. *Photosynth. Res.* **1990**, *25*, 173–185. [CrossRef] [PubMed]
16. Lichtenthaler, H.K.; Wellburn, A.R. Determinations of total carotenoids and chlorophylls a and b of leaf extracts in different solvents. *Biochem. Soc. Trans.* **1983**, *11*, 591–592. [CrossRef]
17. Ábrahám, B.É.; Őri, N.; Szabó, S.; Blaskó, L.; Zsigrai, G.Y. Effects of whitening process on the quality of different grain sorghum hybrids. *Eur. J. Plant Sci. Biotechnol.* **2012**, *6*, 90–93.
18. Szolnoki, Z.; Farsang, A.; Puskás, I. Cumulative impacts of human activities on urban garden soils: Origin and accumulation of metals. *Environ. Pollut.* **2013**, *177*, 106–115. [CrossRef]
19. Stefanovits, P.; Filep, G.; Füleky, G. *Talajtan*; Mezőgazda Kiadó: Budapest, Hungary, 1999; pp. 470p. (In Hungarian)
20. Junsomboon, J.; Jakmunee, J. Determination of potassium, sodium and total alkalies in Portland cement, fly ash, admixtures, and water of concrete by a simple flow injection flame photometric system. *J. Autom. Meth. Manag. Chem.* **2011**, *2011*, 742656. [CrossRef]
21. Kredics, L.; Hatvani, L.; Naeimi, S.; Körmöczi, P.; Manczinger, L.; Druzhinina, I.S. Biodiversity of the genus *Hypocrea/Trichoderma* in different habitats. In *Biotechnology and Biology of Trichoderma*; Gupta, V.K., Schmoll, M., Herrera-Estrella, A., Upadhyay, R.S., Druzhinina, I., Tuohy, M., Eds.; Elsevier Science B.V.: Amsterdam, The Netherlands, 2014; pp. 3–24.
22. Olmedo Monfil, V.; Casas-Flores, S. Molecular mechanisms of biocontrol in *Trichoderma* spp. and their applications in agriculture. In *Biotechnology and Biology of Trichoderma*; Gupta, V.K., Schmoll, M., Herrera-Estrella, A., Upadhyay, R.S., Druzhinina, I., Tuohy, M., Eds.; Elsevier Science B.V.: Amsterdam, The Netherlands, 2014; pp. 429–454.
23. Stewart, A.; Hill, R. Applications of *Trichoderma* in plant growth promotion. In *Biotechnology and Biology of Trichoderma*; Gupta, V.K., Schmoll, M., Herrera-Estrella, A., Upadhyay, R.S., Druzhinina, I., Tuohy, M., Eds.; Elsevier Science B.V.: Amsterdam, The Netherlands, 2014; pp. 415–428.
24. Saldajeno, M.G.B.; Naznin, H.A.; Elsharkawy, M.M.; Shimizu, M.; Hyakumachi, M. Enhanced resistance of plants to disease using *Trichoderma* spp. In *Biotechnology and Biology of Trichoderma*; Gupta, V.K., Schmoll, M., Herrera-Estrella, A., Upadhyay, R.S., Druzhinina, I., Tuohy, M., Eds.; Elsevier Science B.V.: Amsterdam, The Netherlands, 2014; pp. 477–494.

25. Contreras-Cornejo, H.A.; Macías-Rodríguez, L.; López-Bucio, J.S.; López-Bucio, J. Enhanced plant immunity using *Trichoderma*. In *Biotechnology and Biology of Trichoderma*; Gupta, V.K., Schmoll, M., Herrera-Estrella, A., Upadhyay, R.S., Druzhinina, I., Tuohy, M., Eds.; Elsevier Science B.V.: Amsterdam, The Netherlands, 2014; pp. 495–504.
26. Zaidi, N.W.; Dar, M.H.; Singh, S.; Singh, U.S. Trichoderma species as abiotic stress relievers in plants. In *Biotechnology and Biology of Trichoderma*; Gupta, V.K., Schmoll, M., Herrera-Estrella, A., Upadhyay, R.S., Druzhinina, I., Tuohy, M., Eds.; Elsevier Science B.V.: Amsterdam, The Netherlands, 2014; pp. 515–526.
27. Hatvani, L.; Kocsubé, S.; Manczinger, L.; Antal, Z.; Szekeres, A.; Druzhinina, I.S.; Komoń-Zelazowska, M.; Kubicek, C.P.; Nagy, A.; Vágvölgyi, C.; et al. The green mould disease global threat to the cultivation of oyster mushroom (*Pleurotus ostreatus*): A review. In *Science and Cultivation of Edible and Medicinal Fungi: Mushroom Science XVII: Proceedings of the 17th Congress of the International Society for Mushroom Science*; Gruening, M., Ed.; International Society for Mushroom Science: Cape Town, South Africa, 2008; pp. 485–495.
28. Kredics, L.; García Jimenez, L.; Naeimi, S.; Czifra, D.; Urbán, P.; Manczinger, L.; Vágvölgyi, C.; Hatvani, L. A challenge to mushroom growers: The green mould disease of cultivated champignons. In *Current Research, Technology and Education Topics in Applied Microbiology and Microbial Biotechnology*; Méndez-Vilas, A., Ed.; FORMATEX: Badajoz, Spain, 2010; pp. 295–305.
29. Błaszczyk, L.; Siwulski, M.; Sobieralski, K.; Fruzynska-Jozwiak, D. Diversity of *Trichoderma* spp. causing *Pleurotus* green mould diseases in Central Europe. *Folia Microbiol.* **2013**, *58*, 325–333. [CrossRef]
30. Hatvani, L.; Manczinger, L.; Vágvölgyi, C.; Kredics, L. Trichoderma as a human pathogen. In *Trichoderma-Biology and Applications*; Mukherjee, P.K., Horwitz, B.A., Singh, U.S., Mukherjee, M., Schmoll, M., Eds.; CAB International: Wallingford, UK, 2013; pp. 292–313.
31. Viterbo, A.; Landau, U.; Kim, S.; Chernin, L.; Chet, I. Characterization of ACC deaminase from the biocontrol and plant growth-promoting agent *Trichoderma asperellum* T203. *FEMS Microbiol. Lett.* **2010**, *305*, 42–48. [CrossRef]
32. Brotman, Y.; Lisec, J.; Méret, M.; Chet, I.; Willmitzer, L.; Viterbo, A. Transcript and metabolite analysis of the *Trichoderma*-induced systemic resistance response to *Pseudomonas syringae* in *Arabidopsis thaliana*. *Microbiology* **2012**, *158*, 139–146. [CrossRef] [PubMed]
33. Qi, W.; Zhao, L. Study of the siderophore-producing *Trichoderma asperellum* Q1 on cucumber growth promotion under salt stress. *J. Basic Microbiol.* **2013**, *53*, 355–364. [CrossRef] [PubMed]
34. Doni, F.; Isahak, A.; Zain, C.R.C.M.; Yusoff, W.M.W. Physiological and growth response of rice plants (*Oryza sativa* L.) to *Trichoderma* spp. inoculants. *AMB Exp.* **2014**, *4*, 45. [CrossRef]
35. Azarmi, R.; Hajieghrari, B.; Giglou, A. Effect of *Trichoderma* isolates on tomato seedling growth response and nutrient uptake. *Afr. J. Biotechnol.* **2011**, *10*, 5850–5855.
36. Al-Ezerjawi, N.H.; Kadhim, J.H. Effect of two isolates of *Trichoderma harzianum* on total nitrogen, chlorophyll a and b contents and yield of wheat (*Triticum aestivum.* L.) class Eba'a-95. *Int. J. Sci. Res.* **2014**, *3*, 1078–1083.
37. Badar, R.; Qureshi, S.A. Comparative effect of *Trichoderma hamatum* and host specific *Rhizobium* species on growth of *Vigna mungo*. *J. Appl. Pharm. Sci.* **2012**, *2*, 128–132. [CrossRef]
38. Do Vale, L.H.F.; Filho, E.X.F.; Miller, R.N.G.; Ricart, C.A.O.; De Sousa, M.V. Cellulase systems in Trichoderma: An overview. In *Biotechnology and Biology of Trichoderma*; Gupta, V.K., Schmoll, M., Herrera-Estrella, A., Upadhyay, R.S., Druzhinina, I., Tuohy, M., Eds.; Elsevier Science B.V.: Amsterdam, The Netherlands, 2014; pp. 229–244.
39. Błaszczyk, L.; Siwulski, M.; Sobieralski, K.; Lisiecka, J.; Jędryczka, M. *Trichoderma* spp.—Application and prospects for use in organic farming and industry. *J. Plant Prot. Res.* **2014**, *54*, 309–317. [CrossRef]
40. Nosrati, R.; Owlia, P.; Saderi, H.; Olamaee, M.; Rasooli, I.; Akhavian, T.A. Correlation between nitrogen fixation rate and alginate productivity of an indigenous *Azotobacter vinelandii* from Iran. *Iran J. Microbiol.* **2012**, *4*, 153–159.
41. Daiz-Barrera, A.; Soto, E. Biotechnological uses of *Azotobacter vinelandii*: Current state, limits and prospects. *Afr. J. Biotechnol.* **2010**, *9*, 5240–5250.
42. Rubio, E.J.; Montecchia, M.S.; Tosi, M.; Cassán, F.D.; Perticari, A.; Correa, O.S. Genotypic characterization of Azotobacteria isolated from Argentinean soils and plant-growth-promoting traits of selected strains with prospects for biofertilizer production. *Sci. World J.* **2013**, *2013*, 519603. [CrossRef]
43. Cozzolino, A.; Piccolo, A. Polymerization of dissolved humic substances catalyzed by peroxidase. Effects of pH and humic composition. *Org. Geochem.* **2002**, *33*, 281–294. [CrossRef]

44. Seipke, R.F. Strain-level diversity of secondary metabolism in *Streptomyces albus*. *PLoS ONE* **2015**, *10*, e0116457. [CrossRef]
45. Zaburannyi, N.; Rabyk, M.; Ostash, B.; Fedorenko, V.; Luzhetskyy, A. Insights into naturally minimised *Streptomyces albus* J1074 genome. *BMC Genomics* **2014**, *15*, 97. [CrossRef]
46. Nayaka, S.; Vidyasagar, G.M. Development of eco-friendly bio-fertilizer using feather compost. *Ann. Plant Sci.* **2013**, *2*, 238–244.
47. Domenech, J.; Reddy, M.S.; Kloepper, J.W.; Ramos, B.; Gutierrez-Mañero, J. Combined application of the biological product LS213 with *Bacillus*, *Pseudomonas* or *Chryseobacterium* for growth promotion and biological control of soil-borne diseases in pepper and tomato. *BioControl* **2006**, *51*, 245–258. [CrossRef]
48. Nguyen, H.T.; Deaker, R.; Kennedy, I.R.; Roughley, R.J. The positive yield response of field-grown rice to inoculation with a multi-strain biofertiliser in the Hanoi area, Vietnam. *Symbiosis* **2003**, *35*, 231–245.
49. Pierson, E.A.; Weller, D.M. Use of mixtures of fluorescent pseudomonads to suppress take-all and improve the growth of wheat. *Phytopathology* **1994**, *84*, 940–947. [CrossRef]
50. Duffy, B.K.; Simon, A.; Weller, D.M. Combination of *Trichoderma koningii* with fluorescent pseudomonads for control of take-all on wheat. *Phytopathology* **1996**, *86*, 188–194. [CrossRef]
51. Anith, K.N.; Manomohandas, T.P. Combined application of *Trichoderma harzianum* and *Alcaligenes* sp. strain AMB 8 for controlling nursery rot disease of black pepper. *Indian Phytopathol.* **2001**, *54*, 335–339.
52. El-Katatny, H.M. Enzyme production and nitrogen fixation by free, immobilized and coimmobilized inoculants of *Trichoderma harzianum* and *Azospirillum brasilense* and their possible role in growth promotion of tomato. *Food Technol. Biotechnol.* **2010**, *48*, 161–174.
53. Maketon, M.; Apisitsantikul, J.; Siriraweekul, C. Greenhouse evaluation of *Bacillus subtilis* AP-01 and *Trichoderma harzianum* AP-001 in controlling tobacco diseases. *Braz. J. Microbiol.* **2008**, *39*, 296–300. [CrossRef] [PubMed]
54. Rabie, G.H.; Almadini, A.M. Role of bioinoculants in development of salt-tolerance of *Vicia faba* plants under salinity stress. *Afr. J. Biotechnol.* **2005**, *4*, 210–222.
55. Singh, S.; Kapoor, K.K. Inoculation with phosphate-solubilizing microorganisms and a vesicular-arbuscular mycorrhizal fungus improves dry matter yield and nutrient uptake by wheat grown in a sandy soil. *Biol. Fertil. Soils* **1999**, *28*, 139–144. [CrossRef]

9

Vegetal-Derived Biostimulant Enhances Adventitious Rooting in Cuttings of Basil, Tomato and Chrysanthemum via Brassinosteroid-Mediated Processes

Hye-Ji Kim [1,*], Kang-Mo Ku [2], Seunghyun Choi [1] and Mariateresa Cardarelli [3]

[1] Department of Horticulture and Landscape Architecture, Purdue University, West Lafayette, IN 47907, USA; choi510@purdue.edu

[2] Division of Plant and Soil Sciences, West Virginia University, Morgantown, WV 26505, USA; kangmo.ku@mail.wvu.edu

[3] Consiglio per la ricerca in agricoltura e l'analisi dell'economia agraria, Centro di Ricerca Orticoltura e Florovivaismo, 84098 Pontecagnano, Italy; mteresa.cardarelli@crea.gov.it

* Correspondence: hjikim@purdue.edu

Abstract: Plant-derived protein biostimulants exhibit hormone-like activities promoting plant growth and yield, yet detailed investigations on hormonal function have remained limited. This study was conducted to investigate the effects of vegetal-derived-biostimulant on morphological and metabolic changes in cuttings of three herbaceous species demonstrating different rooting ability, basil (*Ocimum basilicum* L.), tomato (*Solanum lycopersicum* L.), and chrysanthemum (*Chrysanthemum indicum* L.), in comparison to auxin. Unrooted cuttings were applied with or without biostimulant (100, 1000, 5000, and 10,000 mg L^{-1}) or auxin [1% indole-3-butyric acid (IBA) plus 0.5% 1-naphthaleneacetic acid (NAA); 100, 200, 300, and 500 mg L^{-1}] as a basal quick-dip, stuck into inert media, and evaluated at 20 days after placement under intermittent mist. Both compounds increased adventitious rooting in all cuttings. Biostimulant required a significantly higher threshold for a series of adventitious rooting responses than auxin, and the maximum effectiveness was achieved at 5000 mg L^{-1} for biostimulant and 100, 200, and 300 mg L^{-1} for auxin in basil, tomato, and chrysanthemum, respectively. Adventitious rooting responses (dry mass and length) to biostimulant showed a gradual logarithmic rise as a function of increasing dosages, which was not in agreement with biphasic dose-response of auxin. Biostimulant significantly increased or tended to increase fine roots in all tested cuttings, which was not consistent with auxin. Relatively high levels of endogenous brassinosteroids (BRs) were present in non-treated cuttings of basil, tomato, and chrysanthemum in decreasing order. Both compounds had no effects or concomitantly increased or decreased BR levels in plant tissues, with fewer effects on basil and tomato, containing high BR levels, but more prominent effects on chrysanthemum, containing relatively low BR levels. Contrasting effects of biostimulant and auxin were found in antioxidant activities, which were promoted by biostimulant but inhibited by auxin either in roots or shoots. These results indicate that the hormonal effects of vegetal-derived biostimulant are primarily exerted by BR-mediated processes while involving interaction with auxin. Both the biostimulant-derived BRs and auxin were suggested to modulate endogenous BR pool via overlapping and interdependent regulatory functions, inducing morphological and metabolic changes during adventitious rooting of cuttings in a plant species-specific manner.

Keywords: stem cuttings; propagation; root morphology traits; indole-3-acetic acid (IAA); indole-3-butyric acid (IBA); gibberellins; phenolic compounds

1. Introduction

Plant-derived biostimulants represent a well-known group of biostimulants and have been proposed as an innovative tool to address the sustainability challenges facing horticulture and to ensure high yield and quality of horticultural commodities [1–3]. Manufactured from plant protein sources using partial hydrolysis, plant biostimulants are considered as a subgroup of growth regulators and bioregulators which are composed of a mixture of polypeptides, oligopeptides, and amino acids [4]. Plant-derived biostimulants are reported to be more effective than animal-derived biostimulants as they contain a higher concentration of amino acids and soluble peptides, with peptides being the principal active compounds [5–7]. Plant-derived biostimulants are defined as materials other than fertilizers that promote plant growth when applied in small quantities or metabolic enhancers [8]. They are available on the market as various forms, including liquid products, soluble powder or in granular form, and were demonstrated to be effective as a seed treatment, foliar spray, and soil drench for crop production [9–11]. When applied as a foliar spray or soil-drench, biostimulants can induce a series of physiological responses in crops changing their phenotypic characteristics and promoting plant growth [10,12].

It has been proposed that such plant responses induced by biostimulants are derived from hormone-like activities and the production of secondary metabolites [13]. Auxin- and gibberellin-like activities were demonstrated in corn coleoptiles and tomato cuttings [5,6], particularly due to the presence of bioactive peptides [14,15]. Peptides are known to be involved in cell differentiation, protease inhibitor induction, cell division, and pollen self-incompatibility response [16,17]. Similar results were reported in degraded soybean meal products, which had promotive effects on root hairs in *Brassica rapa* and tomato cuttings [7].

The positive effects of biostimulant on plant growth and yield have been demonstrated in many studies. The application of biostimulant not only enhanced the growth of corn seedlings [5,6,18] and stem cuttings of tomato [5], but also improved nutrient status, yield, and quality of herbaceous and woody plants, including corn, bean, tomato, sweet yellow pepper, strawberry, banana, papaya, and red grape [5,13,19–22]. It also enhanced tolerance to a wide range of abiotic stresses, such as drought [23], salinity [9], extreme temperatures [24], nutrient deficiency [25], and adverse soil pH [26]. The application of biostimulant increased root morphology, such as root dry mass, total root length, and root surface area, which was associated with improved nitrogen status [5,6]. However, it is not clear how such morphological and physiological changes are induced by the biostimulant.

Adventitious rooting involves significant cellular metabolic activities, leading to the formation of new roots at the base of stem cuttings. Auxin plays a pivotal role in promoting cell growth, cell division, and adventitious root formation in cuttings [27–29] and its mode of action on adventitious rooting is well-elucidated [27,30–32]. Rooting compounds commonly contain indole-3-butyric acid (IBA), 1-naphthaleneacetic acid (NAA), or a combination of the two compounds. The application of auxin to unrooted cuttings promotes adventitious rooting at relatively low doses.

Meanwhile, the mode of action of plant-derived biostimulants on adventitious rooting is largely unknown. An auxin-signaling mediated pathway was proposed to be involved in adventitious rooting of tomato cuttings and their improved nitrogen status, as represented by higher Soil Plant Analysis Development (SPAD) values, following the biostimulant Trainer® treatment [5]. They also found that biostimulant applications increased the shoot elongation rate of dwarf pea plants, prompting the idea of gibberellin involvement in regulating their shoot growth. Unlike auxins, gibberellins are known to inhibit the production of adventitious roots [33–35]. Therefore, stem cuttings provide an ideal experimental system with which mechanistic investigations on hormonal regulations associated with plant-derived-biostimulant can be undertaken. The system eliminates: (1) gibberellic acid as a potential candidate for biostimulant effects due to their antagonistic nature on adventitious rooting and (2) nutritional effects of biostimulant because nutrients are not required for initial stages of adventitious root formation. Nevertheless, carbohydrates play important roles in adventitious rooting, not only by providing energy and carbon chains for biosynthetic processes of new meristems and roots, but also by affecting gene expression, in collaboration with auxin [32].

Recent metabolomic investigations of the hormonal profile on greenhouse melon demonstrated that the application of biostimulant induced upregulation of metabolites related to brassinosteroids (BRs) and their interactions with other phytohormones were postulated to play a critical role in plant growth responses [36]. Similarly, transcriptomic profiles of lateral roots of maize seedlings demonstrated the involvement of BR signal transduction when treated with biostimulant [37]. Meanwhile, the effects of biostimulants were varied by plant species and/or cultivars, growing seasons, and the application method and concentration of the product [38] although the causes for these variations are not clear.

The objectives of the present study were: (i) to examine the hormonal effects of a plant-derived-biostimulant on adventitious rooting in cuttings of three herbaceous plant species with different rooting ability, (ii) to determine dose responses of stem cuttings to biostimulant and auxin, and (iii) to characterize morphological and metabolic changes induced by biostimulant. Cuttings of basil, tomato, and chrysanthemum were treated with biostimulant by a basal quick-dip, and morphological, physiological, and metabolic changes were evaluated to elucidate the hormonal regulation of biostimulant involved in adventitious rooting formation.

2. Materials and Methods

2.1. Plant Materials

Based on our preliminary observations on adventitious root formation of herbaceous plants, cuttings of basil (*Ocimum basilicum* L. cv. Genovese), tomato (*Solanum lycopersicum* L. cv. Washington Cherry), and chrysanthemum (*Chrysanthemum indicum* L. cv. Hollister) were chosen in this study for differences in their relative rooting ability: easy-to-root, moderate-to-root, and difficult-to-root, respectively. In general, herbaceous plant species can produce adventitious roots without application of exogenous auxin; however, auxin application is of commercial importance in cutting propagation, because the endogenous level of auxin is critical to increase the ease during root induction period [34]. The experiment was carried out in summer 2017 to spring 2018, in a glass greenhouse situated at Purdue University, West Lafayette, IN (lat. 40N, long. 86W; altitude 188m above sea level).

Seeds of tomato and basil were acquired from a commercial source (Johnny's Selected Seeds, Albion, ME, USA). The seeds were sown and grown in a growth room for 2 to 3 weeks at Plant Growth Facilities. Meanwhile, unrooted cuttings of chrysanthemum 'Hollister' were obtained from a commercial source (Syngenta Flowers, LLC., Gilroy, CA, USA). Immediately upon receipt, a box of stem cuttings was kept in a refrigerator maintained at 5 °C. The cuttings were applied with auxin within three days and then stuck into the media and propagated as described below to produce stock plants. Uniform seedlings of basil and tomato and rooted cuttings of chrysanthemum were randomly chosen and transplanted into 2 L plastic containers filled with a commercial potting mix (Fafard 2P Mix; Conrad Fafard, Agawam, MA, USA). Plants were fertigated with acidified water supplemented with a combination of two water-soluble fertilizers (3:1 mixture of 15N–2.2P–12.5K and 21N–2.2P–16.6K, respectively; Everris NA Inc., Dublin, OH, USA) to provide the following (in mg L^{-1}): 150 nitrogen (N), 20 phosphorous (P), 122 potassium (K), 38 calcium (Ca), 15 magnesium (Mg), 0.8 iron (Fe), 0.4 manganese (Mn) and zinc (Zn), 0.2 copper (Cu) and boron (B), and 0.1 molybdenum (Mo). Nitrate form was 76% of nitrogen provided. Irrigation water was supplemented with 93% sulfuric acid (Brenntag, Reading, PA, USA) at 0.08 mL L^{-1} to reduce alkalinity to 100 mg L^{-1} calcium carbonate ($CaCO_3$) and pH to a range of 5.8 to 6.2. The stock plants were grown in a glass-glazed greenhouse with exhaust fan and evaporative-pad cooling, radiant hot water heating, and retractable shade curtains controlled by an environmental computer (Maximizer Precision 10; Priva Computers, Vineland Station, ON, Canada). The average day and night temperatures were 23.8 $\pm$ 0.8 and 20.3 $\pm$ 0.9 °C, respectively. The photoperiod was 14-h (0800 to 2200 HR) consisting of natural day lengths with supplemental lighting using high-pressure sodium (HPS) lamps. A supplemental photosynthetic photon flux (*PPF*) was measured using a quantum sensor (LI-250A light meter; LI-COR Biosciences, Lincoln, NE, USA) and was approximately 150 μmol m^{-2} s^{-1} at canopy height. The relative humidity inside the greenhouse ranged from 50% to 70% during the study.

Cuttings were taken from the tips of mature stock plants grown in a greenhouse for 2 to 3 months. The cuttings were prepared to have four apical leaves by removing extra leaves from the basal node and trimmed to be uniform in length.

2.2. Biostimulant and Auxin Treatments and Propagation Conditions

The commercial plant-derived biostimulant Quik-link® (Italpollina S.p.a, Rivoli Veronese, Italy) was used in this study. It contains trace elements (10 g kg^{-1} Fe; 7 g kg^{-1} Mn; 3 g kg^{-1} Zn; 1 g kg^{-1} Cu; 0.2 g kg^{-1} Mo) and organic compounds biologically active like vegetal amino acids and peptides. The aminogram (expressed as percentage of the total amino acids) is: Ala(4.5), Arg(6.7), Asp(12.7), Cys(1.1), Glu(20.2), Gly(4.5), His(3.0), Ile(4.9), Leu(8.3), Lys(6.8), Met (1.5), Phe (5.6), Pro (5.6), Thr (4.1), Trp (1.1), Tyr (4.1), Val (5.3). The product also contains the Root Hair Promoting Peptide (RHPP) which is a signaling peptide stimulating root growth [36].

Quik-link is allowed in organic agriculture according to the Council Regulation (EC) No. 834/2007 of 28 June 2007 [36], and is manufactured by Italpollina USA Inc. (Anderson, IN, USA). The biostimulant was prepared in five concentrations of 0 (control), 1000, 3000, 5000, and 10,000 mg L^{-1}. Meanwhile, a commercial formulation of indole-3-butyric acid (IBA) and 1-naphthaleneacetic acid (NAA) (Dip'N Grow, Inc., Clackamas, OR, USA) was used for auxin treatment, since indole-3-acetic acid (IAA), a naturally occurring compounds, can be easily degraded in the presence of light and is susceptible to destruction in the plant by IAA-oxidase [27–29]. IBA and NAA are more effective than the naturally occurring or synthetic IAA for rooting, and therefore, are the most widely used auxins for rooting stem cuttings [34]. The formulation was prepared in five concentrations of 0, 100, 200, 300, and 500 mg L^{-1}, providing IBA and NAA concentrations at 0, 492 μM IBA + 537 μM NAA, 984 μM IBA + 1074 μM NAA, 1476 μM IBA + 1611 μM NAA, and 2460 μM IBA + 2685 μM NAA, respectively.

Stem base of unrooted cuttings were dipped into a solution of either biostimulant or auxin using a basal quick dip method for 3 s to a depth of 2 cm. The stems were quickly stuck into polystyrene cell packs (300 cm^3 soil volume per cell) filled with inert media (1:1 (v/v) perlite and vermiculite mixture). The cell packs were then placed into polystyrene trays and placed under an intermittent mist, providing bottom heat and overhead mist for 10 s every 20 min during daylight hours with 76 to 98% relative humidity at canopy height for a rooting period of 21 days. The photoperiod was 14-h (0800 to 2200 HR) consisting of natural day lengths with supplemental lighting using high-pressure sodium (HPS) lamps. A supplemental photosynthetic photon flux (*PPF*) was approximately 75 μmol m^{-2} s^{-1} at canopy height and daily maximum/minimum temperatures in the greenhouse were 23.3 ± 0.8 /22.6 ± 0.7 °C.

2.3. Plant Growth Measurements

When the maximum rooting was observed at day 20, stem length was measured from the stem end to the apical growing point and the number of leaves were recorded. The Soil Plant Analysis Development (SPAD) value, an index of chlorophyll content per unit leaf area, was measured using the SPAD chlorophyll meter (Minolta Corporation, Ltd., Osaka, Japan) on three newly expanded leaves and three fully matured leaves separately, and averaged at each group.

At the end of the rooting experiment, plant parts were separated into leaves, stems and roots. The fresh mass of each part was determined immediately after harvest and were dried in a forced-convection oven at 70 °C (Heratherm OMH400, Thermo Scientific Inc., Waltham, MA, USA) for 3 days until a constant weight was reached. Shoot dry mass was calculated as the sum of aerial vegetative parts, and total dry mass was calculated as the shoot and root dry mass. The root-to-shoot ratio was calculated based on the dry mass of roots and shoots. Total plant dry mass was calculated by adding the dry mass of each plant part. Shoot and root dry mass were analyzed by regression as a continuous response to log [concentration] and by analysis of variance to the treatment. The dried plant tissues were ground in a Wiley mill to pass through a 20-mesh screen, and 0.1 mg samples were weighed and subsequently analyzed for the nitrogen content using a Flash EA elemental analyzer (Thermo Scientific, Waltham, MA, USA).

2.4. Measurements of Root Morphological Traits

Cuttings were subjected to root morphological analysis at day 20. The number of adventitious roots were counted manually at harvest when the roots were separated from the stems using a razor blade. Entire roots were carefully rinsed and scanned using the Epson Expression 11000XL scanner (Epson America Inc., Long Beach, CA, USA). The debris removal filter was set to discount objects less than 1 cm^2 with a length/width ratio less than 4. The scanned images were then used to determine root morphological traits, such as total root length, root surface area, average root diameter, and root volume, using WinRHIZO Pro software (Regent Instrument Inc., Quebec City, QC, Canada). After root images were taken, the roots were weighed and dried in an oven set at 75 °C until the samples were completely dry to weigh dry mass. Diameter class length (root length within a diameter class) were generated in the images of adventitious roots acquired from WinRHIZO. The roots were divided into 26 diameter classes at 0.25 mm intervals and root length per each root diameter class was calculated. The root diameter class distribution was computed based on the proportion of the root length in each root diameter class compared to the total root length.

2.5. Primary Metabolite Extraction and Qualitative Analysis from Biostimulant

Primary metabolites were extracted following published protocols with modifications of extraction solvent volume. Quik-link (0.5 mL) were weighed into 2 mL microcentrifuge tubes, followed by the addition of 0.2 mL of water. To fractionate non-polar compounds, 0.375 mL of cold chloroform (−20 °C) and 0.7 mL methanol were added. After vigorous up-and-down mixing by hand (50 times), the extracts were centrifuged at 12,000 × *g* for 4 min, 100 μL supernatant (water soluble metabolites) and organic phase (lipid soluble metabolites) were transferred to 1.5 mL microcentrifuge tubes, respectively. The extracts were dried using Vacufuge concentrator (Eppendorf, Thermo Fisher Scientific, Waltham, MA, USA) with 20 μL of methanol to facilitate water evaporation. For water soluble metabolites, dried extracts were derivatized with 50 μL methoxyamine hydrochloride (40 mg ml^{-1} in pyridine) for 90 min at 37 °C, then with 100 μL MSTFA + 1% TMCS at 50 °C for 20 min. For lipid soluble metabolites, dried extracts were derivatized with 200 μL n,o-bis(trimethylsilyl)trifluoroacetamide with 1% of trimethylchlorosilane at 75 °C for 30 min. Metabolites were analyzed using a gas chromatography-mass spectrometry (GC-MS) (Trace 1310 GC, Thermo Fisher Scientific, Waltham, MA, USA) coupled to an MS detector system (ISQ QD, Thermo Fisher Scientific, Waltham, MA, USA) and an autosampler (Triplus RSH, Thermo Fisher Scientific, Waltham, MA, USA). A capillary column (Rxi-5Sil MS, Restek, Bellefonte, PA, USA; 30 m × 0.25 mm × 0.25 μm capillary column w/10 m Integra-Guard Column) was used to detect polar metabolites. For water-soluble metabolite analysis, after an initial temperature hold at 80 °C for 2 min, the oven temperature was increased to 330 °C at 15 °C min^{-1} and held for 5 min. For lipid-soluble metabolite analysis, after an initial temperature hold at 150 °C for 1 min, the oven temperature was increased to 320 °C at 12 °C min^{-1} and held for 7 min. Injector and detector temperatures were set at 250 °C and 250 °C, respectively. An aliquot of 1 μL was injected with the split ratio of 70:1. The helium carrier gas was kept at a constant flow rate of 1.2 mL min^{-1}. The mass spectrometer was operated in positive electron impact mode (EI) at 70.0 eV ionization energy at m/z 40–500 scan range. Metabolite identification was based on the National Institute of Standards and Technology (NIST) library.

2.6. Quantification of BRs in Plant-Derived Biostimulant and Plant Samples

Campesterol, stigmasterol, and beta-sitosterol were quantified based on GC-MS. Quik-link (0.5 mL) were weighed into 2 mL microcentrifuge tubes, followed by the addition of 0.2 mL of water. To fractionate non-polar compounds, 0.375 mL of cold chloroform (–20 °C) and 0.7 mL methanol were added. After vigorous up-and-down mixing by hand (50 times), the extracts were centrifuged at 12,000× g for 4 min, and 187.5 μL chloroform layer were transferred to 1.5 mL microcentrifuge tubes. For the quantification of BRs from plant samples, 100 μL of organic phase from the primary metabolite analysis above session was used. The extracts were dried using Vacufuge concentrator (Eppendorf, Thermo Fisher Scientific, Waltham, MA, USA). Dried extracts were derivatized with 200 μL n,o-bis(trimethylsilyl)trifluoroacetamide with 1% of trimethylchlorosilane at 75 °C for 30 min. BRs were analyzed using a GC-MS (Trace 1310 GC, Thermo Fisher Scientific, Waltham, MA, USA) coupled to an MS detector system (ISQ QD, Thermo Fisher Scientific, Waltham, MA, USA) and an autosampler (Triplus RSH, Thermo Fisher Scientific, Waltham, MA, USA). A capillary column (Rxi-5Sil MS, Restek, Bellefonte, PA, USA; 30 m × 0.25 mm × 0.25 μm capillary column w/10 m Integra-Guard Column) was used to detect polar metabolites. After an initial temperature hold at 150 °C for 1 min, the oven temperature was increased to 320 °C at 12 °C min^{-1} and held for 7 min. Injector, MS detector temperatures were set at 250 °C, 250 °C, and 300 °C respectively. An aliquot of 1 μL was injected with the splitless mode. The helium carrier gas was kept at a constant flow rate of 1.2 mL min^{-1}. The mass spectrometer was operated in positive electron impact mode (EI) at 70.0 eV ionization energy at m/z 45–600 scan range. Metabolite identification was based on standard compounds in comparison with the mass spectra and retention time. The standard BRs were injected from 25 ng mL^{-1} to 1000 ng mL^{-1} concentrations.

2.7. Amino Acid Quantification of Vegetal-Biostimulant

To quantify the free amino acid content in the sample, EZ:faast free amino acid for GC-MS kit (Phenomenex, Torrance, CA, USA) was utilized to extract and measure the amino acid concentration. 75 mg of the sample was incubated with 1.5 mL water overnight to extract the free amino acid from the sample. After the 24 h incubation, samples were centrifuged at 12,000 × g for 3 min. Amino acid purification and derivatization were conducted on EZ:faast instruction. The analysis of amino acid was carried out in a gas chromatograph (Trace 1310 GC, Thermo Fisher Scientific, Waltham, MA, USA) coupled to a flame ionization detector (FID), and an autosampler (Triplus RSH, Thermo Fisher Scientific, Waltham, MA, USA). A capillary column (ZebronTM EZ-AAA amino acid GC, Phenomenex, Torrance, CA, USA; 10 m, 0.25 mm) was used. The injection ratio was set at 1:15 and the injection temperature was 250 °C. The injection volume was 1.5 μL. The carrier gas was helium and the flow rate was 1.1 mL min^{-1}. The column oven was set at 110 °C and increased 30 °C per minute to 320 °C. FID temperature was set at 220 °C and the air flow 450 mL min^{-1} and the hydrogen flow was 45 mL min^{-1}.

2.8. Total Phenolic Content and Antioxidant Capacity

Total phenolic content and antioxidant capacity were analyzed using methanol extracts that described above primary metabolite analysis based on the published methods [39]. Freeze-dried samples (20 mg) were extracted in 1.4 mL of 100% methanol at 60 °C for 10 min. After centrifuge the supernatants were used for the total phenolic content, 2,2′-azino-bis(3-ethylbenzothiazoline-6-sulphonic acid) (ABTS), and 2,2-diphenyl- 1-picryl-hydrazyl-hydrate (DPPH) antioxidant capacity analyses. Various concentrations of vitamin C were used as standard curves for ABTS and DPPH assays [39]. For the DPPH assay, reaction mixtures containing test samples (10 μL) and 190 μL of a 200 μM DPPH in ethanol were incubated at room temperature for 30 min in 96-well plates. The absorbance of the DPPH free radical was measured at 515 nm using an Epoch 2 plate reader (Biotek Instruments Inc., Winooski, VT, USA). Antioxidant data were expressed as vitamin C equivalent concentration (μg g^{-1} DW). For the ABTS assay, 7 mM ABTS ammonium salt was dissolved in a potassium phosphate buffer (pH 7.4) and treated with 2.45 mM potassium persulfate. The mixture was then allowed to stand at room temperature for 12–16 h for full color development (dark blue). The solution was then diluted with potassium phosphate buffer until absorbance reached 1.0 ± 0.02 at 735 nm using an Epoch 2 plate reader (Biotek Instruments Inc., Power Wave XS, Winooski, VT, USA).

Subsequently, 190 μL of this solution was mixed with 10 μL of the sample extracts. The absorbance was recorded at room temperature after 6 min. Antioxidant data were expressed as vitamin C equivalent concentration (μg g^{-1} DW). For total phenolic content, Folin-ciocalteu reagent was used to determine total phenolic content [39]. Each sample (10 μL) was mixed with (100 μL) of Folin-Ciocalteu reagent (0.2 N) followed by 3 min of incubation at room temperature. Then, 90 μL of sodium carbonate (7.5%) was added. After 60 min of incubation in the dark at room temperature, absorbance was obtained at 735 nm. The total phenolic concentration was determined based on a standard curve of gallic acid.

2.9. Experimental Design and Statistical Analysis

Treatments were arranged in a completely randomized block design. The procedure was repeated at three different time blocks, and each block consisted of 9 treatments and 10 replicates per treatment, amounting to a total of 270 cuttings (90 samples per each plant species) per each time block. All data were subjected to analysis of variance using JMP for Windows, Version 13.2 (SAS Institute Inc., Cary, NC, USA). Polynomial contrasts were used to compare the treatment effects of biostimulant and auxin. Mean separation within each measured parameter was performed by Tukey's honestly significant difference (HSD) test at $p < 0.05$. Regression analysis was carried out to look for trends in response to the concentration for each treatment. Results from the three experiments showed similar trends and the data sets were consistent with each other. However, because the error variance was not homogeneous between experiments, statistical analyses were conducted separately for each experiment, and data from the two trials were pooled and presented here.

3. Results

3.1. The Effects of Biostimulant on Adventitious Rooting in Cuttings of Basil, Tomato, and Chrysanthemum

All cuttings achieved 100% rooting regardless of plant species and treatment. The average number of adventitious roots in untreated cuttings of basil, tomato, and chrysanthemum were 39, 22, and 16, respectively, demonstrating genetic variations in rooting ability. Total root length was higher in the order of tomato, basil, and chrysanthemum (Table 1).

Meanwhile, both biostimulant and auxin increased adventitious rooting in a dose-dependent manner: The number of adventitious roots, root dry mass, and total root length in all plant species increased or showed an increasing trend by higher concentrations of biostimulant and auxin, with exception of overdoses (IBA + NAA0.5) in auxin (Table 1). However, the response level of plant species varied significantly by the treatments. Rooting response increased more prominently by auxin than biostimulant. An optimal level of auxin to induce rooting was highly plant-species specific and maximum root length was achieved nearly at concentrations of 100, 200, and 300 mg L^{-1} in basil, tomato, and chrysanthemum, respectively (Figure 1f). When compared to auxin, biostimulant was required approximately 15- to 50-time higher concentrations to induce the onset of adventitious root formation. In general, the application of biostimulant at a concentration of 5000 mg L^{-1} increased total root length in all tested cuttings (Table 1). An overdose of auxin tended to negatively affect the dry mass of adventitious roots in basil and tomato to the levels of unrooted cuttings but not in chrysanthemum. Such response was contrasting to biostimulant, where higher concentrations of biostimulant tended to increase or gradually increased adventitious rooting, and even the highest concentration at 10,000 mg L^{-1} did not negatively affect root morphological characteristics (Table 1). The root dry mass was positively correlated with the total root length of cuttings treated with either biostimulant or auxin (Figure S1); however, the relationship between dry mass and total root length slightly varied among plant species and between the treatments (Figure S1), indicating that dry mass and/or total root length do not precisely predict response levels of cuttings to biostimulant and auxin applications.

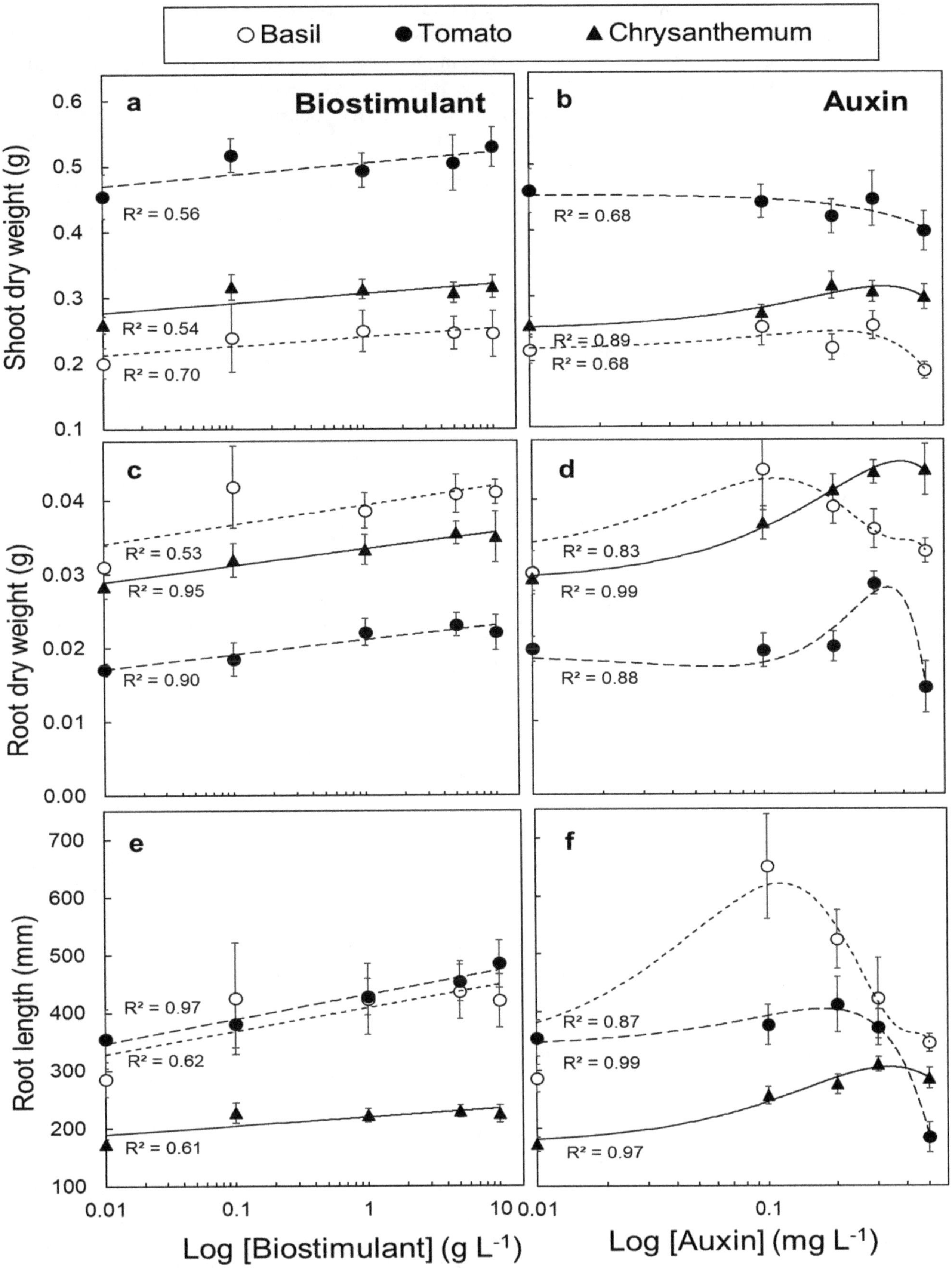

Figure 1. Dose response curves showing the effects of biostimulant or auxin applications on shoot and root dry mass, and total length of adventitious roots in cuttings of basil, tomato, and chrysanthemum. Each data point is the mean ± SE of 20 replicates.

Table 1. Effects of biostimulant (B) or auxin (IBA + NAA) applications on adventitious root characteristics in cuttings of basil, tomato, and chrysanthemum. Cuttings were treated with or without biostimulant (100, 1000, 5000 (B_5), and 10,000 (B_{10}) mg L^{-1}) or auxin (100, 200 (IBA + $NAA_{0.2}$), 300, and 500 (IBA + $NAA_{0.5}$) mg L^{-1}) at the stem base as quick-dip, stuck into inert media, and evaluated at day 20 after placement under intermittent mist. Note that two concentrations per each treatment were presented here.

Treatment	Root Number	Root Dry Mass (g plant^{-1})	Total Root Length (mm)	Root Surface Area (mm^2)	Root Volume (mm^3)	Root Diameter (mm)
			Basil			
Control	39.4	0.033 a	286 b	48.7 b	0.67 b	0.55
B_5	52.0	0.041 a	436 a	69.0 a	0.88 a	0.51
B_{10}	51.9	0.041 a	420 ab	66.4 a	0.84 a	0.52
IBA+$NAA_{0.2}$	42.4	0.034 a	322 ab	60.1 ab	0.89 a	0.60
IBA+$NAA_{0.5}$	49.8	0.033 a	346 ab	60.9 ab	0.87 a	0.57
Significance	ns	ns	*	*	*	ns
			Tomato			
Control	22.2 c	0.017 ab	355 b	41.8 a	0.39	0.38 b
B_5	36.9 a	0.023 a	454 ab	50.2 a	0.44	0.35 bc
B_{10}	36.2 a	0.022 a	484 a	50.7 a	0.42	0.33 c
IBA+$NAA_{0.2}$	28.3 b	0.021 ab	362 ab	42.4 a	0.40	0.38 b
IBA+$NAA_{0.5}$	16.2 c	0.014 b	183 c	28.0 b	0.34	0.48 a
Significance	***	*	***	***	ns	***
			Chrysanthemum			
Control	16.4 c	0.029 d	173 c	35.0 c	0.57 b	0.63
B_5	20.1 bc	0.033 cd	223 b	45.7 b	0.75 a	0.65
B_{10}	21.1 bc	0.036 bc	230 b	47.0 b	0.77 a	0.66
IBA+$NAA_{0.2}$	39.8 a	0.041 ab	275 ab	53.4 ab	0.83 a	0.63
IBA+$NAA_{0.5}$	25.1 b	0.046 a	285 a	57.1 a	0.92 a	0.64
Significance	***	***	***	***	***	ns

ns, *, **, and *** indicate non-significant, or significant at $p < 0.05$, 0.01, and 0.001, respectively. Different letters within each column indicate significant differences according to Tukey's HSD test ($p = 0.05$). Data are means of 20 replicates.

3.2. *The Effects of Biostimulant on Root Diameter Class Distribution*

Average diameters of adventitious roots varied among plant species. Tomato cuttings had relatively fine roots with an average root diameter of 0.38 mm, while the roots of basil and chrysanthemum composed of coarser roots with average root diameters of 0.55 and 0.63 mm, respectively (Table 1). In tomato, most of the roots were in the finer root class ranging from 0.0 to 0.50 mm, accounting for 75% of total root length (Table 2). The roots of basil consisted of a mixture of finer root diameter classes: about 43% of the total roots were in the finer root class (0.0 to 0.50 mm) and about 41% were intermediate root class (0.50 to 0.75 mm) (Table 2). On the other hand, chrysanthemum produced a wide range of root diameter classes (0.0 to 3.0 mm). About 40% of the total roots were in the finer root class, while the rest of the roots were composed of coarser roots (>0.5 mm). Unlike basil and tomato, where the roots thicker than 1 mm were only a small fraction among the roots (less than 3 and 0.5%, respectively) and were primarily proximal near the stem, more than 10% of the total roots of chrysanthemum were in the coarse root class (>1.0 mm).

Root diameter class distribution analyses revealed treatment differences even within the same plant species (Table 2). In basil and tomato, increasing biostimulant concentrations promoted fine roots (0.0 to 0.25 mm). These results were contrasting to auxin-treated cuttings, in which higher auxin concentrations had an increasing trend of promoting coarser roots (0.50 to 1.00 mm). The response of chrysanthemum roots was quite different from those of basil and tomato: Auxin had more pronounced effects on changing root morphological traits in chrysanthemum, and an optimal auxin concentration (IBA + NAA0.2) significantly promoted the formation of fine roots (0 to 0.25 mm) while decreasing coarser roots (>0.75 mm). Likewise, vegetal-biostimulant tended to promote finer roots in chrysanthemum, but to a lesser degree than auxin.

Table 2. Root diameter class (mm) and relative diameter class length (%) of cuttings of basil, tomato, and chrysanthemum. Cuttings were treated with or without biostimulant (100, 1000, 5000 (B_5), and 10,000 (B_{10}) mg L^{-1}) or auxin (100, 200 (IBA + $NAA_{0.2}$), 300, and 500 (IBA + $NAA_{0.5}$) mg L^{-1}) at the stem base as quick-dip, stuck into inert media, and evaluated at day 20 after placement under intermittent mist. Note that two concentrations per each treatment were presented here. Percentage values at each diameter class are given.

Treatment	Root diameter class (mm)				
	0–0.25	0.25–0.50	0.50–0.75	0.75–1.00	>1.00
	Relative root diameter class length (%)				
	Basil				
Control	15.7 b	27.0	41.0	13.8	2.5
B_5	22.7 a	28.5	40.0	11.7	3.8
B_{10}	21.6 a	30.3	38.2	11.4	4.5
IBA+$NAA_{0.2}$	20.6 ab	26.8	33.4	11.3	3.1
IBA+$NAA_{0.5}$	15.5 b	25.3	32.2	15.5	3.6
Significance	**	ns	ns	ns	ns
	Tomato				
Control	35.2 b	40.1	22.4 b	1.2 b	0.3 b
B_5	39.2 ab	42.3	17.3 bc	1.0 b	0.2 b
B_{10}	43.4 a	41.6	13.8 c	1.0 b	0.2 b
IBA+$NAA_{0.2}$	33.1 b	45.7	18.6 bc	1.9 b	0.7 b
IBA+$NAA_{0.5}$	21.6 c	38.1	34.3 a	4.0 a	2.1 a
Significance	***	ns	***	***	***
	Chrysanthemum				
Control	10.4 b	30.1 a	29.9 b	19.1 a	10.4 ab
B_5	11.5 ab	29.8 a	27.2 b	20.1 a	11.4 a
B_{10}	11.9 ab	26.9 ab	30.4 b	19.3 a	11.5 a
IBA+$NAA_{0.2}$	14.0 a	27.2 ab	32.0 b	17.7 ab	8.9 b
IBA+$NAA_{0.5}$	12.8 ab	24.8 b	38.7 a	15.0 b	8.6 b
Significance	*	**	***	**	**

ns, *, **, and *** indicate non-significant, or significant at $p < 0.05$, 0.01, and 0.001, respectively. Different letters within each column indicate significant differences according to Tukey's HSD test ($p = 0.05$). Data are means of 20 replicates.

3.3. The Effects of Biostimulant on Shoot Growth

Consistently with adventitious rooting, response levels of shoots to biostimulant and auxin slightly varied among plant species (Table 3; Tables S1–S3). Increasing concentrations of biostimulant increased or showed an increasing trend of shoot dry mass (Table 3). Biostimulant increased shoot dry mass in cuttings by 10 to 20% at a concentration of 5000 mg L^{-1}, which was somewhat associated with the increase in leaf or stem dry mass of the cuttings. Contrarily, auxin did not affect shoot dry mass of cuttings in basil and tomato with exception of chrysanthemum. SPAD index measured on newly expanded leaves and three fully matured leaves were not significantly different among the treatments, and therefore, pooled for comparisons. The results showed that SPAD index increased only in chrysanthemum when applied with auxin at an optimum level (IBA + NAA0.2), but there were no differences in total nitrogen (N) concentration among treatments (Table 3).

Table 3. Effects of biostimulant (B) or auxin (IBA + NAA) applications on stem length, leaves, stems, and shoot dry mass, the Soil Plant Analysis Development (SPAD) index, total nitrogen (N), and root-to-shoot ratio of basil, tomato, and chrysanthemum cuttings. Cuttings were treated with or without biostimulant (100, 1000, 5000 (B_5), and 10,000 (B_{10}) mg L^{-1}) or auxin (100, 200 (IBA + $NAA_{0.2}$), 300, and 500 (IBA + $NAA_{0.5}$) mg L^{-1}) at the stem base as quick-dip, stuck into inert media, and evaluated at day 20 after placement under intermittent mist. Note that two concentrations per each treatment were presented here.

Treatment	Dry Mass (g plant^{-1})				Stem Length (cm)	SPAD Index	Total N (%)	Root-to-Shoot Ratio
	Total	Shoots	Leaves	Stems				
					Basil			
Control	0.228 ab	0.200 ab	0.145	0.055	6.0	35.1	1.42	0.144
B_5	0.287 a	0.246 a	0.185	0.061	6.3	35.1	1.39	0.170
B_{10}	0.285 a	0.244 a	0.177	0.059	6.6	31.7	1.56	0.177
IBA+$NAA_{0.2}$	0.234 ab	0.200 ab	0.115	0.050	5.8	35.5	1.44	0.175
IBA+$NAA_{0.5}$	0.218 b	0.185 b	0.137	0.049	5.9	32.5	1.45	0.183
Significance	**	*	ns	ns	ns	ns	ns	ns
					Tomato			
Control	0.473 bc	0.457 bc	0.323 ab	0.133 b	7.2 ab	41.5	2.53	0.038 b
B_5	0.526 ab	0.505 ab	0.360 a	0.146 ab	7.5 ab	41.4	2.54	0.046 a
B_{10}	0.551 a	0.530 a	0.358 a	0.173 a	7.8 a	40.8	2.47	0.042 a
IBA+$NAA_{0.2}$	0.414 c	0.399 c	0.283 bc	0.111 b	7.0 b	39.5	2.22	0.046 a
IBA+$NAA_{0.5}$	0.412 c	0.398 c	0.262 c	0.134 b	6.8 b	39.2	2.38	0.036 b
Significance	**	**	**	*	*	ns	ns	ns
					Chrysanthemum			
Control	0.287 b	0.259 b	0.182	0.077 b	6.9 b	34.1 c	3.34	0.104 c
B_5	0.346 a	0.313 a	0.209	0.099 a	7.6 ab	35.4 bc	3.00	0.109 c
B_{10}	0.343 ab	0.307 ab	0.206	0.101 a	8.3 a	34.6 bc	3.02	0.118 bc
IBA+$NAA_{0.2}$	0.356 a	0.315 a	0.212	0.103 a	6.9 b	37.6 a	2.93	0.136 b
IBA+$NAA_{0.5}$	0.343 ab	0.297 ab	0.210	0.087 ab	7.3 ab	36.4 ab	2.97	0.159 a
Significance	*	*	ns	***	***	***	ns	***

ns, *, **, and *** indicate non-significant, or significant at $p < 0.05$, 0.01, and 0.001, respectively. Different letters within each column indicate significant differences according to Tukey's HSD test ($p = 0.05$). Data are means of 20 replicates.

3.4. Tissue Responsiveness to Biostimulant and Auxin

Cuttings of basil, tomato, and chrysanthemum differently responded to biostimulant in comparison to auxin. Biostimulant induced gradual and progressive changes on shoot and root dry mass, as shown in the logarithmic curves (Figure 1a,c), and there were no detrimental effects or phytotoxicity caused by higher doses of biostimulant. This was contradictory to auxin-treated cuttings where higher doses had negative effects on rooting responses, especially in basil and tomato (Figure 1b,d). The rooting responses, as expressed as root dry mass and total root length, were less dramatically influenced by biostimulant than by auxin in all plant species tested.

Basil cuttings were more responsive to a lower concentration of auxin compared to tomato and chrysanthemum, rapidly increasing adventitious roots (Figure 1d,f). At an optimal concentration, auxin-treated cuttings produced similar or higher root biomass relative to biostimulant-treated cuttings (Figure 1c,d). Regression analyses showed that auxin responsiveness of plant species increased in a biphasic manner with increasing concentrations (Figure 1d,f). The response pattern of adventitious rooting to auxin showed that basil and tomato were highly responsive. Basil responds to a lower threshold for rooting followed by a rapid polynomial decay, while tomato required a higher threshold than basil (Figure 1d). Chrysanthemum responded to a lower threshold for rooting, but displayed a gradual polynomial rise to a wide range of auxin, possibly followed by a gradual polynomial fall to a higher concentration of auxin. This rooting response was associated with increased total root length in all the plant species tested (Figure S1). However, a universal scenario of increased root dry mass and/total root length accompanied by biostimulant treatment does not explain the root morphological changes as represented by the proliferation of fine roots, as such subtle changes contribute less to dry mass. Shoots were slightly less responsive to biostimulant applications than roots as characterized by a gentle slope in a logarithmic plot (Figure 1a). Shoot dry mass of basil and tomato did not increase even when a wide range of auxin was applied; however, that of chrysanthemum increased with higher doses of auxin, indicating that chrysanthemum had different responsiveness to auxin from basil and tomato (Figure 1b).

3.5. BRs in Roots and Shoots of Cuttings

Metabolic analyses demonstrated that biostimulant contained precursors of BRs, such as campesterol and stigmasterol (or β-sitosterol) at 11.87 and 28.86 ng mL^{-1} in 5000 mg L^{-1} solution. In addition, a large profile of various compounds, including sugars (ribofuranose, arabinose, and galactose), organic acids (lactic, oxalic acid, glycolic acid, butanoic acid, tartaric acid, and gluconic acid), glucono-1,4-lactone, and fatty acids (palmitic acid and stearic acid) were found to be present as major compounds in the biostimulant (data not shown).

Metabolic profiling of cuttings elucidated that relatively high levels of BRs were present in non-treated cuttings of basil, tomato, and chrysanthemum in decreasing order. Total sterol levels were higher in roots than shoots by 3.4-, 5.3-, and 1.4-times in basil, tomato, and chrysanthemum, respectively) with the highest concentration in roots of basil, tomato, and chrysanthemum in decreasing order, which averaged at 1126, 397, and 213 µg g^{-1} dry weight, respectively (Table 4). There were three major phytosterols present in these plant species: stigmasterol, beta-sitosterol, and campesterol (Table 4). Overall, the combined proportions of stigmasterol and sitosterol were more than 80% of the total sterols, and the proportion of campesterol was less than 20%.

Notably, biostimulant and auxin treatments concomitantly increased or decreased BR levels in plant tissues or had no effects on the levels. Stigmasterol levels in roots tended to be affected by the treatments in all the tested crops ($p \leq 0.12$); however, in a different manner. For example, in basil and chrysanthemum, the optimum level of biostimulant tended to increase stigmasterol levels in roots, while, biostimulant significantly ($p < 0.01$) reduced the levels in tomato. Auxin had similar effects as biostimulant on stigmasterol levels in roots. Correlation relationships were determined between BRs and growth parameters of basil, tomato, and chrysanthemum, i.e., root dry mass, total root length, length of fine roots (0.00 to 0.25 mm) and shoot dry mass. Overall, total sterol levels were not correlated or weakly correlated with root growth parameters in basil and tomato, but moderately correlated (e.g., root dry mass: $r^2 = 0.51$, $p < 0.001$; root length: $r^2 = 0.27$, $p < 0.01$) in chrysanthemum.

Total sterol levels in shoots were the highest in basil, chrysanthemum, and tomato in decreasing order, and averaged at 327, 155, and 75 µg g^{-1} dry weight, respectively (Table 4). Both biostimulant and auxin treatments appeared to have similar increasing or decreasing effects on the levels of BRs as observed in roots. Sitosterol levels were significantly increased in shoots of tomato and chrysanthemum cuttings by biostimulant. Auxin had similar increasing effects on the levels of sitosterol in shoots of those cuttings.

Table 4. Sterol profiles of roots and shoots in cuttings of basil, tomato, and chrysanthemum at day 20 after treatment with or without biostimulant at 5000 (B_5) and 10,000 mg L^{-1} (B_{10}), or auxin at 200 mg L^{-1} (IBA + $NAA_{0.2}$).

Plant Tissue	Treatment	Stigmasterol	ß-Sitosterol	Campesterol	Total
			($\mu g\ g^{-1}$ DW)		
			Basil		
Roots	Control	440 ± 10	454 ± 19	232 ± 10	1126 ± 36
	B_5	474 ± 27	430 ± 17	231 ± 90	1136 ± 51
	B_{10}	410 ± 9	389 ± 32	214 ± 18	1013 ± 52
	IBA+$NAA_{0.2}$	420 ± 13	402 ± 15	217 ± 10	1039 ± 31
	Significance	ns[a]	ns	ns	ns
Shoots	Control	59 ± 40	204 ± 13	64 ± 60	327 ± 23
	B_5	59 ± 60	158 ± 14	55 ± 6	272 ± 25
	B_{10}	47 ± 20	164 ± 40	53 ± 2	264 ± 5
	IBA+$NAA_{0.2}$	47 ± 3	166 ± 16	53 ± 4	261 ± 22
	Significance	ns	ns	ns	ns
			Tomato		
Roots	Control	270 ± 16 a	97 ± 12	30 ± 3 a	397 ± 28 a
	B_5	197 ± 17 b	79 ± 9	22 ± 3 ab	298 ± 27 b
	B_{10}	182 ± 5 b	69 ± 6	20 ± 1 ab	270 ± 90b
	IBA+$NAA_{0.2}$	155 ± 17 b	71 ± 6	17 ± 2 b	243 ± 20 b
	Significance	***	ns	*	**
Shoots	Control	55 ± 8	17 ± 3 b	3.0 ± 0.6	75 ± 12
	B_5	54 ± 40	22 ± 2 ab	5.7 ± 1.8	81 ± 6
	B_{10}	62 ± 90	31 ± 5 a	4.0 ± 1.9	97 ± 15
	IBA+$NAA_{0.2}$	48 ± 50	32 ± 2 a	3.1 ± 1.9	82 ± 7
	Significance	ns	**	ns	ns
			Chrysanthemum		
Roots	Control	83 ± 7	111 ± 9	19.5 ± 2.3	213 ± 18
	B_5	92 ± 5	115 ± 6	20.9 ± 1.5	228 ± 10
	B_{10}	104 ± 6	111 ± 5	20.7 ± 0.8	235 ± 7
	IBA+$NAA_{0.2}$	103 ± 6	114 ± 8	21.0 ± 1.9	238 ± 15
	Significance	ns[b]	ns	ns	ns
Shoots	Control	89 ± 5	65 ± 5 b	1.3 ± 0.5 b	155 ± 10
	B_5	95 ± 2	71 ± 2 ab	1.8 ± 0.2 b	168 ± 40
	B_{10}	99 ± 4	86 ± 4 a	2.0 ± 0.4 b	169 ± 7
	IBA+$NAA_{0.2}$	104 ± 7	78 ± 4 ab	6.5 ± 1.6 a	188 ± 13
	Significance	ns[b]	*	**	ns

ns, *, **, and *** indicate non-significant, or significant at $p < 0.05$, 0.01, and 0.001, respectively. Different letters within each column indicate significant differences according to Tukey's HSD test ($p = 0.05$). Data shown are means ± SE of five replicates. a Significant at $p \leq 0.12$. a Significant at $p \leq 0.1$.

3.6. Antioxidant Capacities and Total Phenolic Content of Cuttings

The radical scavenging activities of roots and shoots in cuttings of basil, tomato, and chrysanthemum were examined as estimated by the DPPH (Figure 2a,d,g) and ABTS assays (Figure 2b,e,h). The DPPH method is one of the most frequently used and inexpensive antioxidant assays; however, pH sensitivity is a major disadvantage of the assay [40]. In order to generate robust results, we used the two different scavenging radical assays in this study. The antioxidant capacities obtained from DPPH assay were in accordance with those obtained from ABTS assays regardless of plant species and tissue type (basil: roots $r^2 = 0.94$, shoots $r^2 = 0.79$; tomato: roots $r^2 = 0.75$, shoots $r^2 = 0.82$; chrysanthemum: roots $r^2 = 0.94$, shoots $r^2 = 0.91$).

The results showed large variations in antioxidant capacities among plant species and tissues. Roots of basil showed the highest antioxidant activities (3.6 mg ascorbic acid equivalent per g DW) followed by chrysanthemum (2.3 mg) and tomato (0.4 mg) (Figure 2). In basil, antioxidant capacities were three-times higher in roots than shoots and were less affected by the treatment (Figure 2a,b), while tomato and chrysanthemum was approximately two-times lower in roots than shoots and were either positively or negatively affected by the treatment (Figure 2d,e,g,h).

Biostimulant significantly increased the scavenging activities of roots of chrysanthemum, and such increases were strongly correlated with the concentrations of total phenolic acids ($r^2 = 0.77$ for DPPH and $r^2 = 0.78$ for ABTS) (Figure 2g–i). Meanwhile, shoots of chrysanthemum behaved differently and demonstrated significantly higher antioxidant activities concomitantly by both biostimulant and auxin compared to control. There was a trend of concentration-dependent increase in radical scavenging activities by biostimulant in shoots and roots of basil and chrysanthemum (Figure 2). Overall, biostimulant had stimulatory effects on antioxidant activities of adventitious roots in cuttings (based on DPPH and ABTS assays), and such results were contradictory to those induced by auxin.

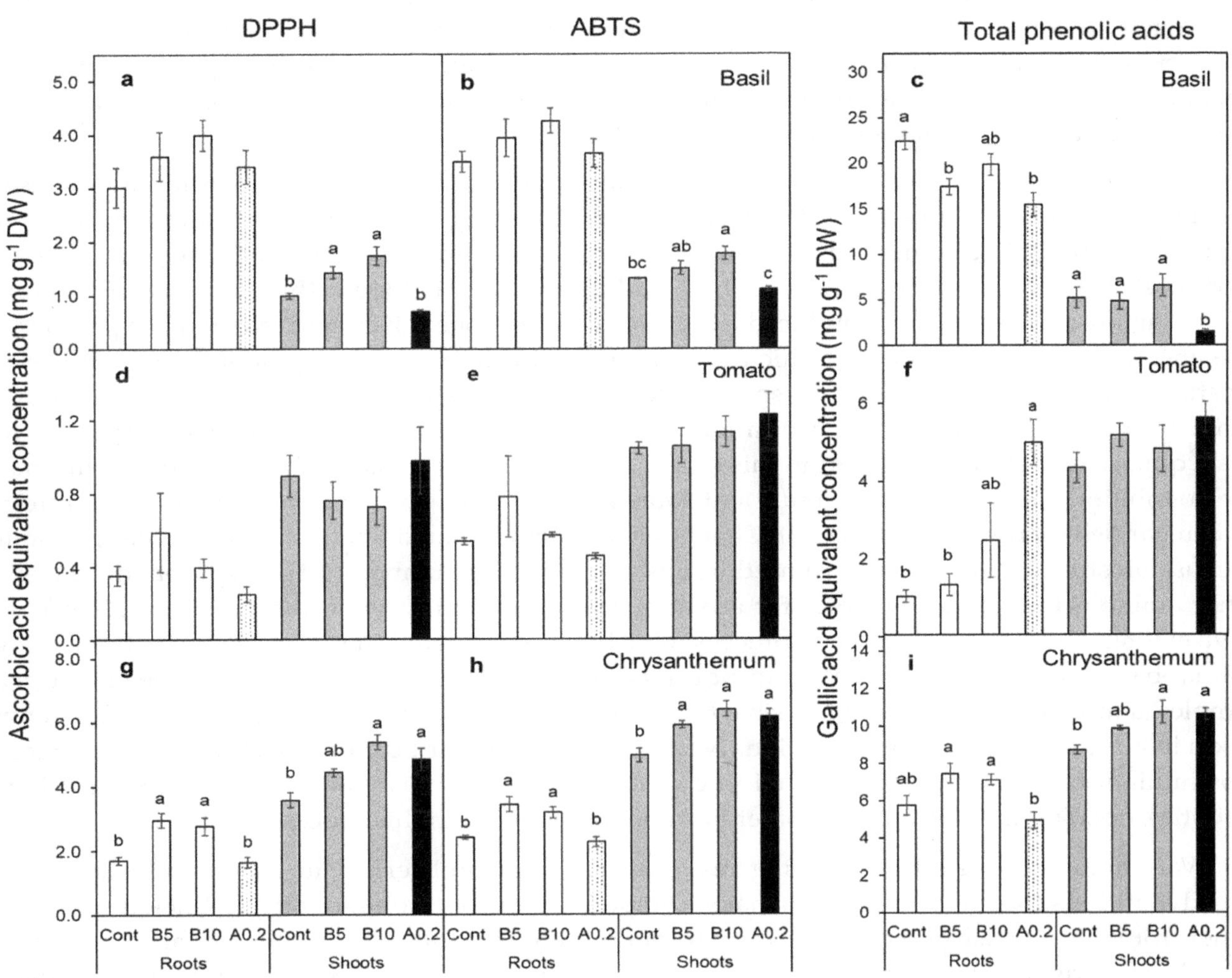

Figure 2. Antioxidant capacity and total phenolic compounds of roots and shoots in cuttings of basil (**a–c**), tomato (**d–f**) and chrysanthemum (**g–i**) at day 20 after treatment with either control (Cont), biostimulant at 5000 (B5) or 10,000 mg L^{-1} (B10), or IAA+NAA at 200 mg L^{-1} (A0.2). Antioxidant capacity was estimated by the DPPH and ABTS assays. The antioxidant capacity and total phenolic compounds of the aqueous extracts are equivalent to indicated concentrations of water-soluble standard antioxidant ascorbic acid (mg g^{-1} DW) and gallic acid (mg g^{-1} DW), respectively. Different letters indicate significant differences within each plant part (roots or shoots) according to Tukey's HSD test ($p = 0.05$). Data shown are means of five replicates.

4. Discussion

4.1. Biostimulant Promotes Adventitious Rooting Responses of Stem Cuttings Similar to Auxin, but to a Lesser Extent

It is well established that auxin promotes the formation of adventitious roots [41,42] and lateral roots [43–45]. Previous studies have reported on root morphological changes induced by biostimulant applications; however, the changes have been focused primarily on the increases in root biomass, total root length, and root surface area [5,6] without sufficient information on root characteristics. Further, no detailed investigations have been made on hormonal effects of biostimulant in promoting plant growth and yield although these aspects have been demonstrated in many studies.

First, we measured morphological responses of stem cuttings, including the number of adventitious roots, root dry mass, total root length, and average root diameter, as a function of the exogenous concentration of either biostimulant or auxin. While most of these variables exhibited plant species-specific responses, partly due to the genetics of differential tissue responsiveness, it was clear that biostimulant was effective in promoting adventitious rooting of basil and tomato, easy-to-root types, as well as chrysanthemum, moderate-to-root type.

Dose-response analyses were used to evaluate the relationship between compound dosage and plant response (Figure 1), and it was found that rooting responses to biostimulant not only were considerably less compared to auxin but also did not fully in agreement with those to auxin. Biostimulant promoted adventitious rooting leading to a gradual logarithmic rise as a function of increasing dosages, while auxin induced a biphasic dose response characterized by rapid polynomial rise and fall in a plant species-specific manner. A significantly higher threshold was required for biostimulant to induce a series of responses compared to auxin. One of the reasons for the mild changes over a wide range of concentrations induced by biostimulant in cuttings is partly due to a basal quick dip method employed in this study. It was confirmed that such an approach eliminates the possibility of biostimulant as a nutrient source, since there were no differences in total nitrogen level regardless of treatments (Table 3). Changes in endogenous auxin pool may be another possibility because biostimulant Quik-link we used in this study contained about 4.1% tryptophan, as well as other amino acids. As a precursor for auxin biosynthesis pathways in plants, tryptophan might have exerted a weak auxin-induced process, which was postulated in maize seedlings treated with animal-based biostimulant [6]. We did not quantify IAA and other auxin derivatives from plant samples, and therefore, it was not possible to determine how biostimulants interact with endogenous auxin in adventitious rooting formation. Nevertheless, a similar but somewhat unique behavior of biostimulant-treated cuttings observed in our study cannot be justified solely by auxin-mediated activities, opening the possibility of other hormonal regulation in this process.

Variations in adventitious rooting responses were also observed in root architectural traits. An adventitious root system has two major components of root: long, relatively thick roots arising either from the cut stem end or the lower part of the stem that forms its framework and shorter, fine lateral roots arising either directly from these framework roots or indirectly as higher-order lateral roots. Since the complete physical separation of lateral roots from adventitious roots was not possible, particularly in basil and tomato due to fibrous nature of their roots, we performed root diameter class distribution analyses to differentiate these root components by carefully manipulating parameters. This method has been proven to be effective in separating different root types [46], and commonly used in root studies. The results revealed that the roots examined in our study are actually classified into very fine (<0.5) to fine (0.5–2 mm) [47] and we further classified them into multiple categories within the range. The adventitious roots of untreated tomato cuttings were composed primarily of finer roots (average root diameter: 0.38 mm) with about 76% of the total roots within 0 to 0.50 mm diameter class, whereas those of basil cuttings consisted primarily of fine to intermediate roots with about 70% of the roots within 0.25 to 0.75 mm diameter class (average root diameter: 0.55 mm). More than 30% of the total roots in chrysanthemum were composed of coarse roots (>0.75 mm) (average root diameter: 0.63 mm) with a wide range of root diameter classes (Table 2).

Interestingly, biostimulant applications significantly increased or tended to increase finer root classes in these plant species, providing direct evidence that biostimulant stimulates proliferation of lateral roots in cuttings. This is in agreement with a recent study in maize seedlings, in which protein hydrolysates increased length and surface area of lateral roots by about 7 and 1.5 times compared to inorganic nitrogen and free amino acids, respectively [48]. Fine roots are considered to be the most permeable part of a root system and play the key role in the acquisition of water and nutrients and root adaptation to extreme environments, particularly in herbaceous plants [49] and such developmental changes may confer significant advantages on long-term plant growth and survival, particularly under suboptimal water and nutrient conditions.

4.2. Biostimulant Induces Adventitious Rooting of Stem Cuttings Primarily via BR-Mediated Processes

In this study, we measured metabolic responses, including BR levels in plant tissues, antioxidant capacities, and total phenolic compounds in roots and shoots of basil, tomato, and chrysanthemum. We found that biostimulant negatively or positively affects BR biosynthesis in plant tissues and increases antioxidant activities and total phenolic compounds in both roots and shoots of cuttings. Consistently with morphological traits, these metabolic responses were not fully in agreement with auxin.

As a group of steroidal plant hormones, BRs are known to mediate modulation of various components of the antioxidant defense system in plants under abiotic stresses, including drought, salinity, and temperature extremes [50]. BRs were reported to be involved in mitigating the adverse effect of high temperature stress on snap bean plants by increasing total free amino acids in leaves and total phenolic acids in the pod [51]. Increases in antioxidative capacities and phenolic compounds were also found in BR-treated *Brassica junica* seedlings under lead toxicity [52]. The BR-mediated antioxidant system was also demonstrated to modulate root growth as the *Arabidopsis det2-9* mutant defective in BR biosynthesis exhibited inhibited root growth and accumulated more reactive oxygen species than the wild type [53]. We found that stigmasterol, sitosterol, and campesterol were the major phytosterols in cuttings of plant species tested. These phytosterols serve as precursors for BR biosynthesis and are integral membrane components which regulate the permeability and fluidity of membranes [54], and phytoserol composition in the plasma membrane affects the proper functioning of auxin transporters [55]. Campesterol influences the level of active BR, and regulates a number of physiological activities in plant development, such as cell elongation, xylem differentiation, and stress tolerance [54].

Herein, we postulate that biostimulant induces adventitious root formation primarily via BR-mediated processes while interacting with auxin-mediated mechanisms and that native BR pool in plant tissues influences adventitious rooting responses to biostimulant. There are at least six pieces of evidence to support this view: (1) endogenous auxin plays the key role in adventitious rooting formation in these cuttings as adventitious roots were produced in cuttings that did not receive any treatment, (2) endogenous BRs also play a critical role in adventitious rooting formation in these cuttings as relatively high levels of native BRs were present in cuttings that did not receive any treatment, (3) both biostimulant and auxin influenced endogenous BR levels in most cuttings, (4) biostimulant exerted weaker effects on adventitious rooting of cuttings than did auxin treatment, (5) adventitious rooting responses to biostimulant was most prominent in chrysanthemum cuttings that have relatively low native BR levels and are less responsiveness to auxin, and (6) antioxidant activities in adventitious roots tended to be increased by biostimulant but decreased by auxin.

As discussed earlier, the extent to which the increased induction and formation of adventitious rooting varied greatly in response to the compound, with more prominent effects by auxin than biostimulant (Figure 2). For example, the optimal levels of auxin and biostimulant increased the dry mass of adventitious roots by 54% and 20% in chrysanthemum, 67% and 26% in tomato, and 42% and 26% in basil, compared to untreated cuttings. Further, major differences between auxin and biostimulant existed not only in the patterns of dose response curve, but also in the absolute amount of the compounds required for promoting adventitious rooting (Figure 1).

Clouse et al. [56] noted that measurable effects on cell elongation induced by BR required much longer treatment time compared with the rapid effects caused by auxin. Nemhauser et al. [57] elucidated that auxin-response element ARFAT is the crucial intersection point of BR and auxin pathways, which is BR responsive and requires BR biosynthesis for normal expression. These findings are consistent with our observations and support our interpretations that the hormonal effects induced by biostimulant is more likely to be related to BRs than auxin and that auxin and BRs interact in controlling BR pool in plant tissues and work coordinately in fine-tuning adventitious rooting responses of cuttings.

4.3. Biostimulant-Induced BRs and Auxin Have Overlapping Functions in Adventitious Root Formation

We found that roots and shoots of cuttings produce relatively high levels of endogenous BRs which were increased or decreased concomitantly by biostimulant and auxin. This similar effect of both compounds demonstrates that their overlapping role in BR biosynthesis. Although auxin was not quantified in this study, there is no doubt that auxin plays an important role in adventitious rooting formation in these tested plant species. Auxin and BRs are two important phytohormones and are known to exert some similar physiological effects exclusively or through their functional interaction, which include cell division and expansion, vascular differentiation, root growth, and senescence [58]. It was reported that a shared auxin and BR pathway is required for seedling growth, and response from one pathway requires the function of the other, and this interdependence occurs at gene expression level [57]. Consistently, auxin-treated cuttings in our study showed increased levels of BRs, indicating that auxin treatments in cuttings also involve BR biosynthesis. The extent of increased response levels and more remarkable effects on rooting responses induced by auxin indicate that auxin triggers cellular and molecular responses of adventitious rooting synergistically and interdependently from BRs. While such synergistic and interdependent interactions of auxin and BRs have been demonstrated in other plant systems and was well reviewed by Tian et al. [59], this is the first time demonstrating the interactions between biostimulant-induced BRs and auxin in adventitious rooting responses of cuttings. The interaction also includes the lateral root formation of an adventitious root system. BRs are required for lateral root development in Arabidopsis and act synergistically with auxin to promote lateral root formation by increasing acropetal auxin transport [58,60]. BRs mainly function at the lateral root primordia initiation while auxin is required for both initiation and emergence stages of lateral root formation [43,58].

Based on this view, various responses of plant species to biostimulant and auxin can be explained by endogenous BR pools of plant species. The application of biostimulant and auxin had negative effects on BR levels in basil and tomato, highly responsive plant species containing a higher level of native BRs, but had positive effects on BR levels in chrysanthemum, less responsive plant species containing lower BR levels (Table 4). We also demonstrated that a high level of auxin has an inhibitory effect on antioxidant capacities and phenolic compounds in chrysanthemum cuttings (Figure 2). A similar inhibitory effect of increased auxin levels on BR-induced growth responses was observed in auxin-overproducing *yucca* mutants [57]. Thus, it is likely that different plant species have a different level of BR-pool which restricts plant growth response to these compounds, and that increased auxin levels saturate the BR-pool, significantly reducing BR-effects on regulatory changes.

The induction phase in cuttings or detached organs, such as leaves, is generally marked by the immediate consequences of the wounding response caused by severance. It encompasses the first hours after cutting removal, with a local increase in jasmonate, phenolic compounds and auxin at the cutting base [32]. Phenolic compounds exert antioxidant properties against oxidative stress [61], and were demonstrated to promote adventitious roots of stem slices from apple microshoots by protecting IAA from decarboxylation and the tissue from oxidative stress caused by wounding [62], contributing to the auxin stability for adventitious root induction [62]. The high positive correlation ($p < 0.001$)

between antioxidant capacities and total phenolic content indicates that phenolic compounds are a major contributor to the antioxidant activities of these plants. Phenolic compounds act as antioxidants protecting auxins from decarboxylation and the tissue from oxidative stress, allowing more auxin is available to induce roots [62].

In addition to BR-related proteins, other plant hormone related proteins were identified when maize seedlings were exposed to protein hydrolysate [37]. Metabolomics studies of greenhouse melons treated with biopolymer-based biostimulant as substrate drench demonstrated that BRs interact with other hormones in the leaves, possibly via translocation from roots, as the compounds related to other hormones were observed in the leaves [36], and this translocation may explain the lower level of BRs in shoot of cuttings observed in our study. These findings suggest that there are cross-talks among hormones during the adventitious rooting process. BRs positively regulate lateral root formation whereas cytokinin and abscisic acid negatively regulate the event, and ethylene has positive and negative roles during lateral root formation [60]. On the contrary, the root growth-stimulating effect of BRs was proposed to be independent of auxin and gibberellin action, in which processes genes related to other phytohormones did not show changes, suggesting that the stimulatory effect of BRs on root growth is an autonomous effect rather than cross-talks with other phytohormones [63].

Relatively little is known about the effects of BRs on growth and development of adventitious roots. There are only a few reports showing that BRs mainly inhibit adventitious root development in cuttings of tomato and mung bean at low concentrations (0.1 μM), but the effects mainly occur on the shoot [64,65]. BRs have shown to be involved in jasmonate signaling and exert a mild negative regulation of jasmonate-induced inhibition of root growth [66]. Increase in adventitious root formation in geranium stem cuttings were observed in the treatment with BRs which also improved shoot growth of coleus cuttings [67,68].

Our results provide evidence that adventitious rooting responses of cuttings treated with biostimulant involve BR biosynthesis and their overlapping function with auxin, leading to the morphological and metabolic changes occurring during adventitious root formation. Due to the short-term investigations on adventitious rooting processes, we did not find subsequent effects of morphological changes occurring in roots and shoots. Yet, it is expected that such developmental changes improve crop performance and resource acquisition under suboptimal water and nutrient environment and confer significant advantages on long-term plant growth and survival, particularly under abiotic stresses.

5. Conclusions

To elucidate the hormonal effects of plant-derived-biostimulant, adventitious rooting responses of cuttings were examined after a basal quick-dip treatment with various concentrations of biostimulant in comparison to auxin. This approach allows detailed investigations on the hormonal function of biostimulant as auxin is known to play a key role in adventitious rooting process and eliminates potential nutrient effects of the compound. Biostimulant exerted similar effects as auxin increasing adventitious rooting responses. Dose-response analyses revealed that biostimulant showed a gradual logarithmic rise as a function of increasing dosages, contrary to a typical biphasic dose response of auxin, and required a significantly higher threshold than auxin. Metabolic profiles showed that BRs were highly present in non-treated cuttings of basil, tomato, and chrysanthemum in decreasing order, and both biostimulant and auxin had fewer effects in basil and tomato, high BR producers, and greater effects in chrysanthemum, less BR producer, indicating that native BR-pools of plant species influence adventitious rooting responses to biostimulant, as well as auxin. Biostimulant promoted antioxidant activities and phenolic compounds in cuttings, particularly in chrysanthemum, while auxin inhibited these metabolic responses. The inhibitory effect of auxin is likely due to the saturation of BR-pool, significantly reducing BR-effects. These provide evidence that biostimulant has overlapping functions with auxin in adventitious root formation, while exerting distinctive and independent contributions. We demonstrate for the first time that biostimulant induces adventitious rooting responses of cutting via BR-mediated processes while interacting with auxin and that there are interdependent effects of BRs and auxin on antioxidant activities of cuttings. Our results provide new insight into the hormonal regulation of biostimulant and a fine-tuning role of BRs in adventitious root formation.

Supplementary Materials
Figure S1. The relationship between root dry mass and total root length of basil, tomato, and chrysanthemum cuttings as affected by biostimulant and auxin applications, Table S1. Polynomial contrasts on the means of adventitious root number, root dry mass, total root length, total root surface area, root volume and average root diameter of basil, tomato, and chrysanthemum cuttings as affected by biostimulant and auxin applications, Table S2. Polynomial contrasts on the means of root diameter class (mm) and relative diameter class length (%) of basil, tomato, and chrysanthemum cuttings as affected by biostimulant and auxin applications. Percentage values at each diameter class are given, Table S3. Polynomial contrasts on the means of on stem length, leaves, stems, and shoot dry mass, SPAD index, and root-to-shoot ratio of basil, tomato, and chrysanthemum cuttings as affected by biostimulant and auxin applications. Percentage values at each diameter class are given.

Author Contributions: H.K. coordinated and supervised the research, provided intellectual inputs for defining the experimental design, data analysis, and interpretation, and wrote the entire manuscript. K.K. performed metabolic analysis and contributed on metabolomics results, S.C. conducted experiments, collected data, assisted in data analysis and interpretation, and implemented the manuscript. M.C. gave support in the experimental design, data analysis, and interpretation, and implemented the manuscript.

Acknowledgments: We are grateful to Rob Eddy, Michael Russell, Nathan Deppe, and Dan Little for their technical assistance; Hye Su Lee and Gaotian Zhu for their help with crop management and data collection; and Hélène Reynaud for the helpful discussions and support on the research.

References

1. Calvo, P.; Nelson, L.; Kloepper, J.W. Agricultural uses of plant biostimulants. *Plant Soil.* **2014**, *383*, 3–41. [CrossRef]
2. Colla, G.; Rouphael, Y. Biostimulants in horticulture. *Sci. Hortic.* **2015**, *196*, 1–2. [CrossRef]
3. du Jardin, P. Plant biostimulants: Definition, concept, main categories and regulation. *Sci. Hortic.* **2015**, *196*, 3–14. [CrossRef]
4. Schaafsma, G. Safety of protein hydrolysates, fractions thereof and bioactive peptides in human nutrition. *Eur. J. Clin. Nutr.* **2009**, *63*, 1161–1168. [CrossRef] [PubMed]
5. Colla, G.; Rouphael, Y.; Canaguier, R.; Svecova, E.; Cardarelli, M. Biostimulant action of a plant-derived protein hydrolysate produced through enzymatic hydrolysis. *Front. Plant Sci.* **2014**, *5*, 448. [CrossRef] [PubMed]
6. Ertani, A.; Cavani, L.; Pizzeghello, D.; Brandellero, E.; Altissimo, A.; Ciavatta, C.; Nardi, S. Biostimulant activity of two protein hydrolyzates in the growth and nitrogen metabolism of maize seedlings. *J. Plant Nutr. Soil Sci.* **2009**, *172*, 237–244. [CrossRef]
7. Matsumiya, Y.; Kubo, M. *Soybean Peptide: Novel Plant Growth Promoting Peptide from Soybean*; InTech Europe: Rijeka, Croatia, 2011; pp. 215–230.
8. Kauffman, G.L.; Kneivel, D.P.; Watschke, T.L. Effects of a biostimulant on the heat tolerance associated with photosynthetic capacity, membrane thermostability, and polyphenol production of perennial ryegrass. *Crop. Sci.* **2007**, *47*, 261–267. [CrossRef]
9. Lucini, L.; Rouphael, Y.; Cardarelli, M.; Canaguier, R.; Kumar, P.; Colla, G. The effect of a plant-derived biostimulant on metabolic profiling and crop performance of lettuce grown under saline conditions. *Sci. Hortic.* **2015**, *182*, 124–133. [CrossRef]
10. Colla, G.; Cardarelli, M.; Bonini, P.; Rouphael, Y. Foliar applications of protein hydrolysate, plant and seaweed extracts increase yield but differentially modulate fruit quality of greenhouse tomato. *Hortscience* **2017**, *52*, 1214–1220. [CrossRef]
11. Colla, G.; Hoagland, L.; Ruzzi, M.; Cardarelli, M.; Bonini, P.; Canaguier, R.; Rouphael, Y. Biostimulant action of protein hydrolysates: Unraveling their effects on plant physiology and microbiome. *Front. Plant Sci.* **2017**, 8. [CrossRef]
12. Rouphael, Y.; Colla, G.; Giordano, M.; El-Nalchel, C.; Kyriacou, M.C.; De Pascale, S. Foliar applications of a legume-derived protein hydrolysate elicit dose-dependent increases of growth, leaf mineral composition, yield and fruit quality in two greenhouse tomato cultivars. *Sci. Hortic.* **2017**, *226*, 353–360. [CrossRef]

13. Ertani, A.; Pizzeghello, D.; Francioso, O.; Sambo, P.; Sanchez-Cortes, S.; Nardi, S. *Capsicum chinensis* L. growth and nutraceutical properties are enhanced by biostimulants in a long-term period: Chemical and metabolomic approaches. *Front. Plant Sci.* **2014**, 5. [CrossRef] [PubMed]
14. Ito, Y.; Nakanomyo, I.; Motose, H.; Iwamoto, K.; Sawa, S.; Dohmae, N.; Fukuda, H. Dodeca-CLE peptides as suppressors of plant stem cell differentiation. *Science* **2006**, *313*, 842–845. [CrossRef] [PubMed]
15. Kondo, T.; Sawa, S.; Kinoshita, A.; Mizuno, S.; Kakimoto, T.; Fukuda, H.; Sakagami, Y. A plant peptide encoded by CLV3 identified by in situ MALDI-TOF MS analysis. *Science* **2006**, *313*, 845–848. [CrossRef] [PubMed]
16. Ryan, C.A.; Pearce, G. Polypeptide hormones. *Plant Physiol.* **2001**, *125*, 65–68. [CrossRef]
17. Ryan, C.A.; Pearce, G.; Scheer, J.; Moura, D.S. Polypeptide hormones. *Plant Cell* **2002**, *14*, S251–S264. [CrossRef]
18. Schiavon, M.; Ertani, A.; Nardi, S. Effects of an alfalfa protein hydrolysate on the gene expression and activity of enzymes of the tricarboxylic acid (TCA) cycle and nitrogen metabolism in *Zea mays* L. *J. Agric. Food Chem.* **2008**, *56*, 11800–11808. [CrossRef]
19. Baglieri, A.; Cadili, V.; Monterumici, C.M.; Gennari, M.; Tabasso, S.; Montoneri, E.; Nardi, S.; Negre, M. Fertilization of bean plants with tomato plants hydrolysates. Effect on biomass production, chlorophyll content and N assimilation. *Sci. Hortic.* **2014**, *176*, 194–199. [CrossRef]
20. Paradikovic, N.; Vinkovic, T.; Vrcek, I.V.; Zuntar, I.; Bojic, M.; Medic-Saric, M. Effect of natural biostimulants on yield and nutritional quality: An example of sweet yellow pepper (Capsicum annuum L.) plants. *J. Sci. Food Agric.* **2011**, *91*, 2146–2152.
21. Parrado, J.; Escudero-Gilete, M.L.; Friaza, V.; Garcia-Martinez, A.; Gonzalez-Miret, M.L.; Bautista, J.D.; Heredia, F.J. Enzymatic vegetable extract with bio-active components: Influence of fertiliser on the colour and anthocyanins of red grapes. *J. Sci. Food Agric.* **2007**, *87*, 2310–2318. [CrossRef]
22. Zodape, S.T.; Gupta, A.; Bhandari, S.C.; Rawat, U.S.; Chaudhary, D.R.; Eswaran, K.; Chikara, J. Foliar application of seaweed sap as biostimulant for enhancement of yield and quality of tomato (*Lycopersicon esculentum* Mill.). *J. Sci. Ind. Res.* **2011**, *70*, 215–219.
23. de Vasconcelos, A.C.F.; Zhang, X.Z.; Ervin, E.H.; Kiehl, J.D. Enzymatic antioxidant responses to biostimulants in maize and soybean subjected to drought. *Sci. Agric.* **2009**, *66*, 395–402. [CrossRef]
24. Botta, A. Enhancing plant tolerance to temperature stress with amino acids: An approach to their mode of action. In *I World Congress on the Use of Biostimulants in Agriculture*; Silva, S.S., Brown, P., Ponchet, M., Eds.; International Society for Horticultural Science: Leuven, Belgium, 2013; Volume 1009, pp. 29–35.
25. Colla, G.; Svecova, E.; Cardarelli, M.; Rouphael, Y.; Reynaud, H.; Canaguier, R.; Planques, B. Effectiveness of a plant-derived protein hydrolysate to improve crop performances under different growing conditions. In *I World Congress on the Use of Biostimulants in Agriculture*; Silva, S.S., Brown, P., Ponchet, M., Eds.; International Society for Horticultural Science: Leuven, Belgium, 2013; Volume 1009, pp. 175–179.
26. Rouphael, Y.; Cardarelli, M.; Bonini, P.; Colla, G. Synergistic action of a microbial-based biostimulant and a plant derived-protein hydrolysate enhances lettuce tolerance to alkalinity and salinity. *Front. Plant Sci.* **2017**, *8*, 131. [CrossRef] [PubMed]
27. Epstein, E.; Ludwigmuller, J. Indole-3-butyric acid in plants—Occurrence, synthesis, metabolism and transport. *Physiol. Plant.* **1993**, *88*, 382–389. [CrossRef]
28. Ludwig-Muller, J. Indole-3-butyric acid in plant growth and development. *Plant Growth Regul.* **2000**, *32*, 219–230. [CrossRef]
29. Nordstrom, A.C.; Jacobs, F.A.; Eliasson, L. Effect of exogenous indole-3-acetic-acid and indole-3-butyric acid on internal levels of the respective auxins and their conjugation with aspartic-acid during adventitious root-formation in pea cuttings. *Plant Physiol.* **1991**, *96*, 856–861. [CrossRef] [PubMed]
30. Husen, A.; Pal, M. Metabolic changes during adventitious root primordium development in Tectona grandis Linn. f. (teak) cuttings as affected by age of donor plants and auxin (IBA and NAA) treatment. *New Forests.* **2007**, *33*, 309–323. [CrossRef]
31. De Klerk, G.J.; Van der Krieken, W.; De Jong, J.C. Review—The formation of adventitious roots: New concepts, new possibilities. *In Vitro Cell. Dev. Biol. Plant.* **1999**, *35*, 189–199. [CrossRef]
32. da Costa, C.T.; de Almeida, M.R.; Ruedell, C.M.; Schwambach, J.; Maraschin, F.S.; Fett-Neto, A.G. When stress and development go hand in hand: Main hormonal controls of adventitious rooting in cuttings. *Front. Plant Sci.* **2013**, 4. [CrossRef]

33. Fabijan, D.; Taylor, J.S.; Reid, D.M. Adventitious rooting in hypocotyls of sunflower (*Helianthus annuus*) seedlings.2. Action of gibberellins, cytokinins, auxins and ethylene. *Physiol. Plant.* **1981**, *53*, 589–597. [CrossRef]
34. Hartmann, H.T.; Kester, D.E.; Davies, F., Jr.; Geneve, R.L. *Plant Propagation: Principles and Practices*, 8th ed.; Prentice-Hall: Upper Saddle River, NJ, USA, 2011.
35. Liu, J.H.; Reid, D.M. Adventitious rooting in hypocotyls of sunflower (*Helianthus annuus*) seedlings.4. The role of changes in endogenous free and conjugated indole-3-acetic-acid. *Physiol. Plant.* **1992**, *86*, 285–292. [CrossRef]
36. Lucini, L.; Rouphael, Y.; Cardarelli, M.; Bonini, P.; Baffi, C.; Colla, G. A vegetal biopolymer-based biostimulant promoted root growth in melon while triggering brassinosteroids and stress-related compounds. *Front. Plant Sci.* **2018**, 9. [CrossRef] [PubMed]
37. Trevisan, S.; Manoli, A.; Ravazzolo, L.; Franceschi, C.; Quaggiotti, S. mRNA-sequencing analysis reveals transcriptional changes in root of maize seedlings treated with two increasing concentrations of a new biostimulant. *J. Agric. Food Chem.* **2017**, *65*, 9956–9969. [CrossRef] [PubMed]
38. Colla, G.; Nardi, S.; Cardarelli, M.; Ertani, A.; Lucini, L.; Canaguier, R.; Rouphael, Y. Protein hydrolysates as biostimulants in horticulture. *Sci. Hortic.* **2015**, *196*, 28–38. [CrossRef]
39. Ku, K.M.; Choi, J.N.; Kim, J.; Kim, J.K.; Yoo, L.G.; Lee, S.J.; Hong, Y.S.; Lee, C.H. Metabolomics analysis reveals the compositional differences of shade grown tea (*Camellia sinensis* L.). *J. Agric. Food Chem.* **2010**, *58*, 418–426. [CrossRef] [PubMed]
40. Huang, D.; Ou, B.; Prior, R.L. The chemistry behind antioxidant capacity assays. *J. Agric. Food Chem.* **2005**, *53*, 1841–1856. [CrossRef] [PubMed]
41. Falasca, G.; Zaghi, D.; Possenti, M.; Altamura, M.M. Adventitious root formation in *Arabidopsis thaliana* thin cell layers. *Plant Cell Rep.* **2004**, *23*, 17–25. [CrossRef] [PubMed]
42. Sorin, C.; Bussell, J.D.; Camus, I.; Ljung, K.; Kowalczyk, M.; Geiss, G.; McKhann, H.; Garcion, C.; Vaucheret, H.; Sandberg, G.; Bellini, C. Auxin and light control of adventitious rooting in Arabidopsis require ARGONAUTE1. *Plant Cell.* **2005**, *17*, 1343–1359. [CrossRef]
43. Casimiro, I.; Marchant, A.; Bhalerao, R.P.; Beeckman, T.; Dhooge, S.; Swarup, R.; Graham, N.; Inze, D.; Sandberg, G.; Casero, P.J.; et al. Auxin transport promotes *Arabidopsis* lateral root initiation. *Plant Cell.* **2001**, *13*, 843–852. [CrossRef]
44. Malamy, J.E.; Benfey, P.N. Down and out in *Arabidopsis*: The formation of lateral roots. *Trends Plant Sci.* **1997**, *2*, 390–396. [CrossRef]
45. Woodward, A.W.; Bartel, B. Auxin: Regulation, action, and interaction. *Ann. Bot.* **2005**, *95*, 707–735. [CrossRef]
46. Gu, D.X.; Zhen, F.X.; Hannaway, D.B.; Zhu, Y.; Liu, L.L.; Cao, W.X.; Tang, L. Quantitative classification of rice (*Oryza sativa* L.) root length and diameter using image analysis. *PloS ONE* **2017**, *12*, e0169968. [CrossRef] [PubMed]
47. Zobel, R.W.; Waisel, Y. A plant root system architectural taxonomy: A framework for root nomenclature. *Plant Biosyst.* **2010**, *144*, 507–512. [CrossRef]
48. Santi, C.; Zamboni, A.; Varanini, Z.; Pandolfini, T. Growth stimulatory effects and genome-wide transcriptional changes produced by protein hydrolysates in maize seedlings. *Front. Plant Sci.* **2017**, 8. [CrossRef] [PubMed]
49. McCully, M.E. Roots in soil: Unearthing the complexities of roots and their rhizospheres. *Ann. Rev. Plant Physiol. Plant Mol. Biol.* **1999**, *50*, 695–718. [CrossRef]
50. Vardhini, B.V.; Anjum, N.A. Brassinosteroids make plant life easier under abiotic stresses mainly by modulating major components of antioxidant defense system. *Front. Environ. Sci.* **2015**, *2*, 1–16. [CrossRef]
51. El-Bassiony, A.M.; Ghoname, A.A.A.; El-Awadi, M.E.; Fawzy, Z.F.; Gruda, N. Ameliorative effects of brassinosteroids on growth and productivity of snap beans grown under high temperature. *Gesunde Pflanz.* **2012**, *64*, 175–182. [CrossRef]
52. Kohli, S.K.; Handa, N.; Sharma, A.; Gautam, V.; Arora, S.; Bhardwaj, R.; Alyemeni, M.N.; Wijaya, L.; Ahmad, P. Combined effect of 24-epibrassinolide and salicylic acid mitigates lead (Pb) toxicity by modulating various metabolites in *Brassica juncea* L. seedlings. *Protoplasma.* **2018**, *255*, 11–24. [CrossRef] [PubMed]
53. Lv, B.; Tian, H.; Zhang, F.; Liu, J.; Lu, S.; Bai, M.; Li, C.; Ding, Z. Brassinosteroids regulate root growth by controlling reactive oxygen species homeostasis and dual effect on ethylene synthesis in *Arabidopsis*. *PLoS Genet.* **2018**, *14*, e1007144. [CrossRef]

54. Clouse, S.D. Brassinosteroids. *The arabidopsis book* **2011**, *9*, e0151. [CrossRef]
55. Men, S.; Boutte, Y.; Ikeda, Y.; Li, X.; Palme, K.; Stierhof, Y.-D.; Hartmann, M.-A.; Moritz, T.; Grebe, M. Sterol-dependent endocytosis mediates post-cytokinetic acquisition of PIN2 auxin efflux carrier polarity. *Nat. Cell. Biol.* **2008**, *10*, 237–244. [CrossRef] [PubMed]
56. Clouse, S.D.; Zurek, D.M.; McMorris, T.C.; Baker, M.E. Effect of brassinolide on gene-expression in elongating soybean epicotyls. *Plant Physiol.* **1992**, *100*, 1377–1383. [CrossRef] [PubMed]
57. Nemhauser, J.L.; Mockler, T.C.; Chory, J. Interdependency of brassinosteroid and auxin signaling in *Arabidopsis*. *PLoS Biol.* **2004**, *2*, 1460–1471. [CrossRef] [PubMed]
58. Bao, F.; Shen, J.J.; Brady, S.R.; Muday, G.K.; Asami, T.; Yang, Z.B. Brassinosteroids interact with auxin to promote lateral root development in Arabidopsis. *Plant Physiol.* **2004**, *134*, 1624–1631. [CrossRef] [PubMed]
59. Tian, H.Y.; Lv, B.S.; Ding, T.T.; Bai, M.Y.; Ding, Z.J. Auxin-BR interaction regulates plant growth and development. *Front. Plant Sci.* **2018**, *8*, 1–8. [CrossRef] [PubMed]
60. Fukaki, H.; Tasaka, M. Hormone interactions during lateral root formation. *Plant Mol. Biol.* **2009**, *69*, 437–449. [CrossRef] [PubMed]
61. Gill, S.S.; Tuteja, N. Reactive oxygen species and antioxidant machinery in abiotic stress tolerance in crop plants. *Plant Physiol. Biochem.* **2010**, *48*, 909–930. [CrossRef] [PubMed]
62. De Klerk, G.J.; Guan, H.Y.; Huisman, P.; Marinova, S. Effects of phenolic compounds on adventitious root formation and oxidative decarboxylation of applied indoleacetic acid in Malus 'Jork 9'. *Plant Growth Regul.* **2011**, *63*, 175–185. [CrossRef]
63. Mussig, C.; Shin, G.H.; Altmann, T. Brassinosteroids promote root growth in Arabidopsis. *Plant Physiol.* **2003**, *133*, 1261–1271. [CrossRef]
64. Guan, M.; Roddick, J.G. Epibrassinolide-inhibition of development of excised, adventitious and intact roots of tomato (*Lycopersicon esculentum*)-comparison with the effects of steroidal estrogens. *Physiol Plant.* **1988**, *74*, 720–726. [CrossRef]
65. Guan, M.; Roddick, J.G. Comparison of the effects of epibrassinolide and steroidal estrogens on adventitious root-growth and early shoot development in mung bean cuttings. *Physiol. Plant* **1988**, *73*, 426–431. [CrossRef]
66. Ren, C.M.; Han, C.Y.; Peng, W.; Huang, Y.; Peng, Z.H.; Xiong, X.Y.; Zhu, Q.; Gao, B.D.; Xie, D.X. A leaky mutation in *DWARF4* reveals an antagonistic role of brassinosteroid in the inhibition of root growth by jasmonate in Arabidopsis. *Plant Physiol.* **2009**, *151*, 1412–1420. [CrossRef] [PubMed]
67. Swamy, K.N.; Seeta Ram Rao, S. Influence of brassinosteroids on rooting and growth of geranium (*Pelargonium* sp.) stem cuttings. *Asian J. Plant Sci.* **2006**, *5*, 619–622.
68. Swamy, K.N.; Seeta Ram Rao, S. Effect of brassinosteroids on rooting and early vegetative growth of Coleus [*Plectranthus forskohlii* (Wild.) Briq.] stem cuttings. *Indian J. Plant Sci.* **2010**, *1*, 68–73.

10

The Influence of Bio-Stimulants and Foliar Fertilizers on Yield, Plant Features and the Level of Soil Biochemical Activity in White Lupine (*Lupinus albus* L.) Cultivation

Alicja Niewiadomska [1], Hanna Sulewska [2], Agnieszka Wolna-Maruwka [1], Karolina Ratajczak [2,*], Zyta Waraczewska [1] and Anna Budka [3]

1 Department of General and Environmental Microbiology, Poznań University of Life Sciences, ul. Szydłowska 50, 60-656 Poznań, Poland; alicja.niewiadomska@up.poznan.pl (A.N.); amaruwka@up.poznan.pl (A.W.-M.); zyta.waraczewska@wp.pl (Z.W.)

2 Department of Agronomy, Poznan University of Life Sciences, ul. Dojazd 11, 60-632 Poznań, Poland; hanna.sulewska@up.poznan.pl

3 Department of Mathematical and Statistical Methods, Poznań University of Life Sciences, ul. Wojska Polskiego 28, 60-637 Poznań, Poland; anna.budka@up.poznan.pl

* Correspondence: karolina.ratajczak@up.poznan.pl

Abstract: The aim of this study is to assess the effect of two biostimulators (Titanit, Rooter) and six foliar fertilizers (Optysil, Metalosate Potassium, Bolero Bo, ADOB 2.0 Zn IDHA, ADOB B, ADOB 2.0 Mo) on white lupine. In addition, we evaluated the enzymatic activity of dehydrogenase, acid, and alkaline phosphatases, catalase, the level of biological nitrogen fixation, yield, plant biometric, chlorophyll fluorescence and chlorophyll content. A field experiment was conducted between 2016 and 2018 at the Gorzyń Experimental and Educational Station, Poznań University of Life Sciences in Poland. The best effects in plant yield were obtained after the application of Optysil or ADOB Zn IDHA. The three years results of dehydrogenase (DHA), alkaline phosphatase (PAL), and the biological index of soil fertility (BIF), show that the bio-stimulants and most of the foliar fertilizers used did not always stimulate the activity of these enzymes and index in the white lupine crops, as compared with the control treatment. Analysis of the results of the acid phosphatase activity (PAC) shows that during the entire white lupine growing season the foliar fertilizers and bio-stimulants decreased the activity of this enzyme. This effect was not observed when the Metalosate potassium foliar fertilizer was applied. The field analyses of biological nitrogen fixation showed that the fertilizers and bio-stimulants significantly stimulated nitrogenase activity under the white lupine plantation. The best effects in plant yield were obtained after application Optysil or ADOB Zn IDHA.

Keywords: soil enzymatic activity; biological index fertility; nitrogenase activity; microelements fertilization (Ti, Si, B, Mo, Zn)

---

1. Introduction

The degradation of the soil environment, excessive use of chemicals, depletion of natural resources, as well as the decreasing biodiversity instigated the European Union to make a decision about the need for integrated crop cultivation and protection [1]. Since 2014 the recommendations concerning integrated protection and cultivation have been in force in Poland. At present we can see the transitional phase between conventional and sustainable agriculture. In order to meet the assumptions of sustainable agriculture it is necessary to diversify the crop structure and minimize the excessive share of cereals. It is also necessary to use integrated methods of agricultural production, so it might be particularly important to restore legume plantations [2].

The significance of legumes in sustainable agriculture is increasing because they improve the physicochemical properties of soil, increase the content of organic matter by leaving large quantities of crop residues, and reduce the need to apply nitrogen fertilizers. White lupine (*Lupinus albus* L.) is one of the most important crops in this group of plants in Poland. It has been the longest known crop species of the *Lupinus* genus. Because of its very high content of protein and fat, especially in seeds, it has been used for human nutrition for thousands of years, despite its high content of bitter alkaloids [3]. It was only in 1930 that low-alkaloid forms were obtained. Because of the introduction of new varieties, the cultivation of white lupine with low alkaloid content became popular in Poland. Between 2005 and 2015 the area of cultivation of large-seeded legumes increased almost four times so that in 2015 they covered an area of 407,000 ha [4].

Lupine species have the largest share in this group of crops. On the other hand, the area of plantations with small-seeded legumes, such as clover and alfalfa, did not fluctuate much in that decade and in 2015 they covered an area of 93,000 ha [5].

Legumes are characterized by the ability to coexist with the nitrogen-fixing diazotrophic bacteria (*Rhizobium*). In order to increase the protein content in plants, which depends on the system developed by the plant and rhizobia, it is necessary to find agents improving the efficiency of this symbiosis.

Scientists are more and more interested in bio-stimulants, which are defined as materials containing one or more active substances and/or microorganisms. They improve the uptake of nutrients by plants, their tolerance to abiotic and biotic stress, and the quality of crops [6]. Bio-stimulants also increase the activity of rhizosphere microorganisms and soil enzymes, as well as they stimulate hormone production and photosynthesis [7]. They also promote the overall plant growth, including increased biomass and crop yields [8]. In the group of synthetic bio-stimulants, there are preparations containing growth regulators, phenolic compounds, inorganic salts, and beneficial nutrients [9,10], which naturally occur in plants in trace amounts (e.g., titanium and silicon). They act mainly by the stimulation of numerous physiological processes, which has a positive effect on plant yield and crop quality. Nutrients assimilable by plants, reduces the impact of stress, which affects the growth and development of plants. They regulate the uptake of macro- and microelements, alleviate the negative effects of periodic water shortage, high salinity, as well as activates the natural immune mechanisms of plants. They also strengthen cell walls and reduce the susceptibility of plants to mechanical damage [11]. Microelements regulate biochemical processes occurring in plants, being part of most enzymes or acting as their activators, therefore their deficit may lead to the inhibition of specific enzymatic reactions, which in turn leads to disorders of many biochemical and physiological processes, adversely affecting the growth and plant development [12,13]. There are many fertilizers that are enriched with amino acids, organic compounds, or surfactants. For example, potassium in fertilizer is in the form of very small molecules complexed with a unique set of natural amino acids. In turn, boron in the fertilizer is in the form of sodium pentaborate decahydrate, and the addition of sorbitol ensures rapid uptake of the fertilizer through the leaves of fertilized plants and high efficiency of the fertilizer. Zinc in modern fertilizers is chelated with the biodegradable IDHA chelating agent, because of which it also gains a form that is very well absorbed by plants. This fertilizer increases the plants' resistance to drought and diseases and increases the germination of seeds. It is produced in the form of microgranules, based on modern microgranulation technology. The manufacturer of molybdenum fertilizer has developed a liquid formula of the fertilizer additionally enriched with biodegradable tensides, which decreases the surface tension of the working liquid and increases the efficiency of covering the leaf blade during spraying increases [14].

Essential plant nutrients are mainly applied to soil and plant foliage in order to achieve maximum economic yields. Soil application is more common and most effective for nutrients that are required in high quantities. However, under certain circumstances, foliar fertilization is more economic and effective. Because of the intensified cultivation foliar fertilization has become an indispensable agrotechnical procedure. Plants exhibit the highest demand for potassium and nitrogen (more than 200 kg in terms of the yield per 1 ha), and the lowest demand for zinc, boron, copper, and molybdenum. Plants need only a few grams of molybdenum in terms of the yield per hectare. This means that foliar fertilization is particularly recommended and effective when it is necessary to supply micronutrients to crops [15].

Each agrotechnical treatment, i.e., the use of fertilizers or bio-stimulants, may cause changes in the soil environment. There have been numerous studies showing various effects of these treatments on the count of selected groups of microorganisms and the amount of soil enzymes they secrete [16].

Measurement of the activity of soil enzymes provides information about the quality of soil. This procedure is important as it indicates the biochemical activity of soil. Enzymes are thought to be good and sensitive indicators because they quickly react to changes in soil caused by natural and anthropogenic factors. Apart from that, it is easy to measure their activity, which affects the main microbiological reactions involving the cycles of nutrients in soil. Studies also showed that agrotechnical procedures influence the enzymatic activity more than other biochemical parameters [17].

The aim of this study is to assess the effect of selected bio-stimulants (Tytanit, Rooter) and foliar fertilizers (Optysil, Metalosate potassium, Bolero Bo, ADOB 2.0 Zn IDHA, ADOB B, ADOB 2.0 Mo) on the yield and plant features, activity of dehydrogenase, acid and alkaline phosphatases, and catalase, as well as the level of biological nitrogen fixation based on the activity of nitrogenase in a white lupine plantation.

2. Material and Methods

2.1. Experimental Design

A field experiment was conducted between 2016 and 2018 at the Gorzyń Experimental and Educational Station, Poznań University of Life Sciences. The GPS coordinates of the experiment are as follows: N-52.56589, E-015.90556, 65 m AMSL. Each year one-factor experiment was conducted as randomized block design in four replications with the following nine factor levels: 1. control treatment—plants not treated with preparations; 2. Tytanit; 3. Optysil; 4. Metalosate Potassium; 5. Rooter; 6. Bolero Mo; 7. ADOB Zn IDHA; 8. ADOB B; 9. ADOB 2.0 Mo. Each fertilizer was applied in a timely manner, according to the manufacturer's recommendations (Table 1).

An experiment was conducted on white lupine (*Lupinus albus* L.) of the Butan cultivar. The lupine seeds were inoculated with the effective strain of *Bradyrhizobium lupinus* root nodule bacteria directly before sowing by using nitragina. Nitragina is a single-component graft, containing a specific bacterial strain for a specific legume plant, in which peat is a carrier. The Butan cultivar can be grown all over Poland, this variety is insensitive to delayed sowing; its growing period is 2–14 days shorter than that of traditional varieties and it is less susceptible to diseases caused by *Fusarium* fungi. The cultivar is more valuable as a feed and it has high content of protein (32–37%) and fat (10–12%), while the content of alkaloids is about 30–40% lower.

The seeds were sown (4 April 2016, 4 April 2017 and 7 April 2018) on plots with an area of 21 m^2, with a distance of rows of 15 cm, and sowing density of 75 seeds per 1 m^2.

According to the FAO/WRB classification [18], the soil in the experimental plots is a typical lessive soil formed from light loamy sands, deposited in a shallow layer on light loam (*Haplic Luvisols*) (Table 2). The soil texture was determined by means of a sieve (sand fraction) for the silt and clay fraction [19].

The agrotechnical and cultivation treatments were carried out in accordance with the principles of good agricultural and experimental practice for this species [20]. In the autumn before winter plowing, basic macronutrients were supplied to the soil in the form of multi-component fertilizer Polifoska 4 in the amount of 350 kg ha^{-1} (N—4%, P—12%, K—32%). Before sowing, urea in the amount of 30 kg ha^{-1} was used.

Table 1. The terms and doses of bio-stimulants and fertilizers applied in the experiment.

Bio-Stimulants/Foliar Fertilizers		Term and Dose of Bio-Stimulant	Fertilizer Characteristics
Bio-stimulants	Tytanit	I: leaf and shoot development (BBCH 13–29)—0.3 dm^3 ha^{-1} II: inflorescence development (BBCH 51–59)—0.3 dm^3 ha^{-1} III: beginning of pod development (BBCH 71)—0.3 dm^3 ha^{-1}	Liquid, mineral stimulant containing titanium (Ti). It increases the yield volume and development of plants, improves yield quality parameters and increases plants' natural resistance to stress factors. Composition: 8.5 g Ti (dm^3)$^{-1}$
	Rooter	BBCH 13–14—1 dm^3 ha^{-1}	Bio-stimulant—it stimulates the growth of the root system, accelerates regeneration and improves the uptake of soil minerals. Composition: P_2O_5 13.0%; K_2O 5.0%
Foliar fertilizers	Optysil	I: leaf and shoot development (BBCH 15–29)—0.5 dm^3 ha^{-1} II: inflorescence development (BBCH 51–55)—0.5 dm^3 ha^{-1} III: beginning of pod development (BBCH 71–73)—0.5 dm^3 ha^{-1}	Liquid, silicon antistressor stimulating the growth and development of plants, activating their natural immune systems and increasing tolerance to unfavorable cultivation conditions. Composition: 200 g SiO_2 (dm^3)$^{-1}$
	Metalosate Potassium	2–3 treatments every 10–14 days during intensive growth—3 dm^3 ha^{-1}	Liquid foliar fertilizer containing an easily absorbable form of potassium, which supplements potassium deficit in plants with amino acids. Composition: K_2O 24%
	Bolero Mo	Before florescence—1.5 dm^3 ha^{-1}	Liquid foliar fertilizer containing boron and molybdenum to supplement the deficit of these elements in plants. Composition: B 8.2%; Mo 0.8%
	ADOB 2.0 Zn IDHA	Before florescence—1 dm^3 ha^{-1}	Foliar fertilizer containing zinc (Zn) fully chelated by biodegradable chelating agent IDHA. Composition: Zn 100 g kg^{-1} (weight percentage content 10, chelated by IDHA)
	ADOB B	I: before florescence—2 dm^3 ha^{-1} II: after florescence on pods—1 dm^3 ha^{-1}	Liquid, highly concentrated foliar fertilizer containing boron that regulates auxin activity and participates in cell division. Composition: N 78 g kg^{-1}; B 150 g kg^{-1}
	ADOB 2.0 Mo	early stages of development—0.15 dm^3 ha^{-1}	Liquid, single-component fertilizer which increases the rate and efficiency of use of nitrogen by plants and improves interaction with iron. Composition: Mo 20%

Table 2. The texture of soil sampled at a depth of 0–25 cm and the soil chemical properties of the 3-year experiment.

Fraction [mm]	Percentage of Soil Fractions			Texture Class
	Sand 2–0.05	**Silt 0.05–0.002**	**Clay <0.002**	
	78	**18**	**4**	**LS**
Soil Chemical Properties				
pH in 1 mol KCl			6.0	
Phosphorus P (mg·kg^{-1})			70.1	
Potassium K (mg·kg^{-1})			99.3	
Magnesium Mg (mg·kg^{-1})			56.7	
Manganese Mn (mg·kg^{-1})			303.4	
Zinc Zn (mg·kg^{-1})			10.9	
Copper Cu (mg·kg^{-1})			2.6	
Iron Fe (mg·kg^{-1})			1525.2	
Boron B (mg·kg^{-1})			>20	
Organic carbon (%)			0.5	
Percent of caries			0.8	

LS—loamy sand.

The agrotechnical procedures were carried out in accordance with the rules adopted for the species used in the test. White lupine was sown in early April. The following products were used for weed control: Afalon Dispersive 450 EC (1.1 L ha^{-1}) in April, Basagran 480 SL (2.6 L ha^{-1}) and Betanal MaxPro 209 OD (1.25 L ha^{-1}) in May. Fusilade Forte 150 EC (1.0 L ha^{-1}) was additionally applied in June. The following products were sprayed to protect the plants from diseases: Gwarant 500 SC (2.0 L ha^{-1}) in May and Korazzo 250 SC (1.0 L ha^{-1}) in mid- and late June.

2.2. Weather Conditions

During the growing seasons in 2016 and 2017 the weather conditions were similar in terms of temperature and rainfall. During the growing season the highest average air temperature was noted in July both in 2016 (19.5 °C) and 2017 (18.9 °C), whereas the lowest temperature was noted in April, i.e., 8.7 °C in 2016 and 7.5 °C in 2017. However, the weather conditions in 2018 were different than in the previous years (Figure 1). The highest average temperature was noted in August (21.2 °C), whereas the lowest was noted in May (12.7 °C). As far as the average monthly temperature from April to September is concerned, 2018 was the warmest—it was 2.9 °C warmer than 2016 and 1.7 °C warmer than 2017. In 2016 there was drought only at the end of the growing season. Likewise, in 2017 there was no rainfall deficit. On the contrary, it was a wet year, especially from June to August. On the other hand, in 2018 rainfall was unevenly distributed and there were droughts that were particularly unfavorable for plants in May, June, and August.

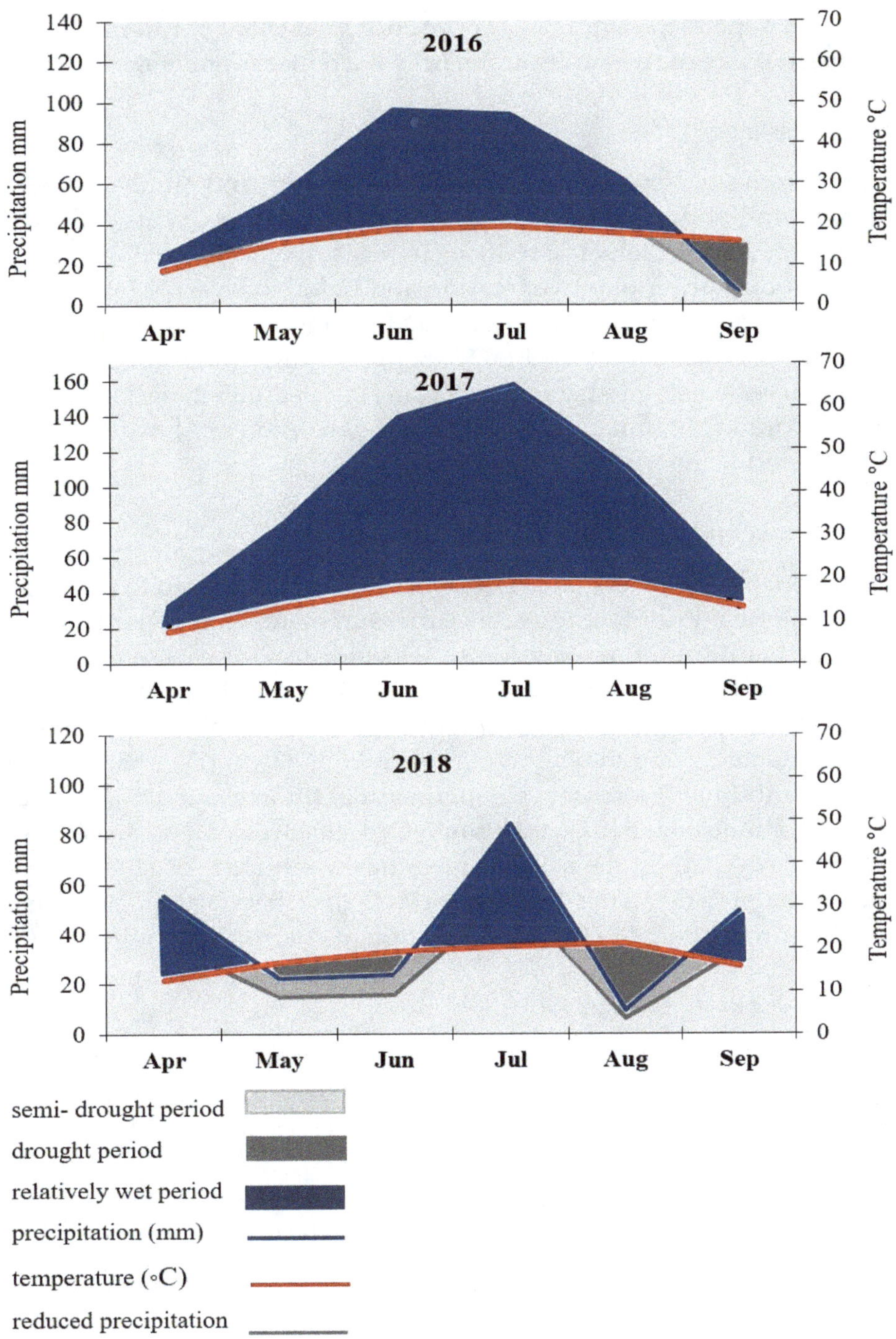

Figure 1. Climate graphs according to Walter [21] characterizing weather conditions in Gorzyń.

2.3. *Influence of Fertilizers on Nitrogenase Activity (Diazotrophy)*

Nitrogenase activity was estimated using the acetylene reduction assay (ARA) at the beginning of the plants' flowering [22]. For this purpose, five plants were randomly selected in plots, in a given experimental treatment and directly were placed tightly in sealed test vials (2000 mL) at the field, purified C_2H_4 was injected to obtain an acetylene concentration of 10% (*v*/*v*) in the gas phase (air). After an hour, 1 mL of the gas phase was taken from inside of the test vessels with a Hamilton gas-tight syringe and stored in small glass vials, which were sealed with rubber septa and aluminum seals. Ethylene concentration was determined using gas chromatograph CHROM 5 (Laboratorni Přistroje, Praha, Czech Republic, 1980). Nitrogenase activity was determined based on the quantity of acetylene reduced to ethylene and expressed in nmolC_2H_4 produced per plant per hour (nMC_2H_4 plant^{-1} h^{-1}). The results are the mean value of five replications from each measurement.

2.4. Plant Biometric Assessment

The plant height (from soil surface to the highest plant point) and number of pods per plant were measured. Shoot, root, and nodule dry mass were determined after drying for 2 days at 70 °C until reaching constant weight. All the biometric traits were measured on 10 randomly selected plants from each object and replication during plant vegetation and before harvest. The total one-sided area of leaf per unit ground surface area expressed by the leaf area index (LAI) and was measured at three randomly selected places of each plot at the BBCH stage 69 using a SunScan Canopy Analysis System type SSI. Lupine was harvested at one stage (BBCH 90–92) with a 1.35-m wide plot combine. The yield of clean seeds was determined in dt·ha^{-1}, given at a standardized (15%) water content and thousand seed weight was measured using a seed counter.

2.5. Chlorophyll Fluorescence and Chlorophyll Content Measurements

A fluorimeter (OS5p; Opti-Sciences, Inc., Hudson, NY, USA) was used to measure the efficiency of the photosynthetic apparatus. Prior to fluorescence measurements, the upper surface of three healthy leaves at the top of one plant from three randomly selected sites for each plot was covered with leaf clips for 30 min. Leaf fluorescence was then measured with a light pulse of 15,000 μmol m^{-2} s^{-1} at a wavelength of 660 nm The assessed parameter was maximum photosynthetic efficiency of PSII (F_v/F_m), which was calculated using the following formula: $F_v/F_m = (F_m - F_0)/F_m$, on the basis of the measured parameters: minimal fluorescence (F_0), maximal fluorescence (F_m), variable fluorescence (F_v) [23].Chlorophyll content meter (CCM-200plus; Opti-Sciences, USA) was used to estimate the chlorophyll content index (CCI) on the same leaves that were used for chlorophyll α fluorescence measurements. CCM-200plus measures the chlorophyll absorbance and calculates the chlorophyll content index, which is proportional to the concentration of chlorophyll in the sample.

2.6. Soil Sampling for Biochemical Analyzes

Soil samples collected from the arable layer (0–20 cm) were used as the research material for biochemical analyses. Each year they were collected at four terms: First term—at the plants' emergence (BBCH 5–10), Second term—at the plants' full growth (BBCH 35–40), third term—at the at the plants' florescence (BBCH 51–59), fourth term—after harvest.

Soil samples were taken from five places of each experimental plot, in four replications for each of the nine treatments of the experiment. In this way, at each analysis term we received 36 samples of soil, each of 1 kg.

2.7. Soil Enzymatic Activity

The analyses of soil enzymatic activity in individual treatments were based on the colorimetric method applied to measure the dehydrogenase activity (DHA), where 1% triphenyl tetrazolium chloride (TTC) was used as the substrate. The activity was measured after 24-h incubation at a temperature of 30 °C and a wavelength of 485 nm and it was expressed as μmol triphenyl formazane TPF 24 h^{-1} g^{-1}dm of soil [24].

Apart from that, the biochemical analyses of soil involved the determination of activities of acid (EC 3.1.3.2) phosphomonoesterases (PAC) and alkaline phosphomonoesterases (PAL) with the method developed by Tabatabai and Bremner [25]. The activities were determined with disodium *p*-nitrophenyl phosphate tetrahydrate used as a substrate after 1 h incubation at 37 °C and at a wavelength of 400 nm. The results were converted into μmol (p-nitrophenol) PNP h^{-1} g^{-1}dm of soil.

Catalase activity was measured by means of permanganometry, according to the method developed by Johnsons and Temple [26], where 0.3% H_2O_2 was the substrate. After 20-min incubation at room temperature (about 20 °C) 0.02 M $KMnO_4$ was titrated to a light pink colour and expressed as μmol H_2O_2 g^{-1} dm min^{-1}.

2.8. Biological Index of Fertility

The biological index of fertility (BIF) was calculated using the dehydrogenase activity (DHA) and catalase activity (CAT) according to the Stefanic method [27] using the following formula: (DHA + kCAT)/2, where k was the factor of proportionality which equaled 0.01.

2.9. Statistical Analyses

The dynamics of changes in the soil enzymatic activity was statistically analyzed. As there were no significant differences between the parameters in the research years, they were treated as replicates and the results were analyzed by two-way ANOVA using Statistica 12.0 software. The fertilization method and the term of analysis were the factors differentiating the traits under study to estimate the soil biochemical activity parameters. Homogeneous subsets of mean were identified via Duncan's test at a significance level of $p = 0.05$. Yield, biometric, physiological traits of plants, and nitrogenase activity were tested once a year for the experiment. Hence, one-way analysis of variance (ANOVA) was used with Duncan's confidence interval, which was applied at a significance level of $p = 0.05$. As there were no significant differences between the parameters in the experimental years, they were treated as replicas.

Principal component analysis (PCA) was used to visualize the multidimensional dependencies between the soil biochemical activity and the types of fertilization [28]. In order to show the existing regularities (correlations) between biometric and physiological parameters of plants in individual years of research, a Pearson correlation matrix was determined, which was illustrated using a heatmap. The colors indicate the correlation coefficient values (from darkest—value −1, to the lightest—value +1). Cluster analysis enables grouping of the studied physiological parameters of the plants in the experiment in such a way that the degree of correlation between parameters within one group was the highest and between groups the smallest [29]. The agglomeration Ward method (Ward Hierarchical Clustering) and the Euclidean distance were used to create a tree diagram.

3. Results

3.1. Yield, Biometric, and Physiological Traits of White Lupine Plants

The studied biostimulators/foliar fertilizers modified the yield and yield components of white lupine. The yield of white lupine seeds was low and ranged from 11.67 dt·ha^{-1} (ADOB B) to 13.88 dt·ha^{-1} (Optysil) and depended significantly on the bio-stimulants or foliar fertilizers that were applied (Table 3). After applying Optysil or ADOB Zn IDHA (13.63 dt·ha^{-1}), the yields were significantly higher when compared to the control plants by 1.82 and 1.57 dt·ha^{-1}, respectively.

Thousand seed weight (TSW) was significantly higher than the control plants when ADOB Zn IDHA (322.7 g) was applied. All tested preparations significantly stimulated the height of white lupine. The strongest stimulation was obtained by Metalosate potassium, which increased the height of white lupine (40.5 cm) by 6.2 cm when compared to the control. Apart from these fertilizers, in the group that most strongly stimulated this trait were: Optysil (39.8 cm), ADOB 2.0 Mo (38.9 cm), and Bolero Mo (38.6 cm).

ADOB Zn IDHA (318.4 pc·m^{-2}) and Tytanit (300.8 pc·m^{-2}) significantly increased the number of pods compared to the control, and the increase was 96.6 and 79 pc·m^{-2}, respectively.

Studies have also shown changes in nodulation and physiological parameters of the plant. Dry mass of root nodules was significantly stimulated after application of ADOB Zn IDHA (0.212 g) by 0.067 g when compared to the control treatment.

Chlorophyll fluorescence (F_v/F_m), showing the level of plant stress, was measured in the BBCH 69 (end of flowering) and BBCH 79 phases (75% of the pods reached typical length). At the end of flowering, the best plant condition, expressed by the F_v/F_m parameter, was obtained after the application of ADOB Zn IDHA (0.815) or Metalosate potassium (0.813) and both values significantly exceeded those obtained both with the control and all other treatments. In the assessment made at a later developmental phase, the tested biostimulators/foliar fertilizers did not significantly differentiate this parameter.

ADOB Zn IDHA application significantly stimulated the content of chlorophyll in leaves, expressed in CCI, which was 50.9 and exceeded the control by 17.4, as well as all other objects. In addition, significantly higher CCI values than in the control object were obtained after using Tytanit (46.7), ADOB B (42.9), or ADOB 2.0 Mo (40.7).

In turn, the significantly highest LAI value in the experiments was obtained after application of Rooter. The LAI value was 2.03 and exceeded the control by 0.62, for which the lowest LAI value was determined.

Table 3. The influence of the bio-stimulants and fertilizers on yield, biometric, and physiological traits of white lupine.

Objects	Seed Yield, $dt \cdot ha^{-1}$	TSW, g	Height, cm	Number of Pods, $pc. \cdot m^{-2}$	Plant Dry Mass, g	Root Nodules Dry Mass, g	F_v/F_m BBCH 69	CCI	LAI
1	12.06 bc	302.4 bc	34.3 e	221.8 bc	5.05	0.145 bc	0.784 cd	33.5 d	1.41 g
2	11.82 bc	295.2 c	37.8 bcd	300.8 a	5.72	0.147 bc	0.796 b	46.7 b	1.73 de
3	13.88 a	301.7 bc	39.8 ab	250.5 b	6.46	0.160 bc	0.792 bc	23.1 f	1.81 c
4	11.96 bc	305.8 bc	40.5 a	189.1 c	5.28	0.170 b	0.813 a	24.2 f	1.70 de
5	13.18 ab	313.9 ab	37.1 cd	235.0 bc	5.36	0.142 bc	0.776 d	25.6 f	2.03 a
6	12.76 abc	306.1 bc	38.6 abc	250.0 b	5.14	0.129 c	0.774 d	30.2 e	1.88 b
7	13.63 a	322.7 a	36.4 d	318.4 a	6.17	0.212 a	0.815 a	50.9 a	1.61 f
8	11.67 c	310.4 b	37.9 bcd	273.4 ab	5.12	0.169 bc	0.779 d	42.9 c	1.68 e
9	11.94 bc	309.0 b	38.9 abc	240.7 b	5.24	0.170 bc	0.797 b	40.7 c	1.76 cd
p-value	0.001	0.000	0.000	0.000	0.236	0.001	0.000	0.000	0.000

1. control—no bio-stimulants or foliar fertilizers applied to the plants; 2. plant + Tytanit; 3. plant + Optysil; 4. plant + Metalosate potassium; 5. plant + Rooter; 6. plant + Bolero Mo; 7. plant + ADOB Zn IDHA; 8. plant + ADOB B; 9. plant + ADOB 2.0 Mo; lack of homogeneous groups means no significant differences at the level of $p < 0.05$, a, b, c, d, e, f, g-homogeneous groups (Duncan's test. $p < 0.05$); TSW-thousand seed weight, F_v/F_m—maximum photosynthetic efficiency of PSII, CCI—chlorophyll content index, LAI—leaf area index.

The results of the experiment showed that foliar fertilizers and bio-stimulants affected the enzymatic activity of the soil and the biological index of fertility (BIF), as well as the nitrogenase activity in the white lupine plantation. The two-way analysis of variance showed that the foliar fertilization/bio-stimulants did not have a significant influence on the enzymatic activity and the soil biological index of fertility (BIF). Only the term of the test (development phase, based on BBCH scale) had a highly significant influence on the enzymatic activity and the biological index of fertility (BIF) of the soil (Table 4). One-way analysis of variance showed that foliar fertilization/bio-stimulants had a significant influence on nitrogenase activity.

Table 4. The test *F* statistics and the significance levels of the two-way analysis of variance for the soil bioactivity. The traits under analysis were affected by two factors, i.e., foliar fertilization and the term of the test.

Parameter	Fertilization	Development Phase	Interaction
	White Lupine *Butan*		
Dehydrogenase	13.393 ^{ns}	159.989 ***	41.123 ^{ns}
Alkaline phosphatase	7.036 ^{ns}	51.672 ***	5.37 ^{ns}
Acid phosphatase	14.907 ^{ns}	116.200 ***	10.116 ^{ns}
Catalase	192.47 ^{ns}	1558.42 ***	121.42 ^{ns}
BIF	2.90 ^{ns}	131.96 ***	2.71 ^{ns}
Nitrogenase	14.08 ***	-	-

F test statistics and significance levels of two-way analysis of variance for activity of enzymes associated with herbicides and terms research fixed factors *** $p = 0.001$, ns—no signification.

3.2. *Biological Fixation of Nitrogen under Lupine Plantation*

The field analyses of the biological fixation of nitrogen showed that the fertilizers and bio-stimulants significantly stimulated the nitrogenase activity in the white lupine plantation (Figure 2). During the three years in all the experimental treatments nitrogenase exhibited higher activity than in the control plot and differences were statistically significant. The highest nitrogenase activity was noted after the application of the ADOB B and ADOB Zn IDHA. The activity of the enzyme was respectively six and four times higher than in the control plot. Apart from the control treatment, the lowest biological fixation of nitrogen was noted after the application of Metalosate potassium.

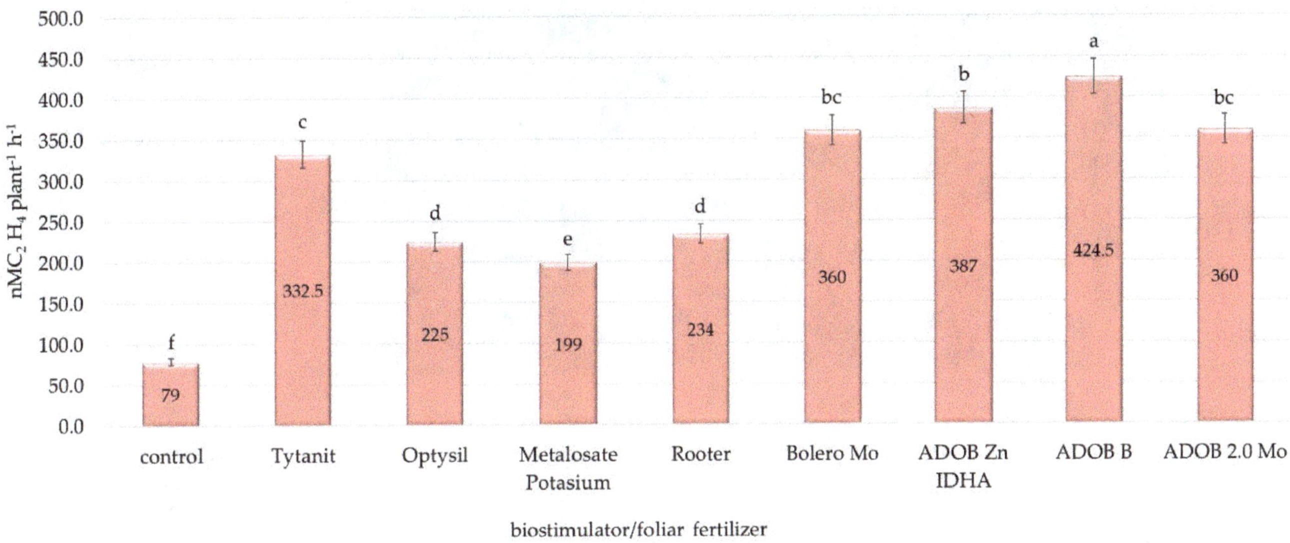

Figure 2. The influence of the bio-stimulants and fertilizers on the level of biological fixation of nitrogen. Abbreviation: means values ± standard errors; a, b, c, d, e, f—homogenous groups according to Duncan's test at level $p = 0.05$.

The heat map presents correlations between all biometric and physiological characteristics of white lupine plants studied (Figure 3). Based on this visualization, relatively higher correlations were found between some features, including: PN (number of pods, pc.$\cdot$m^{-2}), TSW (thousand seed weight), H (height plant), PDM (plant dry mass), Y (seed yield), and PDM, Y, LAI (leaf area index) and Fv/Fm^1 (maximum photosynthetic efficiency of PSII BBCH–69). In turn, BNF (biological nitrogen fixation) and RNDM (root nodules dry mass) are negatively correlated with LAI, Y, PDM, H, TSW, PN, and Fv/Fm^2 (maximum photosynthetic efficiency of PSII BBCH–78). Additionally, based on cluster analysis, groups of related biometric and physiological traits of plants were determined. Three groups have been designated. The first group that is the most distinct from the others contains: RNDM, BNF, CCI, and Fv/Fm^1. The other two groups are: LAI, Y, PDM and H, TSW, PN, Fv/Fm^2.

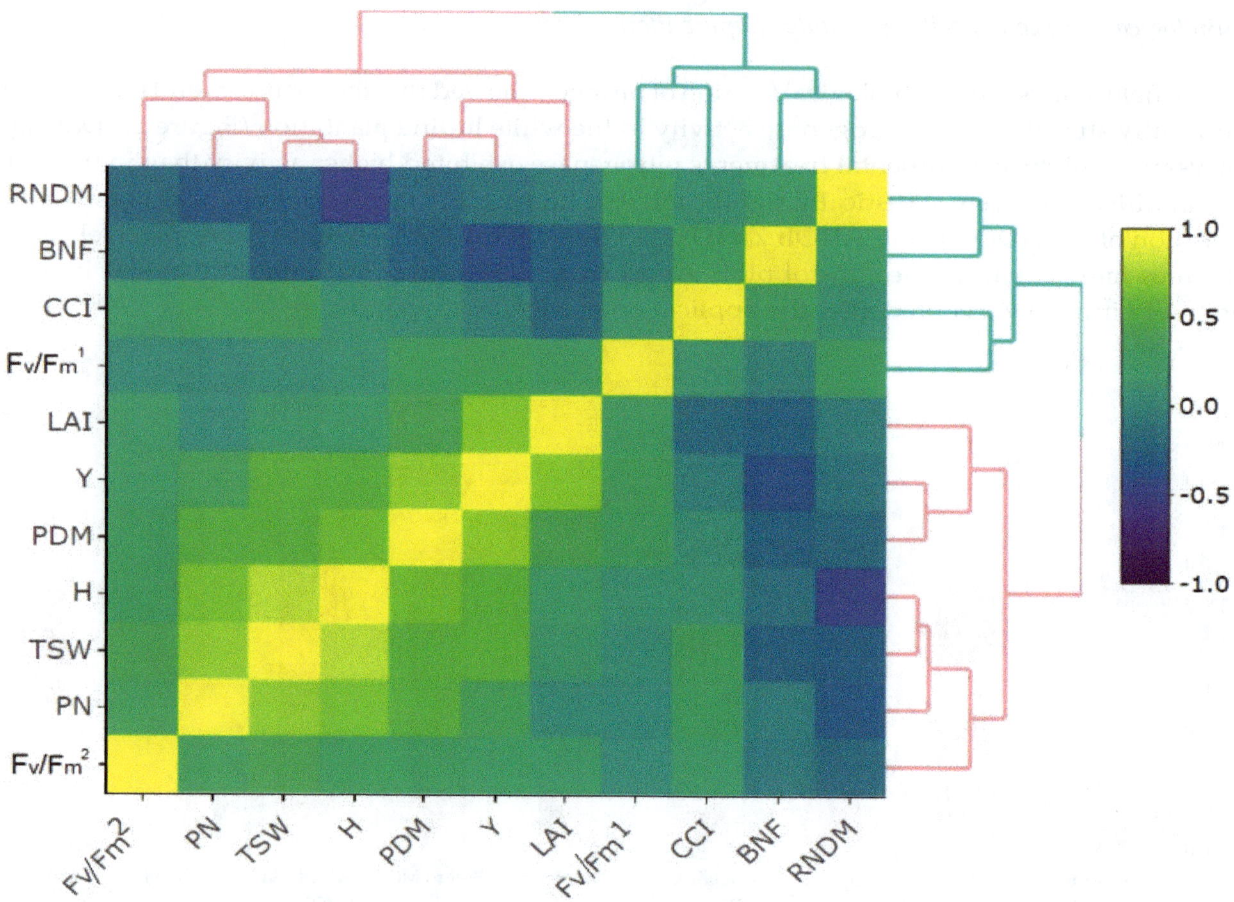

Figure 3. Correlations between all biometric and physiological characteristics of white lupine plants. Abbreviation: RNDM—root nodules dry mass, g—BNF—biological nitrogen fixation, CCI—chlorophyll content index, Fv/Fm^1—maximum photosynthetic efficiency of PSII BBCH–69, LAI—leaf area index, Y—seed yield, PDM—plant dry mass, g, H—height plant, TSW—thousand seed weight, PN—number of pods, pc.·m^{-2}; Fv/Fm^2—maximum photosynthetic efficiency of PSII BBCH–78.

3.3. Analysis of Soil Biochemical Activity

Only the ADOB 2.0 Mo and Metalosate potassium foliar fertilizers stimulated the dehydrogenase activity throughout the growing season, as compared with the control treatment. After the application of the bio-stimulants the level of the enzyme activity was similar to the activity in the control treatment. However, when the Optysil and ADOB B were applied, the activity decreased but not statistically significantly. The experiment also showed that the peak of the dehydrogenase activity significant occurred at the third term of analyses, when the plants began flowering (BBCH 51–59). The results of the analysis of the dehydrogenase activity in the soil under the white lupine plantation are shown in Figure 4.

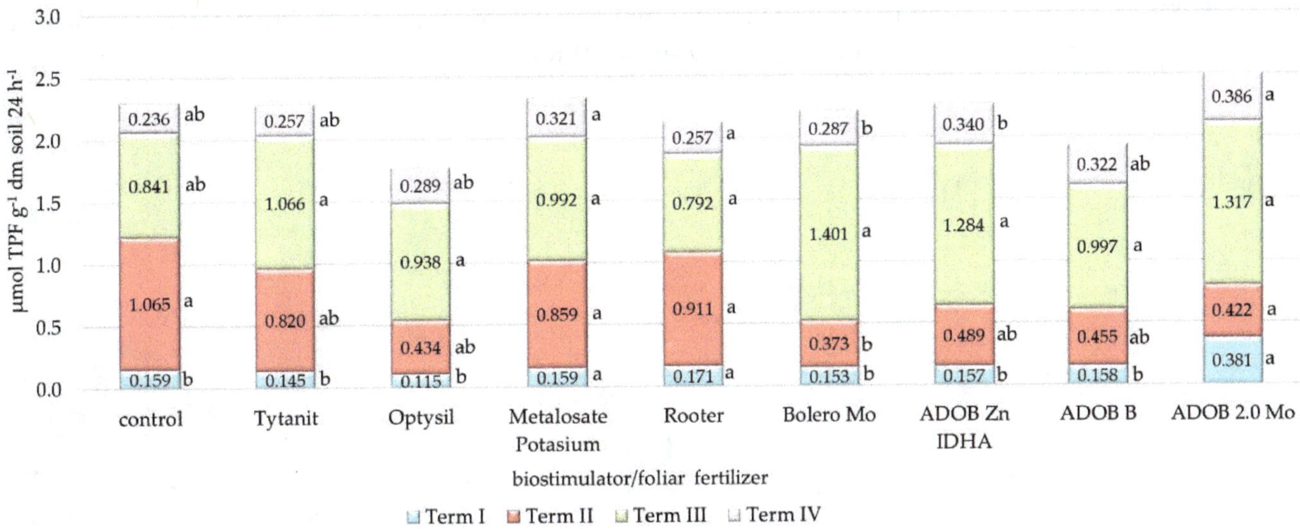

Figure 4. The influence of the bio-stimulants and fertilizers on the dehydrogenase activity. Abbreviation: a, b—homogenous groups according to Duncan's test at level $p = 0.05$; I term—at the plants' emergence (BBCH 5–10), II term—at the plants' full growth (BBCH 35–40), III term—at the at the plants' florescence (BBCH 51–59), IV term—after harvest.

The analysis of the results of the acid phosphatase activity (PAC) shows that during the entire white lupine growing season the foliar fertilizers and bio-stimulants decreased the activity of this enzyme, as compared with the control treatment (Figure 5). This effect was not observed when the Metalosate potassium foliar fertilizer was applied. During the second term of analyses, shortly before flowering, the acid phosphatase activity in all the experimental treatments was higher than in the control treatment. It was very high after the application of the Bolero Mo (0.170 µmol PNP h^{-1} kg^{-1}dm of soil) and ADOB 2.0 Mo (0.171 µmol PNP h^{-1} g^{-1}dm of soil).

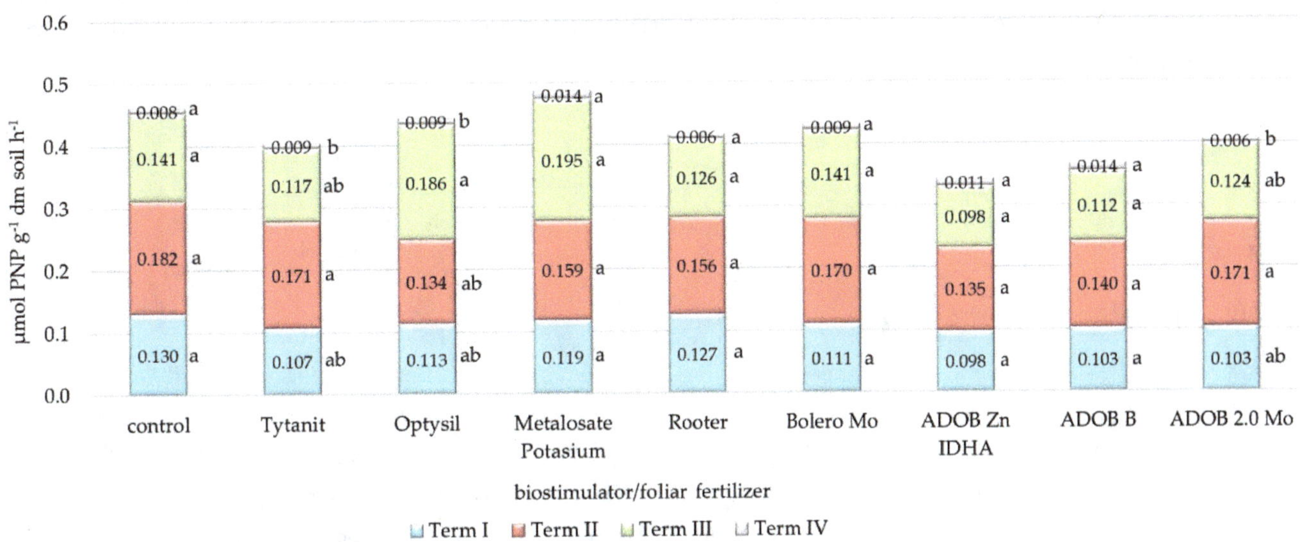

Figure 5. The influence of the bio-stimulants and fertilizers on the acid phosphatase level. Abbreviation: a, b—homogenous groups according to Duncan's test at level $p = 0.05$; I term—at the plants' emergence (BBCH 5–10), II term—at the plants' full growth (BBCH 35–40), III term—at the at the plants' florescence (BBCH 51–59), IV term—after harvest.

The bio-stimulants and most of the foliar fertilizers did not increase the alkaline phosphatase (PAL) activity in the white lupine plantation, as compared with the control treatment (Figure 6). The ADOB 2.0 Mo and Bolero Mo stimulated the activity of this enzyme, which respectively increased by

14% and 5%, as compared with the control treatment. The enzyme exhibited statistically significantly increased activity shortly before they began flowering (II term).

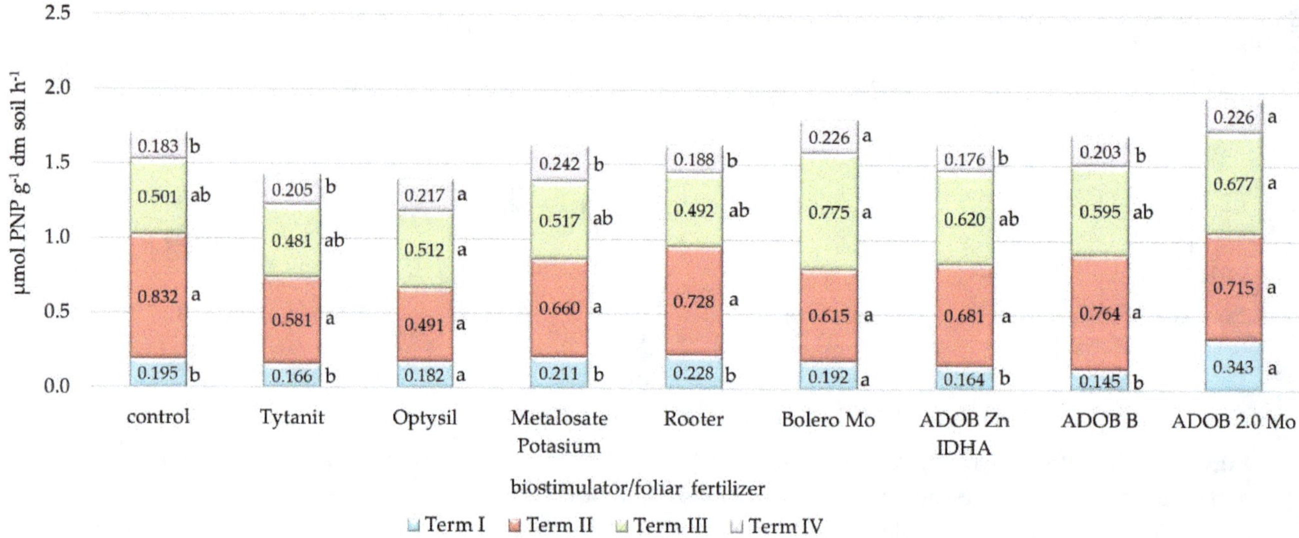

Figure 6. The influence of the bio-stimulants and fertilizers on the alkaline phosphatase level. Abbreviation: a, b—homogenous groups according to Duncan's test at level $p = 0.05$; I term—at the plants' emergence (BBCH 5–10), II term—at the plants' full growth (BBCH 35–40), III term—at the at the plants' florescence (BBCH 51–59), IV term—after harvest.

All the preparations stimulated the catalase activity, as compared with the control treatment (Figure 7), but not significantly. The enzyme significantly exhibited high activity, i.e., when the plants started flowering (III term) in all the experimental treatments. The catalase activity ranged from 98.510 $\mu mol\ H_2O_2g^{-1}$ dm min^{-1} after the application of the Tytanit to 135.819 $\mu mol\ H_2O_2g^{-1}$ dm min^{-1} after the application of the Bolero Mo.

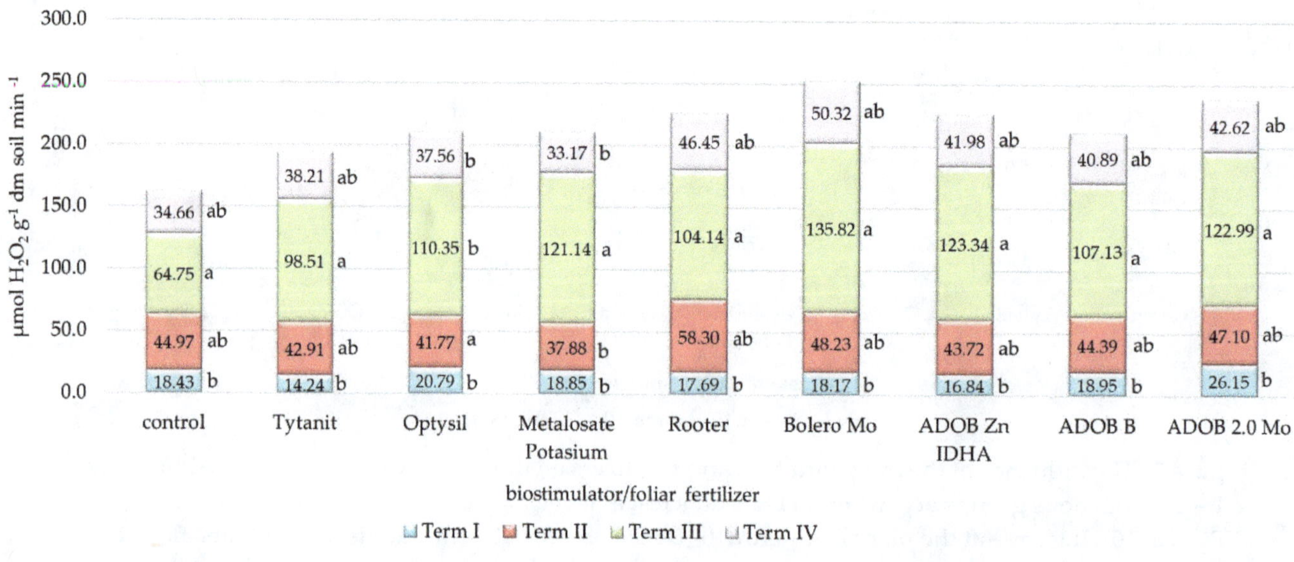

Figure 7. The influence of the bio-stimulants and fertilizers on the catalase activity. Abbreviation: a, b—homogenous groups according to Duncan's test at level $p = 0.05$; I term—at the plants' emergence (BBCH 5–10), II term—at the plants' full growth (BBCH 35–40), III term—at the at the plants' florescence (BBCH 51–59), IV term—after harvest.

The biological index of fertility (BIF), which was calculated on the basis of the dehydrogenase and catalase activity, was not always higher after the application of the bio-stimulants and foliar fertilizers (Figure 8). The highest value of this indicator was noted after the application of the Optysil and the lowest after ADOB Zn IDHA. The BIF was significantly high at the beginning of flowering, as it ranged from 5.17 after the application of ADOB Zn IDHA to 12.34 after the application of ADOB 2.0 Mo. The indicator was also high after the application of the Bolero Mo and Optysil.

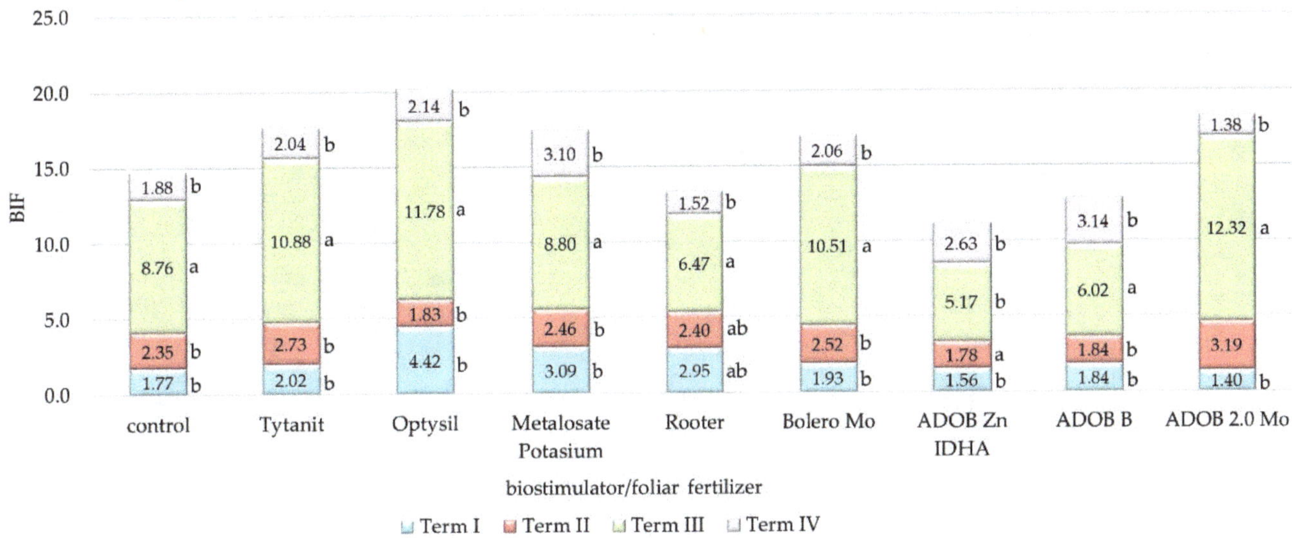

Figure 8. The influence of the bio-stimulants and fertilizers on the BIF. Abbreviation: a, b—homogenous groups according to Duncan's test at level $p = 0.05$; I term—at the plants' emergence (BBCH 5–10), II term—at the plants' full growth (BBCH 35–40), III term—at the at the plants' florescence (BBCH 51–59), IV term—after harvest.

Principal component analysis (PCA) was used to show how the foliar fertilizers and bio-stimulants affected the white lupine plantation. The first two principal components accounted for over 89.2% of the total variation (Figure 9). The parameters of the soil biochemical activity in 2018 differed significantly from 2016 to 2017. This effect may have been caused by the weather conditions (Figure 1). In 2018 the season was the warmest of all the research years. The average temperature difference between 2018 and the previous years was 2.9 °C in August and 1.7 °C in May. As the thermal conditions were very similar in 2016 and 2017, the PCA showed similar dependencies for these two years. In 2016 the fertilizer preparations and bio-stimulants significantly affected the catalytic activity of acid phosphatase (PAC) at all the terms of analyses. This dependency was not observed for the other parameters of soil biochemical activity.

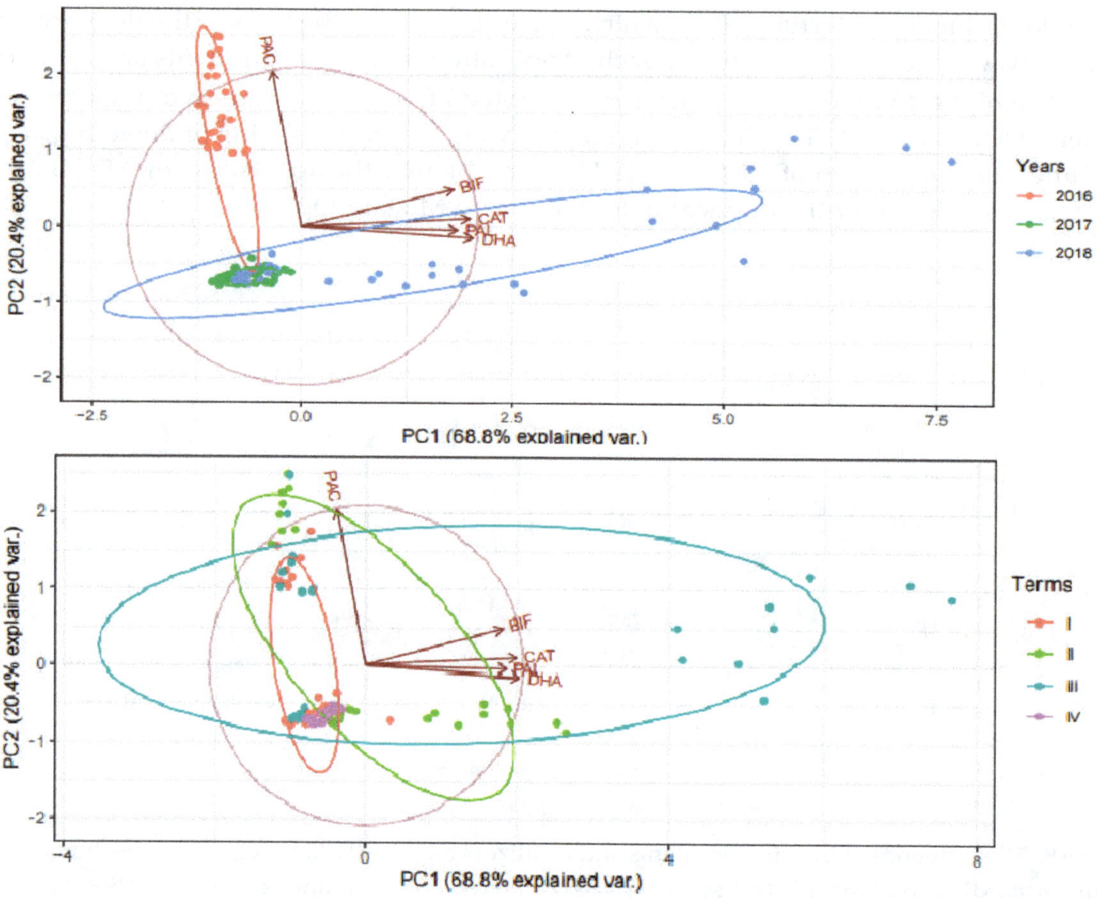

Figure 9. The dependence between the soil enzymatic activity and all treatments with fertilizers and bio-stimulants at the terms of analyses. Abbreviation: I term—at the plants' emergence (BBCH 5–10), II term—at the plants' full growth (BBCH 35–40), III term—at the at the plants' florescence (BBCH 51–59), IV term—after harvest. BIF—index of fertility, CAT—catalase activity, PAC—acid phosphomonoesterases, PAL—alkaline phosphomonoesterases, DHA—dehydrogenase activity.

In 2017 the application of the fertilizers did not cause significant differences in the activity of the soil enzymes or the biological index of soil fertility. In dry 2018 the preparations did not significantly affect the catalytic activity of the test parameters only during plants' emergence (I term). However, at the plants' full growth (II term), the foliar fertilizers and bio-stimulants strongly influenced the catalytic activity of catalase (CAT), dehydrogenase (DHA), alkaline phosphatase (PAL), and the biological index of soil fertility (BIF). Apart from that, the principal component analysis showed that in 2018 the indicators of soil biochemical activity were affected most strongly by foliar fertilizers and bio-stimulants the flowering of the plants (III term) and after the harvest (IV term).

4. Discussion

4.1. Yield, Biometric, and Physiological Traits

Silicon, iron, manganese, boron, copper, molybdenum, and zinc are the basic micronutrients. The silicon content in most plants is comparable to the content of calcium, magnesium, and phosphorus. Many studies have shown the positive effects of silicon on plants, their development, yield, and sensitivity to biotic and abiotic stress [30]. In many tests, silicon has been shown to significantly influence the regulation of nutrient uptake such as: calcium, magnesium, and phosphorus. In other studies [31], silicon fertilization increased the yield of sugar beet roots by 13.7–15.9%, as well as the yield of many other species [11], especially in the form of spraying plants under stress conditions. According to Fageria and Baligar [32] and Duffy [33] Zn is the microelement most limiting crop yield.

Zinc is taken up in small amounts and it participates in all major functions of the plant, increases nitrogen uptake, and activates CO_2 binding in later stages. Hence, it is necessary in plant nutrition and its importance in plant production is growing [13]. Similarly, Kaya et al. [34] obtained the highest common bean plants (*Phaseolus vulgaris* L.), with the largest number of pods and seeds per plant after application of a foliar mixture of zinc.

The preparations used in our study also stimulated the tested biometric parameters of the plants. Plant height was stimulated the most after application of Metalosate potassium (by 8.5%) when compared to the control treatments. In turn, the number of nodules was most strongly stimulated by ADOB Zn IDHA (by 68.4%) and LAI by Rooter (by 69.5%). Other fertilizers containing boron, molybdenum, silicon, and titanite also increased the parameters indicated above. These results are consistent with the results of Raj and Raj [12] regarding the beneficial effects of Zn on plant efficiency, physiological parameters, plant height, and nodulation formation. Our results are also consistent with field studies of Omer et al. [35], in which the treatments of molybdenum application did not modify any of the studied lentil characteristics, except for the height of the plant. Also Rahman et al. [36] showed that the use of molybdenum in its deficiency in soil, stimulates the formation of nodules. Of the physiological traits studied, chlorophyll fluorescence (Fv/Fm) was most strongly stimulated by ADOB Zn IDHA (by 3.9%) and Metalosate potassium (by 3.7%). In turn, the CCI index was most strongly stimulated by ADOB Zn IDHA, whose application resulted in an increase of this parameter by 51.9% when compared to the control treatment. The results of research on *Vigna sinensis* [37] and on *Celosia* [38] showed that Zn spraying on plants caused a significant increase in chlorophyll content. In a study conducted by Artyszak et al. [39], foliar fertilization with silicon increased LAI and effective quantum efficiency of PSII—ΦPSII, as well as positively affected the growth and development of many plant species [40,41].

4.2. Biological Fixation of Nitrogen

The bio-stimulants and foliar fertilizers which improved the biological fixation of nitrogen in the white lupine plantation contained important macro- or microelements. Scientific reports suggest that some elements are particularly significant to the nitrogen fixation process.

Mineral nutrients may influence N_2 fixation in legumes at various stages of the symbiotic process: infection and nodule development, nodule function, and host plant growth. For healthy and vigorous growth, intact plants need to take up relatively large amounts of some inorganic elements: ions of nitrogen (N), potassium (K), calcium (Ca), phosphorus (P) and sulphur (S), and small quantities of other elements: iron (Fe), nickel (Ni), chlorine (Cl), manganese (Mn), zinc (Zn), boron (B), copper (Cu), and molybdenum (Mo). Molybdenum and iron are especially important because they are components of the nitrogenase complex in rhizobia which is required for nitrogen fixation. They are components of nitrogenase—the bacterial enzyme that enables the diazotrophy process. The nitrogenase protein consists of two subunits: the larger one containing the FeMo cofactor and the smaller one containing iron alone [42]. Plants growing on acidic, moist, and poorly buffered soils do not have sufficient supply of molybdenum. When molybdenum is applied in a field to the leaves of legumes, the nitrogen fixation of these plants is more efficient, and the mass of their root nodules and the yield of seeds increase [43,44]. The use of ADOB 2.0 Mo with high molybdenum content in our experiment confirmed this fact. There are small amounts of boron in plants, but this micronutrient plays an important role in various physiological processes. It affects the separation of plant tissues and it is necessary for the optimal growth of plants. Boron-deficient plants have less bacteria of the *Rhizobium* genus and fewer infection threads [44]. The significant increase in the level of biological fixation of nitrogen may have been caused by the application of the foliar fertilizer containing boron (ADOB B). Our research also proved that zinc supplied with the ADOB Zn IDHA foliar fertilizer significantly increased the nitrogenase activity. Although plants absorb moderate amounts of zinc, this element has significant influence on bacteria of the *Rhizobium* genus. The research by Wani et al. [45] showed that higher concentrations of this element in soil stimulate bacteria of the *Rhizobium* genus to produce phytohormones (including indoleacetic acid), which promote the growth of plants by increasing the number of root nodules, their dry mass, and the content of leghemoglobin in the nodules.

Many researchers have studied the role of phosphorus in symbiotic systems. Phosphorus plays a crucial role in the nitrogen fixation process [46,47]. The Rooter bio-stimulant, which contained phosphorus and potassium, stimulated this process considerably. Phosphorus participates in a wide range of molecular and biochemical processes. Apart from that, some phosphate bonds are carriers of the energy used in cells. The presence of phosphorus in soil affects the plant's ability to produce root nodules, especially the weight and the number of nodules [48], which translates into the level of nitrogen fixation.

When the supply of phosphorus is insufficient, plants often suffer from nitrogen deficiency. Sulphur and potassium are less important for symbiotic systems than the aforementioned elements. Nevertheless, potassium ions are very desirable in saline soils because they function as an osmolyte. In view of the fact that nearly half of irrigated soils around the world are considered saline, the addition of potassium helps to maintain the bacteria-plant system [48,49].

4.3. *Biochemical Activity*

The activities of soil enzymes are considered sensitive indicators of important microbial reactions involved in nutrient cycles and they respond to changes in the soil caused by natural or anthropogenic factors. In this regard, soil enzyme activities are often used to evaluate the impact of human activity on soil life [50].

Soil enzymes are a group of catalysts that significantly affect the ecological properties of the pedosphere. These are both extracellular enzymes and the ones that are present in microorganisms (both in proliferating cells and in endospores). Enzymes control the course of all chemical reactions in microbial cells, e.g., the synthesis of proteins, nucleic acids, and carbohydrates [51]. Soil enzymes are involved in the decomposition of organic substances released into the soil during the plant's growth as well as the formation and decomposition of humus in the soil. They release and transfer minerals to plants. In spite of the dynamic nature of the microbiological and biochemical properties, soil enzymes are accurate and significant determinants of soil fertility, and they are important indicators of changes taking place in the soil [52,53].

Dehydrogenases (DHA) are enzymes belonging to the group of oxidoreductases. They are responsible for catalyzing the oxidation of organic compounds. Active dehydrogenases are present only inside living cells and they indicate the presence of physiologically active microorganisms. Dehydrogenases are commonly found in the pedosphere, where they are involved in the decomposition of organic compounds. Measurement of the dehydrogenase activity in soil shows the intensity of respiratory metabolism of soil microorganisms, mainly actinobacteria and bacteria.

Our research showed that only some foliar fertilizers (ADOB 2.0 Mo and Metalosate Potassium) stimulated the dehydrogenase activity in the white lupine plantation, however, the results were not significant.

Dehydrogenase exhibited high activity at the beginning of the plants' flowering phase (BBCH 51–59). It may have been caused by an increased secretion from the root system during that period [54,55]. In consequence, the count of microorganisms increased [56].

Also macro- and microelements applied in the form of foliar fertilizers and biostimulators could affect dehydrogenase activity. Bielińska et al. [57] observed that fertilizing preparations with nitrogen, phosphorus, and potassium increased the content of these enzymes in the soil. There was a similar effect observed in our study after the application of the Metalosate potassium foliar fertilizer. There were analogous results of experiments on similar bio-conditioners conducted by [58] and [53]. According to Bilen et al. [59], boron improves the dehydrogenase activity. Taran et al. [60] showed that molybdenum stimulated the production of these enzymes by the root nodules of legumes. They also observed that the content of titanium might be positively correlated with the soil biochemistry.

The results of the experiment showed that both the bio-stimulants (Tytanit and Rooter) and foliar fertilizers positively affected the acid phosphatase activity, which was lower than in the control treatment. The Metalosate potassium foliar fertilizer did not cause this effect. This shows that the preparations used in our experiment positively influenced the plants' ability to absorb phosphorus. It is necessary to remember that phosphorus-deficient plants are characterized by increased secretion of acid phosphatase through the root system into the soil. Ciereszko et al. [61] found that the deficit of this macroelement stimulated plants' secretion of acid phosphatases. Lemanowicz et al. [62] and Niewiadomska et al. [56] also suggest these relationships in their studies on the effect of the PRP SOL fertilizer containing phosphorus, potassium, zinc, boron, and molybdenum on the lupine plantation. They observed a decrease in the catalytic activity of this enzyme because of the activation of the compounds that were inaccessible to plants. Bielińska and Mocek-Płóciniak [63] made similar observations. Wang et al. [64] also found that these enzymes exhibited higher activity in the experimental treatment without phosphorus fertilization.

The alkaline phosphatase activity increased significantly only after the application of the ADOB 2.0 Mo and Bolero Mo foliar fertilizers. This effect may have been caused by the increased activity of soil microorganisms, which were stimulated by organic phosphorus compounds secreted into the soil by white lupine plants. Waldrip et al. [65] proved that the content of organic forms of phosphorus was correlated with the activity of alkaline phosphatases in the soil.

All the preparations used in the experiment significantly stimulated catalase activity. As early as 1963, Koter [66] found that the catalase activity increased when plants were fertilized with boron. Hu and Zhu [67] observed that the catalase and dehydrogenase activity increased when plants were fertilized with silicon. Such elements as copper and zinc are essential constituents of physiological processes in all living organisms, including microorganisms. Some soils suffer from zinc deficits, which is why they are enriched with fertilizers containing this element to satisfy the nutritional requirements of crops and improved soil activity [68].

The results of the enzymatic analyses of the dehydrogenase and catalase activities enabled the calculation of the biological index of soil fertility (BIF). The treatments with the Optysil and ADOBE 2.0 Mo preparations had influence on the BIF values, as compared with the control sample. The use of the Optysil preparation resulted in particularly high values in the soil samples collected at the beginning of the flowering phase. The BIF value resulted from the significant influence of these fertilizers on the activity of catalase and dehydrogenase. Siwik-Ziomek and Szczepanek [69] indicated that mineral fertilization, which increases the yield of crops, indirectly affects root secretion, and thus increases the biochemical activity of soil at specific phases of plants' development.

5. Conclusions

When non-root fertilization is applied to plants, they take up all necessary elements chiefly through their leaves as well as the stalk and the whole aerial system. A strong stimulating effect on the yield of white lupine plants in comparison with the control object was obtained after the application of silicon (Optysil) or chelated zinc (ADOB Zn IDHA). The use of zinc in foliar fertilizers (ADOB Zn IDHA) in comparison with control treatment stimulated most of the tested features/parameters: TSW, number of pods per 1 m^2, root nodules dry matter, photochemical efficiency of PSII (F_v/F_m), and chlorophyll content (CCI). However, it is noteworthy that this way of "feeding" cannot substitute soil fertilization. It can only be used to quickly supply necessary nutrients to plants at difficult phases so as not to slow down their growth. The bio-stimulants and foliar fertilizers used in our study improved some of the biochemical parameters of soil activity and the nitrogen fixation process in the white lupine plantation. This effect may have been caused by the higher rate of penetration and better uptake of nutrients applied to the plants' leaves. Although macro- and micronutrients differ in their penetration rates, this process can be accelerated up to about a dozen times by non-root fertilization. The downside of foliar fertilization is the fact that only a limited amount of fertilizer can be supplied to plants in this way. Therefore, this method is particularly effective when plants need to be provided with the elements they need in smaller amounts, e.g., iron, boron, and molybdenum. Not only is foliar fertilization a more efficient method of supplying micronutrients, but it is also safer for the environment and the plants. The search for methods that improve the yield and biochemical parameters of the soil environment is in agreement with the sustainable agriculture policy.

Author Contributions: A.N., H.S., A.W.-M., K.R., and Z.W. conceived and designed the experiments; A.N., H.S., A.W.-M., K.R., and Z.W. performed the field experiments and analyzed the data; A.B. statistical analysis; A.N., Z.W., H.S. wrote the paper; A.N., Z.W. revised the manuscript. All authors have read and agreed to the published version of the manuscript.

Acknowledgments: This research was financed by the Ministry of Agriculture and Rural Development of the Republic of Poland in the framework of the Multi-annual Program "Increasing the use of domestic feed protein for the production of high quality animal products in conditions of balanced development" implemented at the Department of Agronomy of the University of Life Sciences in Poznan. The publication was co-financed within the framework of Ministry of Science and Higher Education programme as, Regional Initiative Excellence" in years 2019-2022, Project No. 005/RID/2018/19.

References

1. Directive 2009/128/EC of the European Parliament and of the Council of 21 October 2009 Establishing a Framework for Community Action to Achieve the Sustainable Use of Pesticides:1-16. Available online: https://ec.europa.eu/food/plant/pesticides/sustainable_use_pesticides_en (accessed on 9 October 2019).
2. Małecka-Jankowiak, I.; Blecharczyk, A.; Sawinska, Z.; Waniorek, W. Effect of lupins and pea on the subsequent winter wheat yield depending of tillage and nitrogen fertilization. *Fragm. Agron.* **2018**, *35*, 67–79.
3. Prusinski, J. White lupin (*Lupinus albus* L.)—Nutritional and health values in human nutrition—A review. *Czech J. Food Sci.* **2017**, *35*, 95–105.
4. GUS. *Agriculture in 2015 Year*; GUS: Warsaw, Poland, 2016; p. 177.
5. Dobosz, R.; Krawczyk, R. Comparison of the development ability of the northern root-knot nematode *Meloidogyne hapla* Chitwood, 1949 on selected species and varieties of Fabaceae Lindl. *Prog. Plant Prot.* **2017**, *57*, 159–163.
6. Calvo, P.; Nelson, L.; Kloepper, J.W. Agricultural uses of plant biostimulants. *Plant Soil* **2014**, *383*, 3–41. [CrossRef]
7. Fageria, N.K.; Filho, M.B.; Moreira, A.; Guimarães, C.M. Foliar fertilization of crop plants. *J. Plant Nutr.* **2009**, *32*, 1044–1064. [CrossRef]
8. Kocira, S.; Szparaga, A.; Kocira, A.; Czerwińska, E.; Wójtowicz, A.; Bronowicka-Mielniczuk, U.; Koszel, M.; Findura, P. Modeling biometric traits, yield and nutritional and antioxidant properties of seeds of three soybean cultivars through the application of biostimulant containing seaweed and amino acids. *Front. Plant. Sci.* **2018**, *9*, 388. [CrossRef]
9. Przybysz, A.; Gawronska, H.; Gajc-Wolska, J. Biological mode of action of a nitrophenolates-based biostimulant: Case study. *Front. Plant Sci.* **2014**, *5*, 713. [CrossRef]
10. Du Jardin, P. Plant biostimulants: Definition, concept, main categories and regulation. *Sci. Hortic.* **2015**, *196*, 3–14. [CrossRef]
11. Artyszak, A. Effect of silicon fertilization on crop yield quantity and quality—A literature review in Europe. *Plants* **2018**, *7*, 54. [CrossRef]
12. Raj, A.B.; Raj, S.K. Zinc and boron nutrition in pulses: A review. *J. App. Nat. Sci.* **2019**, *11*, 673–679. [CrossRef]
13. Fageria, N.K. *The Use of Nutrients in Crop Plants*; CRC Press: Boca Raton, FL, USA, 2009.
14. Sulewska, H.; Ratajczak, K. Chemical composition of selected preparations supporting plant development and evaluation of their activity in soybean cultivation. *Chem. Ind.* **2017**, *96*, 1352–1355.
15. Karaca, A.; Cetin, S.C.; Turgay, O.C.; Kizilkaya, R. Soil enzymes as indication of soil quality. In *Soil Enzymology*; Springer: Berlin/Heidelberg, Germany, 2010; pp. 119–148.
16. Nannipieri, P.; Kandeler, E.; Ruggiero, P. Enzyme activities and microbiological and biochemical processes in soil. In *Enzymes in the Environment. Activity, Ecology and Applications*; Marcel Dekker Inc.: New York, NY, USA, 2002; pp. 1–33.
17. Saviozzi, A.; Levi-Minzi, R.; Cardelli, R.; Riffaldi, R. A comparison of soil quality in adjacent cultivated, forest and native grassland soils. *Plant Soil* **2001**, *233*, 251–259. [CrossRef]

18. IUSS Working Group WRB. *World Reference Base for Soil Resources 2006, First Update 2007*; World Soil Resources Reports No.103; FAO: Rome, Italy, 2007.
19. Van Reeuwijk, L.P. *Procedures for Soil Analysis*, 6th ed.; ISRIC: Wageningen, The Netherlands, 2002.
20. Krawczyk, R.; Mrówczyński, M. (Eds.) *Metodyka Integrowanej Ochrony Łubinu Wąskolistnego, Żółtego i Białego dla Producentów*; Instytut Ochrony Roślin-Państwowy Instytut Badawczy: Poznan, Poland, 2012.
21. Walter, H. *Vegetation Zones and Climate*; PWRiL: Warsaw, Poland, 1976; pp. 26–31.
22. Sawicka, A. *The Ecological Aspects of Dinitrogen Fixation*; Dissertations Scientific: 134; University of Agriculture in Poznan: Poznan, Poland, 1983.
23. Kalaji, M.H.; Guo, P. Chlorophyll fluorescence: A useful tool in barley plant breeding programs. In emphPhotochemistry Research Progress; Sánchez, A., Gutierrez, S.J., Eds.; Nova Science Publishers: Hauppauge, NY, USA, 2008; pp. 447–471.
24. Thalmann, A. Zur Methodik der Bestimmung der Dehydrogenase aktivität im Boden mittels triphenytetrazoliumchlorid (TTC). *Landwirtsch Forsch.* **1968**, *21*, 249–258.
25. Tabatabai, M.A.; Bremner, J.M. Use of p-nitrophenyl phosphate for assay of soil phosphatase activity. *Soil Biol. Biochem.* **1969**, *1*, 301–307. [CrossRef]
26. Johnson, J.L.; Temple, K.L. Some variables affecting the measurement of "catalase activity" in soil 1. *Soil Sci. Soc. Am.* **1964**, *28*, 207–209. [CrossRef]
27. Stefanic, F.; Ellade, G.; Chirnageanu, J. Researches concerning a biological index of soil fertility. In *Proceeding of the 5th Symposium of Soil Biology*; Nemes, M.P., Kiss, S., Papacostea, P., Stefanic, C., Rusan, M., Eds.; Romanian National Society of Soil Science: Bucharest, Romania, 1984; pp. 35–45.
28. Abdi, H.; Lynne, J.W. Principal component analysis. *Wiley Interdiscip. Rev. Comput. Stat.* **2010**, *2*, 433–459. [CrossRef]
29. Murtagh, F.; Legendre, P. Ward's hierarchical clustering method: Clustering criterion and agglomerative algorithm. *arXiv* **2011**, arXiv:1111.6285.
30. Sacała, E. Role of silicon in plant resistance to water stress. *J. Elementol.* **2009**, *14*, 619–630. [CrossRef]
31. Artyszak, A.; Gozdowski, D.; Kuchcińska, K. The effect of silicon foliar fertilization in sugar beet—*Beta vulgaris* (L.) ssp. vulgaris conv. crassa (Alef.) prov. altissima (Döll). *Turk. J. Field Crop.* **2015**, *20*, 115–119. [CrossRef]
32. Fageria, N.K.; Baligar, V.C. Growth components and zinc recovery efficiency of upland rice genotypes. *Pesqui. Agropecu. Bras.* **2005**, *40*, 1211–1215. [CrossRef]
33. Duffy, B. Zinc and plant disease. In *Mineral Nutrition and Plant Disease*; Datnoff, L.E., Elmer, W.H., Huber, D.M., Eds.; APS Press: St. Paul, MN, USA, 2007; pp. 155–175.
34. Kaya, M.U.; Atak, M.E.; Khawar, K.M.; Ciftci, C.Y.; Ozcan, S. Effect of pre-sowing seed treatment with zinc and foliar spray of humic acids on yield of common bean (*Phaseolus vulgaris* L.). *Int. J. Agric. Biol.* **2005**, *7*, 875–878.
35. Omer, F.A.; Abbas, D.N.; Khalaf, A.S. Effect of molybdenum and potassium application on nodulation, growth and yield of lentil (*Lens culinaris* MEDIC). *Pak. J. Bot.* **2016**, *48*, 2255–2259.
36. Rahman, M.M.; Bhuiyan, M.M.H.; Sutradhar, G.N.C.; Rahman, M.M.; Paul, A.K. Effect of phosphorus, molybdenum and *Rhizobium* inoculation on yield and yield attributes of mung bean. *Int. J. Sustain. Crop Prod.* **2008**, *3*, 26–33.
37. Hassanein, M.S.; Shalaby, M.A.F.; Rashad, E.M. Improving growth and yield of some faba bean cultivars by using some plant growth promoters in newly cultivated land. *Ann. Agric. Sci.* **2000**, *38*, 2141–2155.
38. Tobbal, Y.F.M. Physiological Studies on the Effect of Some Nutrients and Growth Regulators on Plant Growth and Metabolism. Ph.D. Thesis, Faculty of Science Al-Azhar University, Cairo, Egypt, 2006.
39. Artyszak, A.; Gozdowski, D.; Kucińska, K. Effect of foliar fertilization with silicon on the chosen physiological features and yield of sugar beet. *Fragm. Agron.* **2016**, *33*, 7–14.
40. Guntzer, F.; Keller, C.; Meunier, J.D. Benefits of plant silicon for crops: A review. *Agron. Sustain. Dev.* **2012**, *32*, 201–213. [CrossRef]
41. Marschner, H. *Mineral Nutrition of Higher Plants*, 2nd ed.; Academic Press: London, UK, 2002.
42. Weisany, W.; Raei, Y.; Allahverdipoor, K.H. Role of some of mineral nutrients in biological nitrogen fixation. *Bull. Environ. Pharm. Life Sci.* **2013**, *2*, 77–84.
43. Vieira, R.F.; Cardoso, E.J.B.N.; Vieira, C.; Cassini, S.T.A. Foliar application of molybdenum in common beans. I. Nitrogenase and reductase activities in a soil of high fertility. *J. Plant Nutr.* **1998**, *21*, 169–180. [CrossRef]

44. Brown, P.H.; Bellaloui, N.; Wimmer, M.A.; Bassi, E.S.; Ruiz, J.; Hu, H.; Pfeffer, H.; Dannel, F.; Römheld, V. Boron in plant biology. *Plant Biol.* **2002**, *4*, 205–223. [CrossRef]
45. Wani, P.A.; Khan, S.; Zaidi, A. Impact of zinc-tolerant plant growth-promoting rhizobacteria on lentil grown in zinc-amended soil. *Agron. Sustain. Dev.* **2008**, *28*, 449–455. [CrossRef]
46. Bucher, M. Functional biology of plant phosphate uptake at root and mycorrhiza interfaces. *New Phytol.* **2007**, *173*, 11–26. [CrossRef] [PubMed]
47. Bonilla, I.; Bolaños, L. Mineral nutrition for legume-rhizobia symbiosis: B, Ca, N, P, S, K, Fe, Mo, Co, and Ni. A review. *Sustain. Agric. Rev.* **2009**, *1*, 253–274.
48. Abusuwar, A.O.; Omer, E.A. Effect of intercropping, phosphorus fertilization and rhizobium inoculation on the growth and nodulation of some leguminous and cereal forages. *Agric. Biol. J. N. Am.* **2011**, *2*, 109–124. [CrossRef]
49. Zahran, H.H. Rhizobium-legume symbiosis and nitrogen fixation under severe conditions and in arid climate. *Microbiol. Mol. Biol. Rev.* **1999**, *63*, 968–989. [CrossRef] [PubMed]
50. Karaca, A.; Cetin, S.C.; Turgay, O.C.; Kizilkaya, R. Effects of heavy metals on soil enzyme activities. In *Soil Heavy Metals*; Springer: Berlin/Heidelberg, Germany, 2010; pp. 237–262.
51. Das, S.K.; Varma, A. Role of enzymes in maintaining soil health. In *Soil Enzymology, Soil Biology*; Springer-Verlag: Berlin/Heidelberg, Germany, 2011; pp. 25–42.
52. Bielińska, E.; Mocek-Płóciniak, A. Impact of the tillage system on the soil enzymatic activity. *Arch Environ. Prot.* **2012**, *38*, 75–82. [CrossRef]
53. Panasiewicz, K.; Niewiadomska, A.; Sulewska, H.; Wolna-Maruwka, A.; Borowiak, K.; Budka, A.; Ratajczak, K. The effect of sewage sludge and BAF inoculant on plant condition and yield as well as biochemical and microbial activity of soil in willow (*Salix viminalis* L.) culture as an energy crop. *Peer J.* **2019**, *7*, e6434. [CrossRef]
54. De-la-Peña, C.; Badri, D.V.; Lei, Z.; Watson, B.S.; Brăndao, M.M.; Silva-Filho, M.C.; Sumner, L.W.; Vivanco, J.M. Root secretion of defense-related proteins is development-dependent and correlated with flowering time. *J. Biol. Chem.* **2010**, *285*, 30654–30665. [CrossRef]
55. Badri, D.V.; De-la-Peña, C.; Lei, Z.; Manter, D.K.; Chaparro, J.M.; Guimarães, R.L.; Sumner, L.W.; Vivanco, J.M. Root secreted metabolites and proteins are involved in the early events of plant-plant recognition prior to competition. *PLoS ONE* **2012**, *7*, e46640. [CrossRef]
56. Niewiadomska, A.; Sulewska, H.; Wolna-Maruwka, A.; Ratajczak, K.; Głuchowska, K.; Waraczewska, Z.; Budka, A. An assessment of the influence of co-inoculation with endophytic bacteria and rhizobia, and the influence of PRP SOL and PRP EBV fertilizers on the microbial parameters of soil and nitrogenase activity in yellow lupine (*Lupinus luteus* L.) cultivation. *Polish J. Environ. Stud.* **2018**, *27*, 2687–2702. [CrossRef]
57. Bielińska, E.J.; Futa, B.; Bik-Mołodzińska, M.; Szewczuk, C.; Sugier, D. The impact of fertilizing agents on the enzymatic activity of soils. *J. Res. Appl. Agric. Eng.* **2013**, *58*, 15–19.
58. Swędrzyńska, D.; Zielewicz, W.; Przybył, P.; Starzyk, J. Effect of bio soil conditioner on microbial state and enzymatic activity of soil under *Lolium perenne* cultivation. *Grassl. Sci.* **2013**, *16*, 111–128.
59. Bilen, S.; Bilen, M.; Bardhan, S. The effects of boron management on soil microbial population and enzyme activities. *Afri. J. Biotech.* **2011**, *10*, 5311–5319.
60. Taran, N.Y.; Gonchar, O.M.; Lopatko, K.G.; Batsmanova, L.M.; Patyka, M.V.; Volkogon, M.V. The effect of colloidal solution of molybdenum nanoparticles on the microbial composition in rhizosphere of *Cicer arietinum* L. *Nanoscale Res. Lett.* **2014**, *9*, 289–297. [CrossRef] [PubMed]
61. Ciereszko, I.; Szczygła, A.; Żebrowska, E. Phosphate deficiency affects acid phosphatase activity and growth of two wheat varieties. *J. Plant Nutr.* **2011**, *34*, 815–829. [CrossRef]
62. Lemanowicz, J.; Siwik-Ziomek, A.; Koper, J. Effect of spring barley nitrogen fertilization on the changes in the content of phosphorus and the activity of alkaline and acid phosphatase in soil. *Ecol. Chem. Eng. A* **2012**, *19*, 1497–1507.
63. Bielińska, E.; Mocek-Płóciniak, A. *Phosphatases in the Soil Environment*; UP Pub.: Poznan, Poland, 2009; ISBN 978-83-7160-554-3.
64. Wang, J.B.; Chen, Z.H.; Chen, L.J.; Zhu, A.N.; Wu, Z.J. Surface soil phosphorus and phosphatase activities affected by tillage and crop residue input amounts. *Plant Soil Environ.* **2011**, *57*, 251–257. [CrossRef]
65. Waldrip, H.M.; He, Z.; Erich, M.S. Effects of poultry manure amendment on phosphorus uptake by ryegrass, soil phosphorus fractions and phosphatase activity. *Biol. Fert. Soils* **2011**, *47*, 407–418. [CrossRef]

66. Koter, M. Research on the impact of boron on the growth and development of plants. *Soil Sci. Ann.* **1963**, *13*, 185–211.
67. Hu, X.J.; Zhu, Z.J. Effect of silicon on resistance of powdery mildew and the activities of antioxidative enzymes in leaf apoplast of cucumber. *Acta Agric. Zhejiang* **2008**, *20*, 67–71.
68. Wyszkowska, J.; Borowik, A.; Kucharski, M.A.; Kucharski, J. Effect of cadmium, copper and zinc on plants, soil microorganisms and soil enzymes. *J. Elementol.* **2013**, *18*, 2587–2602. [CrossRef]
69. Siwik-Ziomek, A.; Szczepanek, M. Soil extracellular enzyme activities and uptake of N by oilseed rape depending on fertilization and seaweed biostimulant application. *Agron. J.* **2019**, *9*, 480. [CrossRef]

11

Effect of *Pterocladia capillacea* Seaweed Extracts on Growth Parameters and Biochemical Constituents of Jew's Mallow

Mohamed Ashour [1], Ahmed A. El-Shafei [2,3,*], Hanan M. Khairy [4], Doaa Y. Abd-Elkader [5], Mohamed A. Mattar [2,6,*], Abed Alataway [3,*] and Shimaa M. Hassan [5]

1 Invertebrate Aquaculture Lab., Aquaculture Division, National institute of Oceanography and Fisheries (NIOF), Alexandria Branch, Alexandria 21576, Egypt; microalgae_egypt@yahoo.com
2 Department of Agricultural Engineering, College of Food and Agriculture Sciences, King Saud University, Riyadh 11451, Saudi Arabia
3 Prince Sultan Bin Abdulaziz International Prize for Water Chair, Prince Sultan Institute for Environmental, Water and Desert Research, King Saud University, Riyadh 11451, Saudi Arabia
4 Hydrobiology Lab., Marine Environmental Division, National Institute of Oceanography and Fisheries, Alexandria 21545, Egypt; hanan_khairy@yahoo.com
5 Department of Vegetable, Faculty of Agriculture (El-Shatby), Alexandria University, Alexandria 21545, Egypt; doaa_004@yahoo.com (D.Y.A.-E.); shaymaa.hassan@alexu.edu.eg (S.M.H.)
6 Agricultural Engineering Research Institute (AEnRI), Agricultural Research Centre, Giza 12618, Egypt
* Correspondence: aelshafei1bn.c@ksu.edu.sa (A.A.E.-S.); mmattar@ksu.edu.sa or mohamedmattar@gmail.com (M.A.M.); aalataway@ksu.edu.sa (A.A.)

Abstract: We performed field experiments to evaluate the influence of two extraction treatments, seaweed (*Pterocladia capillacea* S.G. Gmelin) water extraction (WE) and ultrasound-assisted water extraction (USWE) at three concentrations (5%, 10%, and 15%), as well as control NPK traditional mineral fertilizer on the growth, yield, minerals, and antioxidants of Jew's Mallow (*Corchorus olitorius* L.) during the two seasons of 2016 and 2017 in Egypt. Plant height, number of leaves, and fresh weight of WE10 treatment were the highest ($p < 0.05$) as 59.67 cm, 10.67 and 2.41 kg m^{-2} in 2016, respectively, and 57.33 cm, 11.00 and 2.32 kg m^{-2} in 2017, respectively. WE10 and USWE5 treatments produced the highest dry matter (17.07%) in 2016 and (16.97%) in 2017, respectively. WE10 plants had an increased water productivity of 41.2% relative to control plants in both seasons. The highest chlorophyll 'a' was recorded after the WE10 treatment in 2016 and 2017 (17.79 μg g^{-1} and 17.84 μg g^{-1}, respectively). The highest levels of total antioxidant capacity, total phenolics, and total flavonoids were also recorded after the WE10 treatment. Application of WE10 boosted growth, yield, minerals, and antioxidants of Jew's Mallow. The CROPWAT model was used to estimate the evapotranspiration, irrigation water requirements, and yield response to irrigation scheduling. Our data showed a yield reduction in the initial growth stage if a limited amount of water was provided. Therefore, irrigation water should be provided during the most important stages of crop development with the choice of effective irrigation practices to avoid water losses, as this helps to maximize yield.

Keywords: seaweed extract; ultrasound-assisted water; foliar spray; *Pterocladia capillacea*; bio-fertilizer; growth parameters; antioxidants; Jew's Mallow; CROPWAT model

1. Introduction

Vegetables and their products contain non-enzymatic antioxidants and micro-nutrients that stabilize free radicals and in turn, increase the capacity of the plant to fight against pathogens that may affect humans and animals [1–4]. For example, the antioxidant compounds in vegetables and fruits could help prevent oxidative stress, diabetes, neurodegenerative disorders, cardiovascular disease, and cancer [5,6]. Jute (*Corchorus olitorius* L.), or Jew's mallow, belongs to the *Tiliaceae* family. *C. olitorius* thought to have originated from South China, from where it was introduced to India and Pakistan. However, a wild variety has been discovered in many areas in India, China, Australia, and Africa, particularly in Southeastern Nigeria. Jute leafy vegetable is commonly used in the preparation of soup [7]. The young shoot tips can be consumed raw or cooked and contain elevated concentrations of protein and vitamin C [8,9]. Jute is generally suggested for pregnant and nursing women because it is thought to be rich in iron [10].

Chemical fertilizers have been used in large quantities to compensate the nutrients deficiency in the soil. It has been observed that this use affects soil, plants, and human health. Their potential carcinogenicity and toxicity have been demonstrated, particularly after the reduction of nitrate to nitrite, or just reacting with amines and/or amides in the formation of N-nitroso compounds, N-nitrosamines, and other nitrogen compounds with high levels of nitrate [11]. Screening of native algal species must be considered to achieve a successful commercial and biotechnological potential of native algal species [12]. Seaweeds are the most promising plants from marine ecosystems and are used as a source of food and medicine. The coast of Egypt has a wide range of wild seaweed available throughout the year, even the Mediterranean coast [13] or the Red Sea coast [14]. Along the Egyptian Mediterranean coast, especially near Alexandria, red algae (*Pterocladia capillacea*; Rhodophyta) are the most dominant native seaweeds. Khairy and El-Shafay [13] studied the seasonal variations (spring, summer and autumn 2010) of biochemical composition of *P. capillacea* collected from Abu Qir Bay, Mediterranean Coast of Alexandria, Egypt. In 2010 spring season, *P. capillacea* achieved the highest significant protein (23.72%) and lipid (2.71%), while in 2010 summer season, *P. capillacea* achieved the highest significant carbohydrate (50.96%), ash (15.81%), and moisture (10.19%). Total fatty acids (248–515 μg/g), total saturated fatty acids (189–360 μg/g), total mono-unsaturated fatty acids (29–77 μg/g), total poly-unsaturated fatty acids (30–78 μg/g), total amino acids (2836–3924 μg/g), total essential amino acids (1136–1445 μg/g), and total non-essential amino acids (1700–2445 μg/g) of seasonally collected *P. capillacea* species were observed. Moreover, Khairy and El-Sheikh [15] observed the mineral composition and antioxidant activities of *P. capillacea* species collected seasonally (spring, Summer and autumn 2010) from Abu-Qir Bay, Mediterranean coast of Alexandria, and they concluded that this species is a rich in carotenoids, phenolic compounds, DPPH free radicals and minerals, therefore, this species can be used as potential source of health food in human diets and may be of use to food industry. In general, marine algae are a rich source of protein, lipids, carbohydrates, polysaccharides, minerals, antioxidants, and other bioactive compounds that can serve in multiple biological activities related to different industries [15,16].

Seaweed extracts can be used as fertilizer for flowering plants, vegetables, and grain crops [17–19]. Furthermore, they have been marketed as fertilizer additives, which are better than other fertilizers [20,21]. Using such extracts (bio-fertilizers) in cultivation many protect the soil and improve crop quality. Therefore, applying them to seeds or adding them to the soil stimulates plant growth [22]. Liquid extracts obtained from seaweeds have gained popularity as foliar sprays for many crops. These extracts contain cytokines, growth promoting hormones, elements, vitamins, and amino acids [23,24]. Some unknown bioactive component in seaweed acts to illicit the plant's own production of plant hormones through internal metabolic pathways [25].

Booth [20] reported that the efficacy of seaweeds as extracts was due to the presence of several metabolites and trace elements. The green seaweed *Enteromorpha* has a high potential for commercial exploitation because of its abundant and varied chemical composition, quality, and concentration of basic nutrients [26]. *Enteromorpha* sp. contains 28 times more calcium than spinach, 26 times more than nopal, and 13 times more than quelite [27]. *Ulva lactuca* and *Enteromorpha intestinalis* are used as seaweed liquid extract for many crops [21,28]. Rama Rao [29] reported good yields of *Zizyphus rugosa* fruits, when leaves sprayed with seaweed liquid extracts obtained from *Sargassum*. Seaweed extracts are now available commercially as Maxicrop (Sea-Born), Algifert (Marinure), Goemar GA14, Kelpak 66, Seaspray, Seasol, Cytex, and Seacrop. It has been reported that seaweed extracts are better than other extracts [21,23].

Traditional methods employed for extracting bioactive compounds are time consuming and have low extraction efficiencies. To overcome these disadvantages, novel technologies for extraction of bioactive compounds from marine algae have been investigated including the use of microwaves [30], enzymes [31], and super-critical fluids [32]. Recently, ultrasonic technologies have been used to enhance the extraction efficiencies of bioactive compounds (total phenolics, fucose, and uronic acid) from brown seaweed *Ascophyllum nodosum* [33–35] and starch from microalgae *Chlamydomonas fasciata* Ettl NIES-437 [36]. Moreover, ultrasonic assisted extraction was utilized in various industrial fields including phenolic compounds from citrus peel [37], lycopene from tomatoes [38], and anthocyanins from raspberries [39]. Ultrasound-assisted extraction is a simple and employed to improve extraction of bioactive compounds from seaweed [34–36,40].

Full irrigation should be practiced to maximize water productivity [41]. Weekly skipping of irrigation during seed filling may substantially reduce seed yield and water productivity. Skipping during seed germination may be a viable option when water is scarce and land is not limiting. Economic evaluation will provide guidance to policy makers at basin scales for formulating improved and efficient water management plans under all varying weather conditions. CROPWAT is a software for irrigation planning and management [42,43]. Its main functions are: To calculate reference evapotranspiration (ET_o), crop water requirements, and crop irrigation requirements, which may be used to develop irrigation schedules under multiple management conditions and water supply schemes, to estimate rain-fed production and drought effects, and to evaluate the efficiency of irrigation practices. The CROPWAT model has been validated in previous studies for estimating the dynamics main components of soil water balance [44–47]. We undertook this study to investigate the effect of *P. capillacea* seaweed liquid extracts on the growth, yield, minerals, and antioxidants of Jew's Mallow (*C. olitorius* L.).

2. Materials and Methods

2.1. Seaweed

2.1.1. Sampling

P. capillacea seaweed was collected in spring 2016 from the submerged rocky site near Boughaz El-Maadya, Abu-Qir Bay, Alexandria (31.3000° N and 30.1667° E) in Egypt. Harvested samples were transferred to the Microalgae and Invertebrates Aquaculture Laboratories, National Institute of Oceanography and Fisheries (NIIOF), Alexandria, Egypt. Epiphytes were removed from samples, and the seaweed samples were cleaned, washed, and air-dried in shadow. Dried seaweed samples were powdered and stored at room temperature in plastic bag for further analysis and utilization.

2.1.2. Biochemical Composition

Protein, lipid, carbohydrates, and ash of identified seaweed *P. capillacea*, collected in spring 2016, were determined. Total proteins were extracted according to Rauch [48] and determined according to Hartree [49]. Total carbohydrates were extracted according to Myklestad and Haug [50] and determined according to Dubois et al. [51]. Total lipid was calculated according to Bligh and Dyer [52]. Fatty acids and amino acids were extracted and estimated as described by El-Shenody et al. [14].

2.1.3. Seaweed Liquid Extracts Preparation

In this study, seaweed crude liquid extracts of *P. capillacea* were prepared using two extraction methods: three treatments using water extract (WE) and the other three treatments using

Ultrasound-Assisted Water Extraction (USWE), as shown in Figure 1. For WE, 100 g seaweed powder was soaked in 1 L distilled water in a 60 °C water bath for 60 min (extraction phase I). The residual filtrate was filtered and soaked in 1 L distilled water (1:10, w/v) in a 60 °C water bath for 60 min (extraction phase II) and this process was repeated a third time (extraction phase III). Each extraction phase was filtered through Whatman No. 3 filter paper and the supernatants of the three phases (I, II, and III) were combined to the final WE volume of 3 L and stored at −20 °C. For USWE, the three extraction phases were prepared as described above for WE, but after each phase the mixture was subjected to ultrasonication. The USWE extraction was performed at 60 °C for 5 min and 99% amplitude of 20 kHz; these conditions were adjusted and stable for all three extraction phases. After three USWE extraction phases, the supernatants were combined to a final volume of 3 L, and stored at −20 °C. The final combined supernatants, both WE and USWE, were considered to be a 100% crude extract that was utilized as seaweed foliar spray.

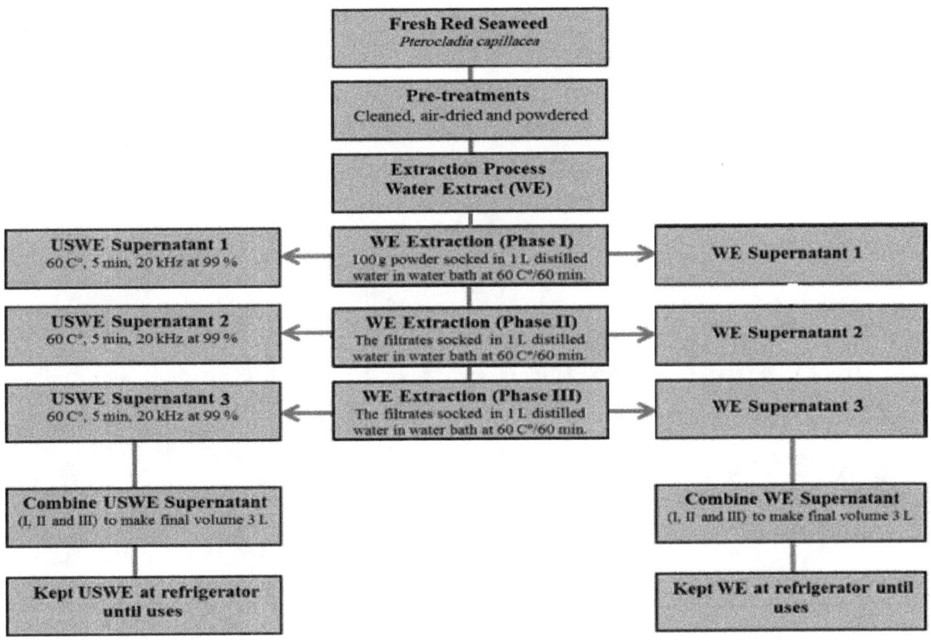

Figure 1. Procedures for water extraction (WE) and ultrasound-assisted water extraction (USWE) of pre-treated *Pterocladia capillacea*.

2.2. Experimental Design

The field experiment with Jew's Mallow (*C. olitorius* cv. Balady) was conducted for two successive growing seasons (2016–2017) at Abeis Experimental Farm, Alexandria University, Alexandria (31.2001° N and 29.9187° E) in Egypt. Before sowing, soil samples were collected (0–30 cm depth) to determine physical and chemical properties following Page [53] (Table 1). Climatic data, such as maximum and minimum air temperature (T_{max} and T_{min}), relative humidity (RH), wind speed (u_2), and rainfall (P), were collected at a meteorological station near the experimental field location (Figure 2) to calculate daily ET_o using the Penman–Monteith FAO-56 equation [54]. Evapotranspiration was estimated during growth using crop coefficient (Kc) values [54] multiplied by ET_o. The experimental area of 220.5 m^2 was divided into three replicate blocks separated by 2-m buffer zones. Each block consisted of seven plots including one traditional fertilizer and six seaweed extract treatments. Each plot covered an area of 10.5 m^2 (3 × 3.5 m). A randomized complete block design (RCBD) was used. Commercial seeds were sown on March 20, 2016 and March 22, 2017 at the rate of 28 kg ha^{-1} [55]. The site was irrigated five times during the first 20 days after sowing to allow germination and establishment before the application of extract treatments. After that, irrigation was carried out every six to seven days for all treatments. The first dose (0.5 m^3 ha^{-1}) of growth fertilizer or seaweed extract was applied 10 days after sowing (DAS), the second one (0.75 m^3 ha^{-1}) was applied 18 DAS, and the third dose (1 m^3 ha^{-1}) was adding 26 DAS. Harvesting included two cuttings at 45 and 70 DAS.

Table 1. Soil physical and chemical properties.

Soil Physical Properties					
Season	**2016**	**2017**	**Season**	**2016**	**2017**
Sand (%)	43.3	42.8	Saturation moisture (m^3 m^{-3})	0.49	0.52
Silt (%)	25.5	23.5	Field capacity (m^3 m^{-3})	0.40	0.41
Clay (%)	31.2	33.7	Wilting point (m^3 m^{-3})	0.17	0.17
Soil texture	Clay loam	Clay loam	Total available moisture (m m^{-1})	0.22	0.24
Bulk density (g cm^{-3})	1.48	1.3	Infiltration rate (mm h^{-1})	3.44	3.20
Soil chemical properties					
pH	8.45	8.88	Total Nitrogen (%)	0.19	0.15
E.C. (dS m^{-1})	3.01	3.0	Phosphorus (ppm)	0. 41	0.44
Soluble cations (meq L^{-1})			**Soluble anions (meq L^{-1})**		
Ca^+	2.08	1.97	CO_3^{--}	0.0	0.0
Mg^{++}	1.98	1.88	HCO_3^-	1.43	1.28
Na^+	2.47	2.39	Cl^-	2.05	1.95
K^+	0.40	0.37	SO_2^{--}	3.46	3.37

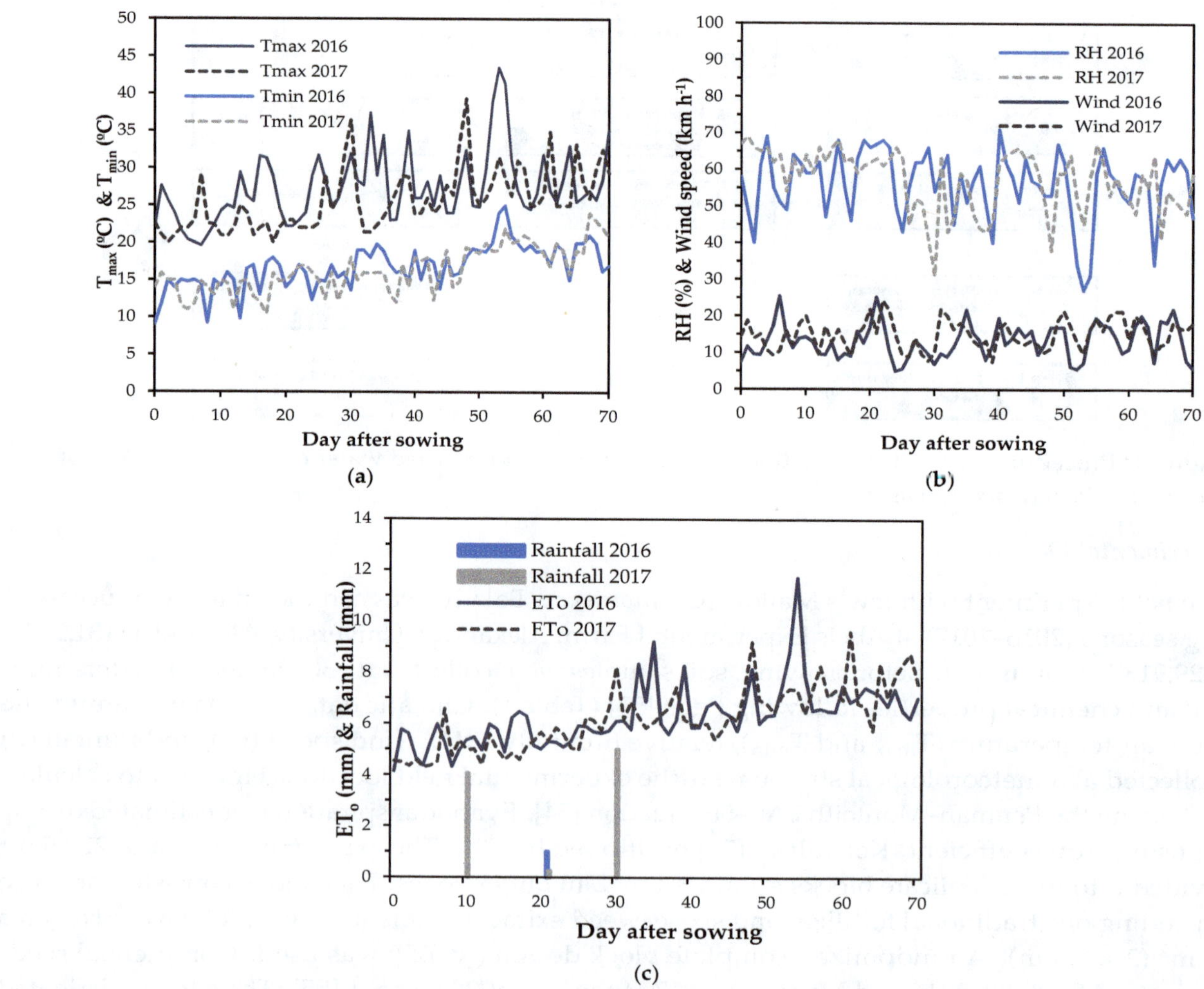

Figure 2. Daily climate parameters in the 2016 and 2017 experimental periods during the Jew's Mallow growing season. (**a**) daily maximum and minimum air temperature (T_{max} and T_{min}, °C), (**b**) relative humidity (RH, %) and wind speed (km h^{-1}), and (**c**) reference evapotranspiration (ET_o, mm) and rainfall (mm).

2.3. Treatments

The following seven treatments were used: mineral NPK fertilizer (control); water extracted seaweed at 5%, 10%, and 15% (WE5, WE10, and WE15); and ultrasound-assisted water extraction seaweed at 5%, 10%, and 15% (USWE5, USWE10, and USWE15).

NPK fertilization was carried out according to the recommendations for commercial production of Jew's Mallow plant. The NPK treatment dose consisted of ammonium nitrate NH_4NO_3 (33%N) at the rate of 300 kg ha^{-1}, calcium superphosphate ($Ca(H_2PO_4)_2.H_2O$ (15% P_2O_5); 525 kg ha^{-1}); and potassium sulphate (K_2SO_4 (48% K_2O); 125 kg ha^{-1}). Nitrogen fertilizer was applied thrice at 7, 15, and 21 DAS. Phosphorus fertilizer was mixed during soil preparation. Potassium fertilizer was applied at 15 DAS.

2.4. Measurements

2.4.1. Agronomic and Physiological

Plants were harvested (cut) twice, at 45 DAS and 70 DAS, from the center of each plot (treatment) per season to determine leaf and stem fresh weight (kg m^{-2}). Five plants were randomly chosen from each plot to measure plant height and; number of leaves. Ratios between dry leaf weight and fresh leaf weight were determined after drying 70 °C in a forced-air oven until reaching a constant weight. Water productivity (WP, kg m^{-3}) was used to evaluate treatments, calculated by dividing total fresh weight (kg m^{-2}) at harvest by the amount of water applied (supplemental irrigation plus rainfall, m^3 m^{-2}) to the crop. Chlorophyll 'a', Chlorophyll 'b', and total carotene (μg g^{-1}) as described by Dere, et al. [56].

2.4.2. Nutrient Contents

Plant nutrient content (N, P, and K) was analyzed and expressed as percentage on leaf dry weight basis. Total N and P contents were determined calorimetrically using a spectrophotometer at 662 and 650 nm, following the methods of Evenhuis [57]. K was quantified by atomic absorption spectrometry as described by Cottenie, et al. [58].

2.4.3. Antioxidant Activities

Antioxidant activities of crude extracts of WE, USWE and Jew's Mallow were observed. Free radical scavenging activity against DPPH (2,2-diphenyl-1-picrylhydrazy) was determined as described by Suresh Kumar, et al. [59]. The total antioxidant content (TAC; mg g^{-1}) was determined with a Phosphomolybdate assay using ascorbic acid as the standard [60]. The total phenolic content (TPC; mg g^{-1}) was determined by using the Folin–Ciocalteu method as modified by Suresh Kumar, et al. [59]. Total flavonoid content (TVC; μg g^{-1}) was determined according to the method of Chang, et al. [61] with Quercetin as the standard.

2.5. CROPWAT Model

Before Jew's Mallow cultivation, the water application depth and irrigation timing intervals were calculated using the CLIMWAT 2.0 and CROPWAT models. CLIMWAT 2.0 is climatic software [62] presenting the monthly agro-climatic data of over 5000 stations worldwide, including the Alexandria-Nouzha agroclimatic station, which was the nearest to the experimental site (4 km). The CROPWAT model was used for calculation of crop water requirements and the development of irrigation schedules using the option to irrigate at critical depletion and refill soil to field capacity. Irrigation times and the amounts were estimated based on the efficiency of the basin irrigation system and applied for both growth seasons (Figure 3). At the end of each season, the CROPWAT model with the options of user defined application depth and irrigation at user defined intervals were used to evaluate the irrigation schedule. The input data for the CROPWAT version 8.0 model [63] required the following data:

- The daily climatic (T_{max}, T_{min}, RH, daylight hours, and u_2) and P data for the seasons of 2016 and 2017 were accessed from the Meteorological Data of Central Laboratory for Agricultural Climate (Figure 1).
- A cropping pattern consisting of the crop type, planting date, growing stage (20, 20, 25, and 8 days for initial, development, mid-season, and late-season stages, respectively), *Kc* (0.7 for initial, 1.15 for mid-, and 0.95 for late-season stage) and critical depletion fraction; *P* (0.3 for initial and development, 0.45 for mid-season stages, and 0.5 for late season stage), rooting depth; Z_r (0.18 m for initial stage and 0.5 m for maximum (mid- and late-season)), and yield response factor; k_y (0.8 for initial, 0.4 for development, 1.2 for mid-, and 1 for late-season). The crop values were assumed as data for a small vegetable according to Allen, et al. [54].
- Soil type: Total available soil moisture, maximum infiltration rate and initial soil moisture depletion were obtained from measured data (Table 1).

The output of CROPWAT model consists of daily root zone depletion ($D_{r,i}$, Equation (1)), deep percolation (DP_i), actual water use by crop $(ET_c)_{actual}$, efficiency of the irrigation schedule (*EIS*, Equation (2)), deficiency of the irrigation schedule (*DIS*, Equation (3)) and yield reduction (Y_R, Equation (4)) were collected and analyzed using the following equation:

$$D_{r,i} = D_{r,i-1} + (ET_{c,i})_{actual} - P_i - I_i + RO_i + DP_i \tag{1}$$

where $D_{r,i}$, and $D_{r,i-1}$ are at days *i* and *i*−1, P_i is total rainfall over day *i*, I_i is net irrigation on day *i*, RO_i is water loss by runoff from the soil surface on day *i*, in our study the *RO* is equal to zero, and DP_i is water loss by deep percolation on day *i*.

$$EIS = \frac{\sum(I_i - DP_i)}{\sum I_i} \times 100 \tag{2}$$

$$DIS = \frac{Sesaonal\ (ET_c)_{\text{potential}} - Sesaonal\ (ET_c)_{\text{actual}}}{Sesaonal\ (ET_c)_{\text{potential}}} \times 100 \tag{3}$$

$$Y_R = K_y\left(1 - \frac{(ET_c)_{\text{actual}}}{(ET_c)_{\text{potential}}}\right) \tag{4}$$

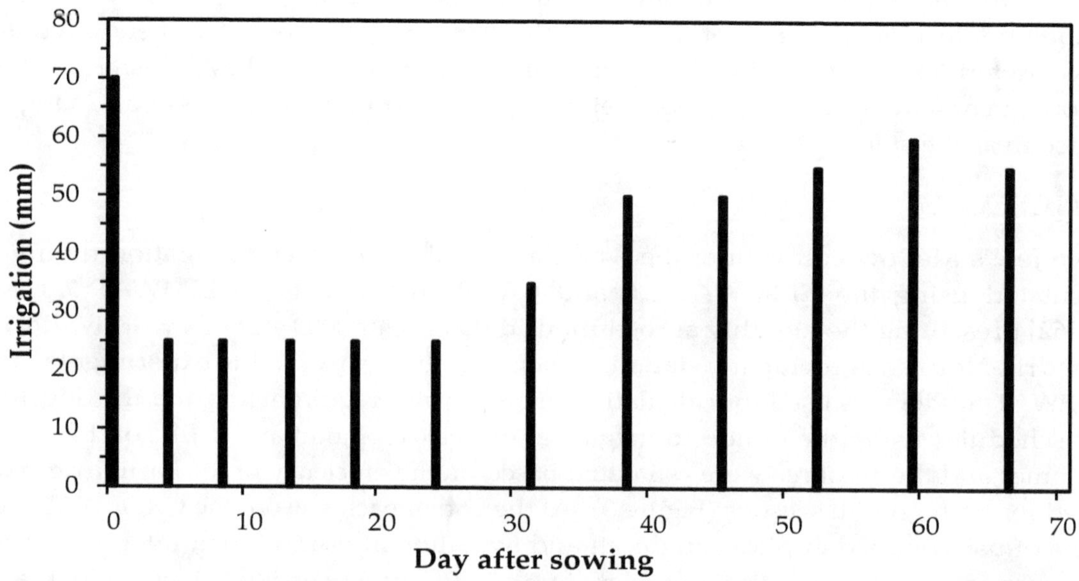

Figure 3. Amounts of irrigation water (mm) applied during the Jew's Mallow growing season.

2.6. Statistical Analyses

Analysis of variance (ANOVA) with RCBD was performed on data obtained from both growing seasons (2016 and 2017) using the IBM SPSS Version 23 software to determine the significance of differences among treatments. Standard errors (SE) were presented for the mean of data from both growing seasons. Differences among means of replicates were measured using Duncan method at $p \leq 0.05$ [64].

3. Results

3.1. Biochemical Composition Seaweed P. Capillacea

Nutritional compositions of the red seaweed species *P. capillacea* collected during spring season of 2016 were investigated. Lipid, protein, carbohydrate, and ash percentages, based on dry weight, were 2.46%, 18.47%, 51.36%, and 13.71%, respectively. Moreover, total fatty acids, total saturated fatty acids, total mono-unsaturated fatty acids, total poly-unsaturated fatty acids, total amino acids, total essential amino acids, and total non-essential amino acids were 247.6, 188.6, 29.1, 29.9, 2836, 1136.3, and 1700 μg g^{-1}, respectively. The antioxidant activities result of WE and USWE crude extracts observed that no significant differences ($p < 0.05$) were found in total antioxidant content (22.48 and 22.33 mg g^{-1}) and total phenolic content (17.79 and 16.85 mg g^{-1}) in WE and USWE, respectively, while USWE achieved a significant difference ($p < 0.05$) in total flavonoid content (45.68 μg g^{-1}) and total carotene (2.03 μg g^{-1}) in comparing to WE (34.77 μg g^{-1} and 1.29 μg $g^{-1,}$ respectively).

3.2. Agronomic Traits

Table 2 shows the significant differences in in plant height ($p < 0.05$). WE10 treatment produced the tallest plants, followed by WE5 in both seasons. In WE10 and WE5 treated plants, height increased by 39.8% and 28.1%, respectively, compared with mineral fertilizer treated plants (control treatment) in 2016 while plant height increased by 28.3% and 23.2% in 2015. The USWE10 treatment had the smallest effect on height in both years, with increases of 4.6% and 0.7% in 2016 and 2017, respectively, relative to control. Statistically significant differences ($p < 0.05$) were found between treatments for leaf number, where WE5 and WE10 treatments had the highest values in both seasons.

Table 2. Effects of water extract (WE) and ultrasound-assisted Water Extraction (USWE) on growth characteristics of Jew's Mallow during 2016 and 2017 growing seasons.

Treatment	Plant Height (cm)		Leaf Number (Plant^{-1})		Fresh Weight (kg m^{-2})		Dry Matter (%)	
	2016	2017	2016	2017	2016	2017	2016	2017
Control	42. 7 ± 0.5 ^{d}	44. 7 ± 0.6 ^{d}	9.3 ± 1.1 ^{ab}	9.0 ± 1.0 ^{b}	1.64 ± 0.1 ^{d}	1.65 ± 0.1 ^{e}	13.9 ± 1.1 ^{c}	14.1 ± 1.6 ^{b}
WE5	54.7 ± 2.0 ^{b}	55.0 ± 2.3 ^{ab}	10.7 ±0.5 ^{a}	11.0 ± 0.2 ^{a}	2.01 ± 0.1 ^{c}	2.07 ± 0.1 ^{bc}	16.4 ± 0.5 ^{ab}	16.3 ± 0.4 ^{a}
WE10	59. 7 ± 1.5 ^{a}	57.3 ± 3.2 ^{a}	10.6 ±1.1 ^{a}	11.0 ± 1.0 ^{a}	2.41 ± 0.1 ^{a}	2.32 ± 0.1 ^{a}	17.1 ± 1.0 ^{a}	16.6 ± 0.4 ^{a}
WE15	52.0 ± 2.0 ^{c}	52.5 ± 2.2 ^{abc}	9.0 ±0.1 ^{b}	9.0 ± 0.1 ^{b}	1.91 ± 0.1 ^{c}	1.84 ± 0.1 ^{de}	16.7 ± 0.9 ^{ab}	16.2 ± 0.4
USWE5	44. 7 ± 0.5 ^{d}	47.7 ± 1.8 ^{cd}	9.3 ±0.6 ^{ab}	9.3 ± 0.5 ^{b}	2.09 ± 0.0 ^{bc}	2.15 ± 0.2 ^{ab}	16.1 ± 1.0 ^{ab}	16.9 ± 1.1 ^{a}
USWE10	44.6 ± 1.5 ^{d}	45.0 ± 2.2 ^{d}	9.0 ±0.2 ^{b}	9.0 ± 0.1 ^{b}	2.22 ± 0.1 ^{ab}	2.16 ± 0.1 ^{ab}	16.4 ± 0.2 ^{ab}	16.6 ± 0.9 ^{a}
USWE15	50.0 ± 1.0 ^{c}	49.9 ± 0.4 ^{bcd}	9.7 ±0.6 ^{ab}	9.6 ± 0.6 ^{b}	1.88 ± 0.2 ^{c}	1.90 ± 0.1 ^{cd}	14.9 ± 1.5 ^{bc}	15.1 ± 1.5 ^{b}

Control: NPK fertilization; WE5, WE10, and WE15: water extracted seaweed at 5%, 10%, and 15%, respectively; USWE5, USWE10, and USWE15: ultrasound-assisted water extraction seaweed at 5%, 10%, and 15%, respectively. Data are means ± SE. Different superscript letters in each column indicate significant differences ($p \leq 0.05$).

There was a significant difference ($p < 0.05$) in fresh weight and dry matter between the treatments in 2016 and 2017. Fresh weight was the highest in WE10 treated plants in both seasons, followed by the USWE10 treatment. For WE10 and USWE10 treatments, the fresh weight increased by 47% and 35.4% in 2016, respectively; then 40.6% and 30.9% in 2017, compared to control treatment. Across all treatments, WE15 and USWE15 treatments reduced the fresh weight in both seasons. There were significant differences ($p < 0.05$) between bio- and mineral fertilizer-treated plants dry matter in 2016 and 2017 (Table 2). The highest dry matter value was recorded in plants that received the WE10 and USWE5 treatments, which was 23% and 19.9%higher, respectively, than control in 2016.

3.3. *Water Productivity*

Water requirement varied from 3.8 to 9.1 and 3.5 to 8.5 mm day^{-1} from the early stage to the peak demand period (mid-season) for 2016 and 2017, respectively. Water productivity (WP) values determined for treatments in 2016 and 2017 are shown in Figure 4. In both seasons, there were significant differences ($p < 0.05$) between WP values. The highest WP values were recorded with the WE10 treatment 41.2% higher than control. Among extract treatments, USWE15 and WE15 had the lowest WP in 2016 and 2017, which was lower by 23% and 20.7% than WE10, respectively.

3.4. *Physiological Traits*

Water extraction treatments had the highest content of chlorophyll 'a' in both seasons (average, 17.49 μg g^{-1}), while the lowest chlorophyll 'a' content was observed with the control treatment (average, 9.4 μg g^{-1}, Table 3). WE10 and WE15 treated plants showed significant increases in chlorophyll 'b' content compared to control treatment in both seasons. The chlorophyll 'b' content for the WE10 and WE15 treated plants (average, 13 μg g^{-1} and 13.3 μg g^{-1}, respectively) was two-fold higher than the content in the control treatment in 2016 and 2017. Conversely, USWE5 and USWE10 application resulted in the lowest chlorophyll 'b' content in both growing seasons, 16.8% and 26.8% lower than control, respectively. The lowest carotene content was achieved by control treatment in 2016 and 2017 (2.9 and 2.8 μg g^{-1}, respectively; Table 3). The highest carotene content in 2016 and 2017 was measured with the USWE10 treatment (71.2% and 72% higher than control, respectively), followed by the USWE15 treatment (53.7% and 54.3% higher than control, respectively).

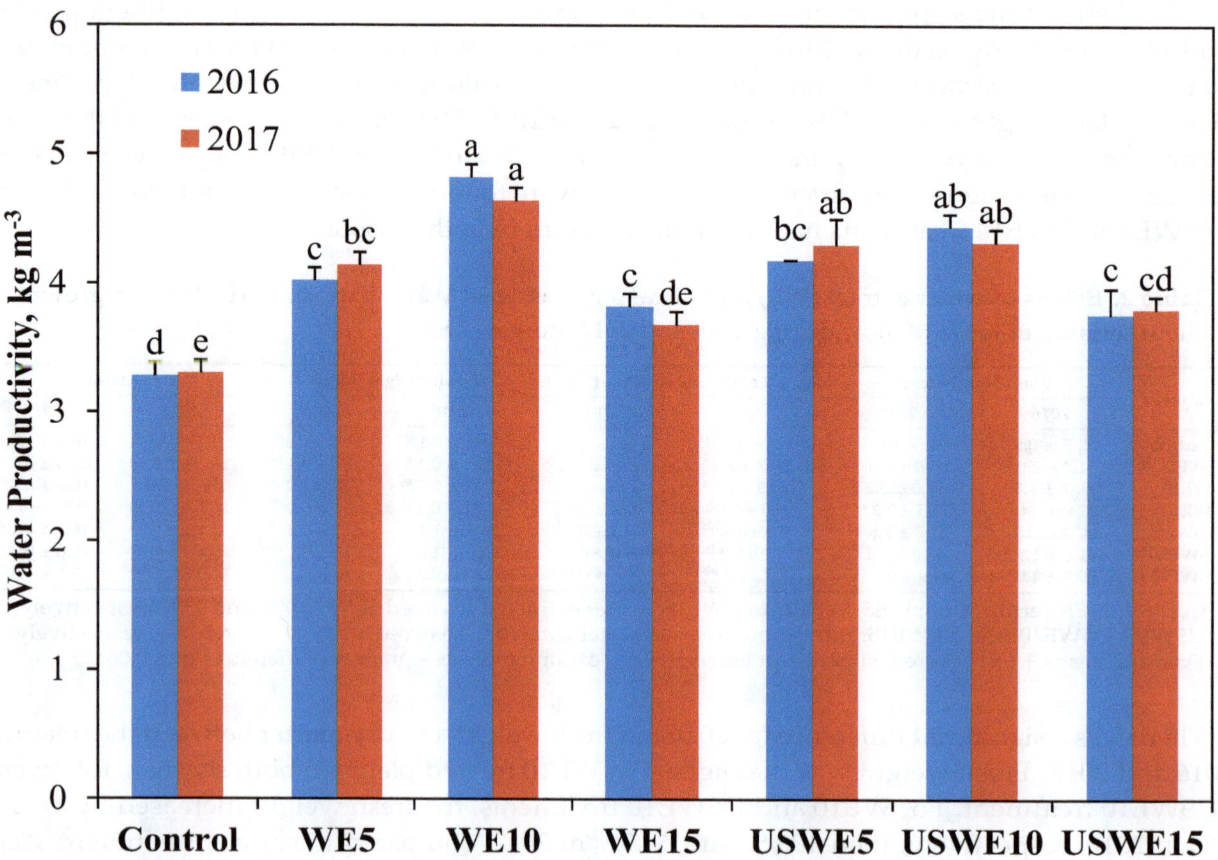

Figure 4. Water productivity (kg m^{-3}) of Jew's Mallow as a function of water extract (WE) and ultrasound-assisted water extract (USWE) treatments in the 2016 and 2017 growing seasons. Different letters (a, b, etc.) above bars indicate a significant difference among treatments in each season.

Table 3. Effects of water extract (WE) and Ultrasound-Assisted Water Extraction (USWE) on chlorophyll 'a', chlorophyll 'b', and carotene concentrations in Jew's Mallow during the 2016 and 2017 growing seasons.

Treatment	Chlorophyll 'a' ($\mu g\ g^{-1}$)		Chlorophyll 'b' ($\mu g\ g^{-1}$)		Carotene ($\mu g\ g^{-1}$)	
	2016	2017	2016	2017	2016	2017
Control	9.4 ± 1.9 ^{d}	9.4 ± 1.9 ^{d}	6.5 ± 0.2 ^{bc}	6.5 ± 0.2 ^{bc}	2.9 ± 0.2 ^{d}	2.8 ± 0.2 ^{d}
WE5	17.0 ± 0.7 ^{a}	17.1 ± 0.7 ^{a}	6.2 ± 0.0 ^{bc}	6.2 ± 0.0 ^{c}	3.1 ± 0.3 ^{cd}	3.1 ± 0.3 ^{cd}
WE10	17.8 ± 0.2 ^{a}	17.8 ± 0.3 ^{a}	13.0 ± 1.2 ^{a}	12.9 ± 1.1^{a}	3.0 ± 0.2 ^{d}	3.1 ± 0.2 ^{cd}
WE15	17.7 ± 1.7 ^{a}	17.5 ± 1.6 ^{a}	13.3 ± 0.8 ^{a}	13.3 ± 0.8 ^{a}	3.1 ± 0.2 ^{cd}	3.0 ± 0.2 ^{cd}
USWE5	12.3 ± 0.1 ^{c}	12.4 ± 0.0 ^{c}	5.4 ± 2.3 ^{c}	5.4 ± 2.3 ^{c}	3.9 ± 0.3 ^{bc}	3.8 ± 0.2 ^{bc}
USWE10	10.3 ± 0.7 ^{cd}	10.3 ± 0.7 ^{cd}	4.8 ± 0.6 ^{c}	4.7 ± 0.7 ^{c}	4.9 ± 0.8 ^{a}	4.9 ± 0.8 ^{a}
USWE15	14. 7 ± 1.7 ^{b}	14.8 ± 1.7 ^{b}	8.5 ± 1.9 ^{b}	8.5 ± 1.8 ^{b}	4.4 ± 0.6 ^{ab}	4.4 ± 0.6 ^{ab}

Control: NPK fertilization; WE5, WE10, and WE15: water extracted seaweed at 5%, 10%, and 15%, respectively; USWE5, USWE10, and USWE15: ultrasound-assisted water extraction seaweed at 5%, 10%, and 15%, respectively. Data are means ± SE. Different superscript letters in each column indicate significant differences ($p \leq 0.05$).

3.5. N, P, and K

Nutrient content (i.e., N, P, and K) of Jew's Mallow plant treated with different seaweed extracts in comparison to control treatment are presented in Table 4. Control treatment had the highest N content in 2016 and 2017 (1.78% and 1.71%, respectively), while WE10 had the lowest N content (1.20% and 1.33%), respectively. USWE10 treatment had the highest P content in 2016 and 2017 (0.74% and 0.77%, respectively). The highest K content (1.90%) was achieved by WE15 treatment in both seasons.

Table 4. Effects of water extract (WE) and ultrasound-assisted Water Extraction (USWE) on N, P, and K content in Jew's Mallow during the 2016 and 2017 growing seasons.

Treatment	N (%)		P (%)		K (%)	
	2016	2017	2016	2017	2016	2017
Control	1.78 ± 0.20 ^{a}	1.71 ± 0.25 ^{a}	0.67 ±0.06 ^{ab}	0.70 ±0.07 ^{ab}	1.40 ±0.01 ^{c}	1.40 ±0.02 ^{b}
WE5	1.33 ± 0.13 ^{b}	1.40 ±0.04 ^{b}	0.64 ±0.01 ^{b}	0.66 ±0.04 ^{b}	1.80 ±0. 10 ^{b}	1.90 ±0.04 ^{a}
WE10	1.20 ±0.10 ^{b}	1.33 ±0.12 ^{b}	0.70 ±0.06 ^{ab}	0.77 ±0.01 ^{a}	1.80 ±0.02 ^{b}	1.80 ±0.01 ^{a}
WE15	1.41 ±0.04 ^{b}	1.49 ±0.23 ^{ab}	0.71 ±0.05 ^{ab}	0.73 ±0.06 ^{ab}	1.90 ±0.03 ^{a}	1.90 ±0.01 ^{a}
USWE5	1.39 ±0.10 ^{b}	1.37 ±0.02 ^{b}	0.66 ±0.06 ^{ab}	0.69 ±0.07 ^{ab}	1.80 ±0.04 ^{b}	1.80 ±0.02 ^{a}
USWE10	1.24 ±0.12 ^{b}	1.32 ±0.11 ^{b}	0.74 ±0.03 ^{a}	0.77 ±0.02 ^{a}	1.20 ±0. 10 ^{d}	1.30 ±0.03 ^{b}
USWE15	1.41 ±0.04 ^{b}	1.39 ±0.06 ^{b}	0.68 ±0.03 ^{ab}	0.74 ±0.03 ^{ab}	1.20 ±0. 11 ^{d}	1.30 ±0.02 ^{b}

Control: NPK fertilization; WE5, WE10, and WE15: water extracted seaweed at 5%, 10%, and 15%, respectively; USWE5, USWE10, and USWE15: ultrasound-assisted water extraction seaweed at 5%, 10%, and 15%, respectively. Data are means ± SE. Different superscript letters in each column indicate significant differences ($p \leq 0.05$).

3.6. Antioxidant Activity

The highest DPPH percentage was achieved by WE10 in 2016 and 2017 (40.78% and 40.74%, respectively). The lowest DPPH percentage was recorded in USWE15 in 2016 and 2017 (8.75% and 8.74%, respectively; Figure 5). The highest TAC was recorded in WE10 in both seasons (43.97 and 44.22 mg g^{-1}, respectively), followed by the USWE10 treatment ((35.69 and 36.38 mg g^{-1}; Table 5). The lowest TAC was recorded with control (26.30 mg g^{-1}) in 2017 and USWE15 treatment (26.00 mg g^{-1}) in 2018. In both seasons, the highest significant TPC was obtained with WE10 treatment (116.28 and 115.81 mg g^{-1}, respectively), while the lowest TPC was obtained with the USWE15 treatment (49.62 and 49.61 mg g^{-1}, respectively). Although USWE5 had a higher TVC value, significant differences between extracts treatments were not observed, except for USWE15 treatment.

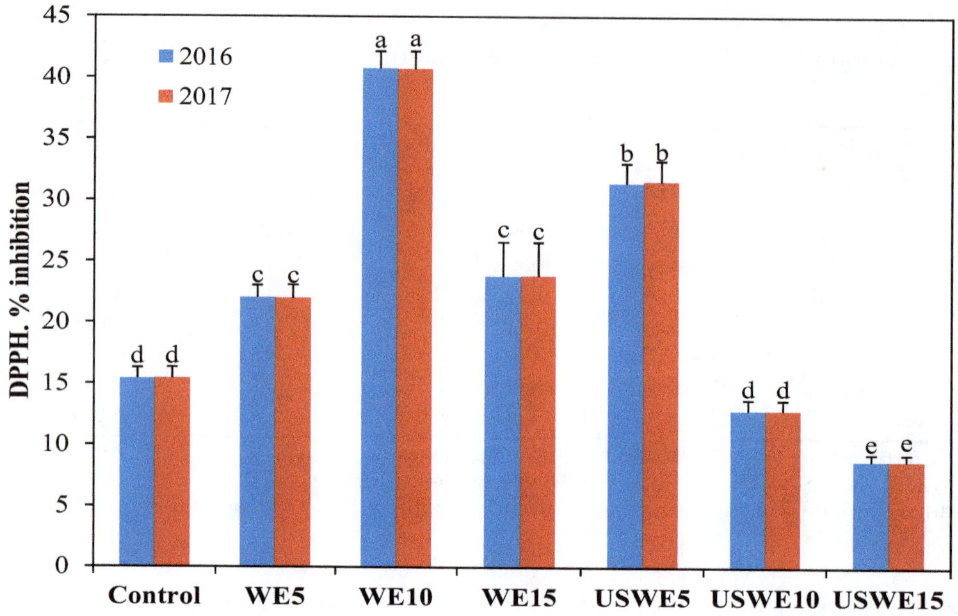

Figure 5. DPPH inhibition (% inhibition) of Jew's Mallow with water extract (WE) and ultrasound-assisted water extract (USWE) in the 2016 and 2017 growing seasons. Different letters (a, b, etc.) above bars indicate a significant difference among treatments in each season.

Table 5. Effects of water extract (WE) and ultrasound-assisted Water Extraction (USWE) on total antioxidant content (TAC), total phenolic content (TPC), and total flavonoid content (TVC) of Jew's Mallow during the 2016 and 2017 growing seasons.

Treatment	TAC (mg g^{-1})		TPC (mg g^{-1})		TVC (µg g^{-1})	
	2016	2017	2016	2017	2016	2017
Control	26.30 ± 3.43 ^{c}	26.22 ± 0.40 ^{c}	97.15 ± 11.78 ^{b}	98.15 ± 10.92 ^{b}	1168 ± 47.4 ^{ab}	1170 ± 49.2 ^{ab}
WE5	34.65 ± 3.36 ^{b}	32.69 ± 1.27 ^{b}	75.22 ± 2.19 ^{cd}	75.40 ± 2.43 ^{cd}	1193 ± 23.9 ^{a}	1194 ± 20.7 ^{a}
WE10	43.97 ± 1.04 ^{a}	44.22 ± 2.40 ^{a}	116.28 ± 6.59 ^{a}	115.81 ± 7.77 ^{a}	1208 ± 26.4 ^{a}	1204 ± 14.7 ^{a}
WE15	27.28 ± 1.37 ^{c}	28.13 ± 2.53 ^{c}	77.92 ± 11.12 ^{cd}	78.33 ± 10.70 ^{cd}	1191 ± 20.9 ^{a}	1187 ± 29.7 ^{a}
USWE5	35.35 ± 1.55 ^{b}	35.52 ± 1.32 ^{b}	66.56 ± 1.69 ^{d}	65.46 ± 1.47 ^{d}	1244 ± 3.1 ^{a}	1232 ± 7.3 ^{a}
USWE10	35.69 ± 2.41 ^{b}	36.38 ± 2.53 ^{b}	49.62 ± 6.54 ^{e}	49.61 ± 751 ^{e}	1205 ± 80.8 ^{a}	1209 ± 75.1 ^{a}
USWE15	26.78 ± 2.28 ^{c}	26.00 ± 2.54 ^{c}	80.20 ± 0.83 ^{c}	79.69 ± 0.56 ^{c}	1113 ± 31.1 ^{b}	1105 ± 30.1 ^{b}

Control: NPK fertilization; WE5, WE10, and WE15: water extracted seaweed at 5%, 10%, and 15%, respectively; USWE5, USWE10, and USWE15: ultrasound-assisted water extraction seaweed at 5%, 10%, and 15%, respectively. Data are means ± SE. Different superscript letters in each column indicate significant differences ($p \leq 0.05$).

3.7. CROPWAT Model

Figure 6 shows the depletion curve before and after each irrigation event during the 2016 and 2017 growth seasons. The highest values of depletion were 52 and 48 mm in the mid-season for each year, respectively. The depletion values were between those of field capacity and readily available moisture except for the initial 20 days of both seasons. Thus, there was no water stress. In the initial stage, there was a maximum *DP* of 28 mm at the first irrigation event in both seasons and after that it decreased to 2.5 and 8.5 mm, respectively, in the 2016 and 2017 growing seasons. The effective rainfall means modeled for 2016 and 2017 were 9.4 and 1 mm, respectively leaving deficits of 387 and 401.7 mm to be made up from irrigation. Thus, effective rainfall showed an ineffective pattern across the growth stages. By applying the basin irrigation system, the application water efficiencies were 78% and 79%, respectively, in 2016 and 2017. Hence, the values of *EIS* were 95.1% and 98.6%, respectively, in 2016 and 2017. The values of $(ET_c)_{actual}$ were 393.2 and 399.8 mm, respectively in 2016 and 2017; the values of *DIS* were 0.8% and 0.7%. On the other hand, there were no yield reductions across the growth stages, with the maximum values of 3.7% and 3.1% in the initial stage, respectively in 2016 and 2017.

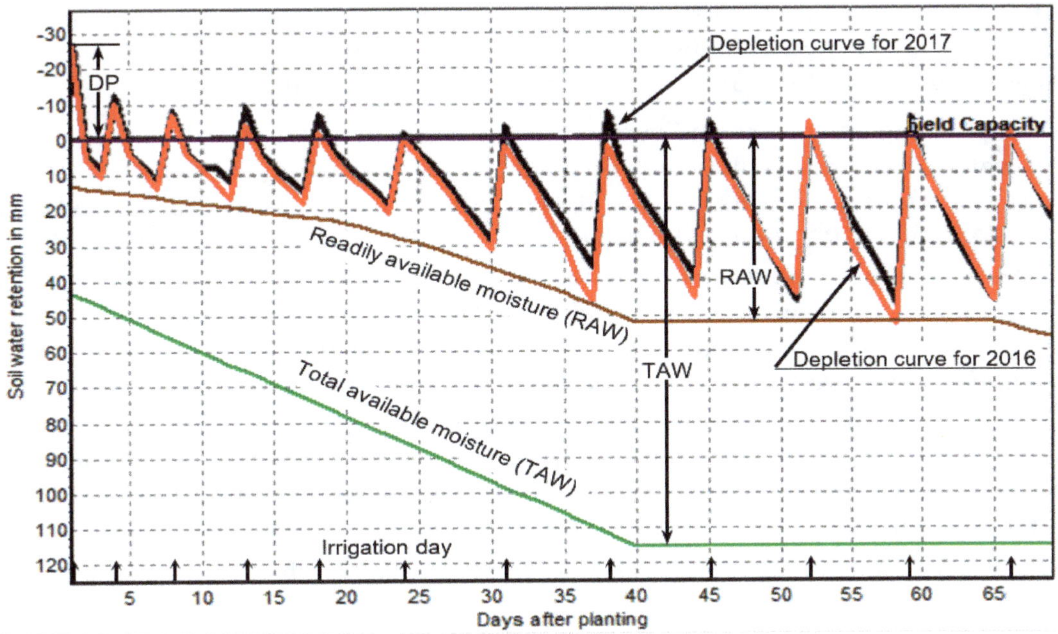

Figure 6. Soil water depletion during the 2016 and 2017 growing seasons.

4. Discussion

Marine algae are considered very important bioindicator for the marine ecosystem [13–15]. Many studies have reported that the constituents, diversity, and communities of marine algae are affected by variations in environmental parameters and nutrient limitation [65–70]. During the last few years, the attention on scientific and commercial interest to biotechnological applications of algae as a sustainable source and global commercial for aquaculture [71–75], biofuel [12,76], extracts [77,78], food supplement, pharmaceuticals, and cosmetics were increased [12].

In current study, data of biochemical composition (protein, lipid and carbohydrate) of *P. capillacea* showed that the large component is carbohydrate (51.36%), followed by protein (21.49%) and lipid (2.06%). The presented data may be act as an indicator for related bioactive secondary metabolites of *P. capillacea* liquid extract. However, our data is in the same line of the results presented by Khairy and El-Shafay [13] who found that, during spring season of 2010, the highest component is carbohydrate (50.49%), followed by protein (23.72%) and lipid (2.71%). Many authors reported that the biochemical constituents of marine algae are affected by variations in environmental conditions and nutrient availability [65–70].

The nutrient contents of seaweed *P. capillacea* used in current study were investigated previously by Khairy and El-Sheikh [14], at the same collected location of current study too, who observed that mineral were potassium (50.9 mg/100g), calcium (68.4 mg/100g), magnesium (22.1 mg/100g), cupper (0.5 mg/100g), ferrous (18.37 mg/100g), and zinc (0.19 mg/100g). In current study, although the applied seaweed extract is a rich source of nutrient, it not characterized as a nutrient fertilizer because of many consecrations like its constituent of bioactive compound which act as a plant growth promoting. Interestingly, the *P. capillacea* seaweed species is reported as a potential source for human healthy food because its constituent of bioactive compounds *like* protein, lipid, carbohydrate, fatty acids (saturated, mono-unsaturated and poly-unsaturated fatty acids), amino acids (essential and non-essential), carotenoids, phenolic compounds, and DPPH [13,14]. Hence, *P. capillacea*, collected from the same study location, is reported as a rich source of alkaloids, flavonoids, steroids, terpenoids, phlobatannins and many other phytochemicals and secondary metabolites [79].

Moreover, *P. capillacea* as a red alga is characterized as a rich source of different phytohormones [40,80,81]. It well known that some unknown bioactive component in seaweed acts to illicit the plant's own production of plant hormones through internal metabolic pathways [25].

Seaweeds and its extracts are becoming of increasing importance because of their bioactive compounds and their potential application in different industries. Liquid seaweed extract is commonly used as commercial agricultural biostimulants because of many considerations.

In current study, to enhance the efficiency of seaweed liquid extract, we evaluate two extraction methods; (1) water (WE); and (2) ultrasound-assisted water extraction (USWE). The effect of different seaweed extracts as a foliar spray on quantity (growth and yield) and quality (minerals and antioxidants activity) of Jew's Mallow (*C. olitorius* L.), comparing to NPK traditional fertilizers were observed. In general, Jew's Mallow (*C. olitorius* L.) treated with liquid seaweed extract (either WE or USWE) achieved the highest significant quantity (yield) and quality (antioxidant activity, P %, and K %), comparing to NPK traditional fertilizers, which only achieved the highest significant N %. Jew's Mallow (*C. olitorius* L.) treated with WE10 and USWE10 were achieved the highest significant yield (fresh weight), and P %. The highest significant Chlorophyll a and b; total antioxidant activity and total phenolic compounds were achieved by WE10, while the highest significant carotene and total flavonoid compounds were achieved by USWE10. In general, in the present study, it was observed that the seaweed liquid extract prepared from *P. capillacea* presented to Jew's Mallow gave better results in all aspects of growth to yield when compared to NPK traditional fertilizers. Using ultrasound-assisted water extraction (USWE) method was significantly improved the total flavonoid and carotene content in *P. capillacea* USWE crude extract, which is positively reflected on these compounds of Jew's Mallow (*C. olitorious* L.), when comparing to WE. Carotenes are indispensable to plants and act as precursors for the biosynthesis of phytohormones and strigolactones, improve the plant development and responses to unstable environmental, and serve as a source of pro-vitamin A [82].

In the present study, WE10-treated plants showed the best response in plant height and leaf number. Similarly, Stephenson [40] reported that seaweed liquid extract prepared from *Ascophyllum* and *Laminaria* accelerated maize growth. Blunden and Wildgoose [83] reported a marked increase in lateral root development in potato plants as a result of treatment with seaweed extract. Similar results were obtained with *Padina* biofertilizer, which induced maximum growth in *Cajanus cajan* [84]. Thirumaran, et al. [85] reported similar findings 20% seaweed liquid extract from brown algae *Rosenvingea intricate* had an increased growth of *Cyamopsis tetragonoloba*. Similarly, Whapham, et al. [86] observed that the application of seaweed *Ascophyllum nodosum* liquid extract increased the chlorophyll content in cucumber cotyledons, tomato, and guar plants [83].

Seaweed liquid extracts can be an effective way to some crop plants to increase both the nutrient content of the soil and crop yield. Hence, seaweeds play a vital role in agriculture, where irrational use of chemical fertilizer and pesticides is a cause of concern. Extensive regional trials with the product are needed to determine the environmental limitations of biological activity and to monitor the survival and dispersal of the inoculate [87]. Hence, use of modern agriculture in conjunction with traditional farming practices is the sustainable solution for the future. The expansion of nature source of other manures, seaweed extract application will be useful in enriching the production in the place of costly chemical fertilizer. The use of seaweed liquid extracts helps to avoid environmental pollution by high doses of chemical fertilizer. The beneficial effects of seaweed extract on terrestrial plants are improving the overall growth, yield and the ability to with stand adverse environmental conditions [88].

From the outputs of the CROPWAT model for 2016 and 2017 growing seasons, it appeared that additional irrigation was required to meet the daily crop water requirements as rainfall had minor effects or none. This high irrigation requirement may be attributed to the low rainfall during the growing seasons. Our data indicate that irrigation is crucial in the initial growth stage of Jew's Mallow due to high *DP* caused by basin irrigation system. To avoid yield reductions in Jew's Mallow cultivation, large quantities of water should be applied during the initial stage. In areas where water is a restricting factor in crop production, a well-designed irrigation schedule can improve water productivity when full irrigation is not plausible. However, a certain yield reduction should be expected due to the relationship between ET_c and yield of some crops [44,89–91].

5. Conclusions

Seaweeds are one of the most important marine resources for food, industrial raw materials, therapeutic and botanical applications. In current study, to enhance the efficiency of seaweed liquid extract, we evaluate two extraction methods; (1) water (WE); and (2) ultrasound-assisted water extraction (USWE). The effect of different seaweed extracts as a foliar spray on quantity (growth and yield) and quality (minerals and antioxidants activity) of Jew's Mallow *C. olitorious* L., comparing to NPK traditional fertilizers were observed. The present study observed that the seaweed liquid extract prepared from *P. capillacea* (either WE or USWE) presented to Jew's Mallow *C. olitorious* L. gave better results in all aspects of quantity and quality when compared to NPK traditional fertilizers. No significant differences of quantity (yield) of *C. olitorious* L. treated with WE10 and USWE10. Water extraction (WE) method improves the Chlorophyll 'a' and 'b'; total antioxidant activity and total phenolic compounds of Jew's Mallow *C. olitorious* L. While, using ultrasound-assisted water extraction (USWE) method improves the carotene and total flavonoid compounds of *P. capillacea* USWE crude extract which positively reflected on the contents of these compounds in Jew's Mallow *C. olitorious* L., when comparing to WE. Carotenes are indispensable to the plants and act as precursors for the biosynthesis of phytohormones and strigolactones, improve the plant development and responses to unstable environmental, and serve as a source of pro-vitamin A. Thus, USWE is an attractive novel technology enhancing the efficiency of seaweed liquid extract on Jew's Mallow. The CROPWAT model has shown that an adequate amount of water is vital, especially during the initial growth stage of Jew's Mallow, but also in other stages. Therefore, it is important to adopt efficient irrigation practices to maximize yields while reducing adverse effects on water resources.

Author Contributions: M.A., conceptualization, funding acquisition, investigation, methodology, writing—original draft, project administration; A.A.E.-S., formal analysis, investigation, data curation, methodology, writing—original draft, writing—review and editing; H.M.K., visualization, methodology, writing—original draft; D.Y.A.-E., supervision, visualization, methodology, writing—original draft; M.A.M., project administration, methodology, writing—review and editing, writing—original draft; A.A., Funding acquisition, project administration, resources, supervision, writing—original draft; S.M.H., methodology, investigation, supervision, project administration, writing—original draft. All authors have read and agree to the published version of the manuscript.

Acknowledgments: This project was financially supported by the Vice Deanship of Research Chairs at King Saud University.

References

1. Butnariu, M.; Butu, A. Chemical Composition of Vegetables and Their Products. In *Handbook of Food Chemistry*; Cheung, P.C.K., Mehta, B.M., Eds.; Springer: Berlin/Heidelberg, Germany, 2015; pp. 627–692.
2. Tang, L.; Hamid, Y.; Sahito, Z.A.; Gurajala, H.K.; He, Z.; Feng, Y.; Yang, X. Evaluation of variation in essential nutrients and hazardous materials in spinach (Spinacia oleracea L.) genotypes grown on contaminated soil for human consumption. *J. Food Compos. Anal.* **2019**, *79*, 95–106. [CrossRef]
3. Hassan, S.M.; Abd Alhafez, Z.A. Efficiency of plastic cover types on essential oil composition and vegetative growth of dill (*Anethum graveolens* L.). In Proceedings of the ISHS Acta horticulturae, VIII International Symposium on Light in Horticulture, East Lansing, MI, USA, 22 May 2016.
4. Oke, J.M.; Hamburger, M.O. Screening of some nigerian medicinal plants for antioxidant activity using 2, 2, diphenyl-picryl-hydrazyl radical. *Afr. J. Biomed. Res.* **2002**, *5*, 77–79.
5. Devasagayam, T.P.A.; Tilak, J.C.; Boloor, K.K.; Sane, K.S.; Ghaskadbi, S.S.; Lele, R.D. Free radicals and antioxidants in human health: Current status and future prospects. *JAPI* **2004**, *52*, 794–804. [PubMed]
6. Yoshikawa, T.; Toyokuni, S.; Yamamoto, Y.; Naito, Y. *Free Radicals in Chemistry, Biology and Medicine*; OICA International: London, UK, 2000.
7. Oboh, G.; Raddatz, H.; Henle, T. Characterization of the antioxidant properties of hydrophilic and lipophilic extracts of Jute (*Corchorus olitorius*) leaf. *Int. J. Food Sci. Nutr.* **2009**, *60*, 124–134. [CrossRef] [PubMed]

8. Shittu, T.A.; Ogunmoyela, O.A. Water blanching treatment and nutrient retention in some Nigerian green leafy vegetables. In Proceedings of the 25th Annual Conference of the Nigerian Institute of Food Science and Technology, Lagos, Nigeria, 5–9 November 2001; pp. 64–65.
9. Helaly, A.; Alkharpotly, A.E.-B.A.; Mady, E.; Craker, L.E. Characterization of Four Molokhia (*Corchorus olitorius*) Landraces by Morphology and Chemistry. *J. Med. Act. Plants* **2017**, *5*, 1–6.
10. Oyedele, D.J.; Asonugho, C.; Awotoye, O.O. Heavy metals in soil and accumulation by edible vegetables after phosphate fertilizer application. *Electron. J. Environ. Agric. Food Chem.* **2006**, *5*, 1446–1453.
11. Brkić, D.; Bošnir, J.; Bevardi, M.; Bošković, A.G.; Miloš, S.; Lasić, D.; Krivohlavek, A.; Racz, A.; Mojsović–Ćuić, A.; Trstenjak, N.U. Nitrate in leafy green vegetables and estimated intake. *Afr. J. Tradit. Complement. Altern. Med.* **2017**, *14*, 31–41.
12. Ashour, M.; Elshobary, M.E.; El-Shenody, R.; Kamil, A.-W.; Abomohra, A.E.-F. Evaluation of a native oleaginous marine microalga *Nannochloropsis oceanica* for dual use in biodiesel production and aquaculture feed. *Biomass Bioenergy* **2019**, *120*, 439–447. [CrossRef]
13. Khairy, H.M.; El-Shafay, S.M. Seasonal variations in the biochemical composition of some common seaweed species from the coast of Abu Qir Bay, Alexandria, Egypt. *Oceanologia* **2013**, *55*, 435–452. [CrossRef]
14. El-Shenody, R.A.; Ashour, M.; Ghobara, M.M.E. Evaluating the chemical composition and antioxidant activity of three Egyptian seaweeds: *Dictyota dichotoma, Turbinaria decurrens*, and *Laurencia obtusa*. *Braz. J. Food Technol.* **2019**, *22*, 623–630. [CrossRef]
15. Khairy, H.M.; El-Sheikh, M.A. Antioxidant activity and mineral composition of three Mediterranean common seaweeds from Abu-Qir Bay, Egypt. *Saudi J. Biol. Sci.* **2015**, *22*, 623–630. [CrossRef] [PubMed]
16. Elshobary, M.; El-Shenody, R.; Ashour, M.; Zabed, H.M.; Qi, X. Antimicrobial and antioxidant characterization of bioactive components from *Chlorococcum minutum*, a newly isolated green microalga. *Food Biosci.* **2020**, in press.
17. Abetz, P. Seaweed extracts: Have they a place in Australian agriculture or horticulture? *J. Aust. Inst. Agric. Sci.* **1980**, *46*, 23–29.
18. Ferreira, M.I.; Lourens, A.F. The efficacy of liquid seaweed extract on the yield of canola plants. *S. Afr. J. Plant Soil* **2002**, *19*, 159–161. [CrossRef]
19. Stirk, W.A.; Arthur, G.D.; Lourens, A.F.; Novák, O.; Strnad, M.; van Staden, J. Changes in cytokinin and auxin concentrations in seaweed concentrates when stored at an elevated temperature. *J. Appl. Phycol.* **2004**, *16*, 31. [CrossRef]
20. Booth, E. The Manurial Value of Seaweed. *Bot. Mar.* **1965**, *8*, 138–143. [CrossRef]
21. Gandhiyappan, K.; Perumal, P. Growth promoting effect of seaweed liquid fertilizer (*Enteromorpha intestinalis*) on the sesame crop plant (*Sesamum indicum* L.). *Seaweed Resour. Util* **2001**, *23*, 23–25.
22. Blunden, G. Agricultural uses of seaweeds and seaweed extracts. In *Seaweed Resources in Europe: Uses and Potential*; Guiry, M.D., Blunden, G., Eds.; John Wiley and Sons: Hoboken, NJ, USA, 1991; pp. 65–81.
23. Ajmal, M.; Ali, H.I.; Saeed, R.; Akhtar, A.; Tahir, M.; Mehboob, M.Z.; Ayub, A. Biofertilizer as an alternative for chemical fertilizers. *J. Agric. Allied Sci.* **2018**, *7*, 1–7.
24. Bokil, K.K.; Mehta, V.C.; Datar, D.S. Seaweeds as manure: II. Pot culture manurial experiments on wheat [India]. *Phykos* **1974**, *13*, 1–5.
25. Craigie, J.S. Seaweed extract stimuli in plant science and agriculture. *J. Appl. Phycol.* **2011**, *23*, 371–393. [CrossRef]
26. Chapman, V.J.; Chapman, D.J. Seaweed as Animal Fodder, Manure and for Energy. In *Seaweeds and Their Uses*; Chapman, V.J., Chapman, D.J., Eds.; Springer: Berlin/Heidelberg, Germany, 1980; pp. 30–61. [CrossRef]
27. Aguilera-Morales, M.; Casas-Valdez, M.; Carrillo-Domínguez, S.; González-Acosta, B.; Pérez-Gil, F. Chemical composition and microbiological assays of marine algae *Enteromorpha* spp. as a potential food source. *J. Food Compos. Anal.* **2005**, *18*, 79–88. [CrossRef]
28. Arunkumar, K.; Puhazhendi, T.; Subramanian, S.; Thangaraj, N.; Rengaswamy, R. Alleviating effect of seaweed liquid fertilizer on water stressed black gram *Vigna mungo* (L.) Hepper. *Seaweed. Res. Utlin* **2002**, *24*, 151–158.
29. Rama Rao, K. Effect of seaweed liquid fertilizer on the yield of *Zizyphus rugosa* (Bores). In Proceedings of the International Symposium on Marine Algae, Bhavnagar, India, 9–12 January 1979; p. 27.

30. Wijesinghe, W.A.J.P.; Jeon, Y.-J. Enzyme-assistant extraction (EAE) of bioactive components: A useful approach for recovery of industrially important metabolites from seaweeds: A review. *Fitoterapia* **2012**, *83*, 6–12. [CrossRef] [PubMed]
31. Rodriguez-Jasso, R.M.; Mussatto, S.I.; Pastrana, L.; Aguilar, C.N.; Teixeira, J.A. Microwave-assisted extraction of sulfated polysaccharides (fucoidan) from brown seaweed. *Carbohydr. Polym.* **2011**, *86*, 1137–1144. [CrossRef]
32. Klejdus, L.; Lojková, L.; Vlcek, J. Hyphenated Solid Phase Extraction/Supercritical Fluid Extraction Methods for Extraction of Phenolic Compounds from Algae. *Curr. Anal. Chem.* **2014**, *10*, 86–98. [CrossRef]
33. Kadam, S.U.; Tiwari, B.K.; Smyth, T.J.; O'Donnell, C.P. Optimization of ultrasound assisted extraction of bioactive components from brown seaweed Ascophyllum nodosum using response surface methodology. *Ultrason. Sonochem.* **2014**, *23*, 308–316. [CrossRef]
34. Kadam, S.U.; Tiwari, B.K.; O'Donnell, C.P. Application of novel extraction technologies for bioactive from marine algae. *J. Agric. Food Chem.* **2013**, *61*, 4667–4675. [CrossRef]
35. Kadam, S.U.; Tiwarim, B.K.; Smyth, T.J.; O'Connell, S.; O'Donnell, C.P. Effect of Ultrasound Pretreatment on the Extraction Kinetics of Bioactives from Brown Seaweed (Ascophyllum nodosum). *J. Sep. Sci. Technol.* **2015**, *50*, 670–675. [CrossRef]
36. Asada, C.; Doi, K.; Sasaki, C.; Nakamura, Y. Efficient extraction of starch from microalgae using ultrasonic homogenizer and its conversion into ethanol by simultaneous saccharification and fermentation. *Nat. Resour.* **2012**, *3*, 175–179. [CrossRef]
37. Ma, Y.-Q.; Chen, J.-C.; Liu, D.; Ye, H.X.-Q. Simultaneous extraction of phenolic compounds of citrus peel extracts: Effect of ultrasound. *Ultrason. Sonochem.* **2009**, *16*, 57–62. [CrossRef]
38. Lianfu, Z.; Zelong, L. Optimization and comparison of ultrasound/microwave assisted extraction (UMAE) and ultrasonic assisted extraction (UAE) of lycopene from tomatoes. *Ultrason. Sonochem.* **2008**, *15*, 731–737. [CrossRef] [PubMed]
39. Chen, F.; Sun, Y.; Zhao, G.; Liao, X.; Hu, X.; Wu, J.; Wang, Z. Optimization of ltrasound-assisted extraction of anthocyanins in red raspberries and identification of anthocyanins in extract using high-performance liquid chromatography–mass spectrometry. *Ultrason. Sonochem.* **2007**, *14*, 767–778. [CrossRef] [PubMed]
40. Stephenson, W.A. Seaweeds in agriculture and horticulture. In *Ratequer, Perumai Valley*, 3rd ed.; Faber & Faber: London, UK, 1974; p. 241.
41. Adeboye, O.B.; Schultz, B.; Adekalu, K.O.; Prasad, K. Crop water productivity and economic evaluation of drip-irrigated soybeans (Glyxine max L. Merr.). *Agric. Food Secur.* **2015**, *4*, 10. [CrossRef]
42. Surendran, U.; Sushanth, C.M.; Joseph, E.J.; Al-Ansari, N.; Yaseen, Z.M. FAO CROPWAT Model-Based Irrigation Requirements for Coconut to Improve Crop and Water Productivity in Kerala, India. *Sustainability* **2019**, *11*, 5132. [CrossRef]
43. Clarke, D.; Smith, M.; El-Askari, K. CropWat for Windows: User Guide. Available online: http://www.fao.org/land-water/databases-and-software/cropwat/en/ (accessed on 1 March 2016).
44. Moseki, O.; Murray-Hudson, M.; Kashe, K. Crop water and irrigation requirements of Jatropha curcas L. in semi-arid conditions of Botswana: Applying the CROPWAT model. *Agric. Water Manag.* **2019**, *225*, 105754. [CrossRef]
45. Tsakmakis, I.D.; Zoidou, M.; Gikas, G.D.; Sylaios, G.K. Impact of Irrigation Technologies and Strategies on Cotton Water Footprint Using AquaCrop and CROPWAT Models. *Environ. Process.* **2018**, *5*, 181–199. [CrossRef]
46. Studer, C.; Spoehel, S. Potential and Actual Water Savings through Improved Irrigation Scheduling in Small-Scale Vegetable Production. *Agronomy* **2019**, *9*, 888. [CrossRef]
47. Flores-Velazquez, J.; Rojano, F.; Aguilar, E. ETo estimation to schedule irrigation in a green wall using Cropwat. In Proceedings of the 2019 ASABE Annual International Meeting, Boston, MA, USA, 7–10 July 2019.
48. Rauch, T. The estimation of microalgal protein content, its meaning the evolution of algal biomass, comparison method for extracting for protein. *Hydrobiologia* **1981**, *78*, 237. [CrossRef]
49. Hartree, E.F. Determination of protein: A modification of the Lowry method that gives a linear photometric response. *Anal. Riochem.* **1972**, *48*, 422–427. [CrossRef]
50. Myklestad, S.; Haug, A. Production of carbohydrates by the marine Chaetoceros affinis var. willei (Gran.) Hustedt. I. Effect of the concentration of nutrients in the culture medium. *J. Exp. Mar. Biol. Ecol.* **1972**, *9*, 125–136. [CrossRef]

51. Dubois, M.; Giles, K.A.; Hamilton, J.K.; Rebers, P.A.; Smith, F. Calorimetric method for determination of sugars and related substances. *Anal. Chem.* **1956**, *28*, 350–356. [CrossRef]
52. Bligh, E.G.; Dyer, W.J. A rapid method of total lipid extraction and purification. *Can. J. Biochem. Physiol.* **1959**, *37*, 911–917. [CrossRef]
53. Page, A.L. *Methods of Soil Analysis. Part 2. Chemical and Microbiological Properties*; American Society of Agronomy, Soil Science Society of America: Madison, WI, USA, 1982.
54. Allen, R.G.; Pereira, L.S.; Raes, D.; Smith, M. *Crop Evapotranspiration-Guidelines for Computing Crop Water Requirements-FAO Irrigation and Drainage Paper No. 56*; FAO—Food and Agriculture Organization of the United Nations: Rome, Italy, 1998.
55. Ghoneim, I.M.; El-araby, S.M. Effect of organic manure source and biofertilizer type on growth, productivity and chemical composition of Jew's Mallow (*Corchorus Olitorious* L.) plants. *J. Agric. Env. Sci. Alex. Univ. Egypt* **2003**, *2*, 88–105.
56. Dere, Ş.; Güneş, T.; Sivaci, R. Spectrophotometric determination of chlorophyll-A, B and total carotenoid contents of some algae species using different solvents. *Turkish J. Bot.* **1998**, *22*, 13–18.
57. Evenhuis, B. *Simplified Methods for Foliar Analysis Vol. Communication 70*; Department of Agriculture Research Product, Research Division, Koninklijk Institut voor den Tropen: Amsterdam, NL, USA, 1978.
58. Cottenie, A.; Verloo, M.; Kiekens, L.; Velghe, G.; Camerlynck, R. *Chemical Analysis of Plants and Soils, Laboratory of Analytical and Agrochemistry*; State University of Ghent: Ghent, Belgium, 1982.
59. Kumar, G.S.; Antony Muthu Prabhu, A.; Seethalashmi, P.G.; Bhuvanesh, N.; Kumaresan, S. Self-catalyzed syntheses, structural characterization, DPPH radical scavenging-, cytotoxicity-, and DFT studies of phenoxyaliphatic acids of 1,8-dioxo-octahydroxanthene derivatives. *J. Mol. Struct.* **2014**, *1059*, 51–60. [CrossRef]
60. Ahmad, M.Z.; Ali, M.; Mir, S.R. Anti-diabetic activity of *Ocimum sanctum* L. roots and isolation of new phytoconstituents using two-dimensional nuclear magnetic resonance spectroscopy. *J. Pharmacogn. Phytother.* **2012**, *4*, 75–85. [CrossRef]
61. Chang, C.-C.; Yang, M.-H.; Wen, H.-M.; Chern, J.-C. Estimation of total flavonoid content in propolis by two complementary colorimetric methods. *J. Food Drug Anal.* **2002**, *10*, 178–182.
62. Swennenhuis, J. *CROPWAT Version 8.0*; FAO—Food and Agriculture Organization of the United Nations: Rome, Italy, 2009.
63. Grieser, J. *Climwat 2.0 for CROPWAT*; FAO—Food and Agriculture Organization of the United Nations: Rome, Italy, 2006.
64. Snedecor, G.W.; Cochran, W.G. *Statistical Methods*, 8th ed.; Iowa State University Press: Ames, IA, USA, 1989.
65. Ashour, M.; Abo-Taleb, H.A.; Abou-Mahmoud, M.M.; El-Feky, M.M.M. Effect of the integration between plankton natural productivity and environmental assessment of irrigation water, El-Mahmoudia Canal, on aquaculture potential of *Oreochromis niloticus*. *Turk. J. Fish. Aquat. Sci.* **2018**, *18*, 1163–1175. [CrossRef]
66. Ashour, M.; Kamel, A. Enhance growth and biochemical composition of *Nannochloropsis oceanica* cultured under nutrient limitation using commercial agricultural fertilizers. *J. Mar. Sci. Res. Dev.* **2017**, *7*, 233. [CrossRef]
67. Nour, M.A.; Zaki, M.A.; Ashour, M.; Kamel, A.; Taha, S.M.; Omar, E. Effect of different *Nannochloropsis oculata* diets (cultured under different nitrogen and phosphorus regime) on growth and biochemical composition of rotifers *Brachionus plicatilis*. In Proceedings of the 12th International Congress on Microbial Interaction and Applications of Beneficial Microbes, Munich, Germany, 17–18 July 2017.
68. Omar, E.; Zaki, M.; Ashour, M.; Kamel, A.; Taha, S.M.; Nour, M.A. Effect of nutrient limitations on growth and biochemical composition of outdoor cultured *Chaetoceros calcitrans*. In Proceedings of the 12th International Congress on Microbial Interaction and Applications of Beneficial Microbes, Munich, Germany, 17–18 July 2017.
69. Nour, A.M.; Ashour, M.; Kamil, A.; Taha, S.M.; Zaki, M.A.; Omar, E. The use of artificial fertilizers for culture of marine microalgae: 1. Growth and biochemical constituents of *Nannochloropsis oculata*. World Congress and Expo on Applied Microbiology, Munich, Germany. *J. Microb. Biochem. Technol.* **2015**, *5*, 120.
70. Nour, A.M.; Ashour, M.; Kamil, A.; Taha, S.M.; Zaki, M.A.; Omar, E. The use of artificial fertilizers for culture of marine microalgae: 2. Growth and biochemical constituents of *Chaetoceros calcitrans*. World Congress and Expo on Applied Microbiology, Munich, Germany, 2015. *J. Microb. Biochem. Technol.* **2015**, *5*, 119.

71. Sharawy, Z.Z.; Ashour, M.; Abbas, E.; Ashry, O.; Helal, M.; Nazmi, H.; Kelany, M.; Kamel, A.; Hassaan, M.; Rossi, W.J.; et al. Effects of dietary marine microalgae, *Tetraselmis suecica* on production, gene expression, protein markers and bacterial count of Pacific white shrimp *Litopenaeus vannamei*. *Aquac. Res.* **2020**, in press.
72. Soliman, H.A.; Fatma, A.A.; AbouZeid, A.E.; Ashour, M. Optimum growth conditions of three isolated diatoms species; *Skeletonema costatum*, *Chaetoceros calcitrans* and *Detonula confervacea* and their utilization as feed for marine penaeid shrimp larvae. *EJAR* **2010**, *36*, 161–183.
73. Soliman, H.A.; Fatma, A.A.; AbouZeid, A.E.; Ashour, M. Population growth rate, fecundity, filtration and ingestion rate of marine rotifer *Brachionus plicatilis* fed with motile and immotile microalgae. *EJAR* **2008**, *34*, 426–439.
74. Heneash, A.; Ashour, M.; Matar, M. Effect of Un-live Microalgal diet, *Nannochloropsis oculata* and *Arthrospira* (*Spirulina*) platensis, comparing to yeast on population of rotifer, *Brachionus plicatilis*. *Mediterr. Aquac. J.* **2015**, *7*, 48–54. [CrossRef]
75. Abo-Taleb, H.; Ashour, M.; El-Shafei, A.; Alataway, A.; Maaty, M.M. Biodiversity of Calanoida Copepoda in Different Habitats of the North-Western Red Sea (Hurghada Shelf). *Water* **2020**, *12*, 656. [CrossRef]
76. Ashour, M. *Marine Microalgae: Aquaculture and Biodiesel Production*, 1st ed.; LAP Lambert Academic Publisher: Saarbrücken, Germany, 2015; pp. 1–240. ISBN 978-3659691683.
77. Ashour, M. Sustainable development and the future of marine natural products in Egypt: SMIAF-Prototype project. In Proceedings of the 4th International Conference and Exhibition on Natural Products Medicinal Plants & Marine Drugs, Rome, Italy, 11 June 2018.
78. Hassan, S.M.; Ashour, M.; Soliman, A.A.F. Anticancer activity, antioxidant activity, mineral contents, vegetative and yield of *Eruca sativa* using foliar application of autoclaved cellular extract of *Spirulina platensis* extract, comparing to N-P-K fertilizers. *J. plant Prod. Mansoura Univ.* **2017**, *8*, 529–536. [CrossRef]
79. Mohy El-Din, S.M.; El-Ahwany, A.M.D. Bioactivity and phytochemical constituents of marine red seaweeds (*Jania rubens*, *Corallina mediterranea* and *Pterocladia capillacea*. *J. Taibah Univ. Sci.* **2016**, *10*, 471–484. [CrossRef]
80. Tarakhovskaya, E.R.; Maslov, Y.I.; Shishova, M.F. Phytohormones in algae. *Russ J. Plant Physiol.* **2007**, *54*, 163. [CrossRef]
81. Nelson, W.R.; van Staden, J. The Effect of Seaweed Concentrate on Wheat culms. *J. Plant Physiol.* **1984**, *115*, 433–437. [CrossRef]
82. Sun, T.; Yuan, H.; Cao, H.; Yazdani, M.; Tadmor, Y.; Li, L. Carotenoid Metabolism in Plants: The Role of Plastids. *Mol. Plant.* **2018**, *11*, 58–74. [CrossRef] [PubMed]
83. Blunden, G.; Wildgoose, P.B. The effects of aqueous seaweed extract and kinetin on potato yields. *J. Sci. Food Agric.* **1977**, *28*, 121–125. [CrossRef]
84. Mohan, V.R.; Venkataraman Kumar, V.; Murugeswari, R.; Muthuswami, S. Effect of crude and commercial seaweed extracts on seed germination and seedling growth in *Cajanus cajan* L. *Phykos* **1994**, *33*, 47–51.
85. Thirumaran, G.; Arumugam, M.; Arumugam, R.; Anantharaman, P. Effect of seaweed liquid fertilizer on growth and pigment concentration of Cyamopsis tetrogonolaba (L) Taub. *Am. Eurasian J. Agron.* **2009**, *2*, 50–56.
86. Whapham, C.A.; Blunden, G.; Jenkins, T.; Hankins, S.D. Significance of betaines in the increased chlorophyll content of plants treated with seaweed extract. *J. Appl. Phycol.* **1993**, *5*, 231. [CrossRef]
87. Kannan, L.; Selvan, T. Effect of seaweed biofertilizer on green gram. In Proceedings of the International Symposium on Phycology, Madras, India, 25 May–15 June 1987; p. 19.
88. Subramanian, S.K. Studies on the Effect of Seaweed Liquid Fertilizer [SLF]. In Proceedings of the International Symposium on Phycology, Madras, India, 25 May–15 June 1987; p. 33.
89. Davidson, R.J. EEG Measures of Cerebral Asymmetry: Conceptual and Methodological Issues. *Int. J. Neurosci.* **1988**, *39*, 71–89. [CrossRef]
90. Bennett, D.R.; Harms, T.E. Crop yield and water requirement relationships for major irrigated crops in southern Alberta. *Can. Water Resour. J.* **2011**, *36*, 159–170. [CrossRef]
91. Fereres, E.; Garda-Vila, M. Irrigation management for efficient crop production. *Crop Sci.* **2019**. [CrossRef]

12

Melatonin as a Chemical Substance or as Phytomelatonin Rich-Extracts for use as Plant Protector and/or Biostimulant in Accordance with EC Legislation

Marino B. Arnao * and Josefa Hernández-Ruiz

Department of Plant Physiology, University of Murcia, 30100-Murcia, Spain; jhruiz@um.es
* Correspondence: marino@um.es

Abstract: Melatonin (*N*-acetyl-5-methoxytryptamine) is a ubiquitous molecule present in animals and plants, and also in bacteria and fungi. In plants, it has an important regulatory and protective role in the face of different stress situations in which it can be involved, mainly due to its immobility. Both in the presence of biotic and abiotic stressors, melatonin exerts protective action in which, through significant changes in gene expression, it activates a stress tolerance response. Its anti-stress role, along with other outstanding functions, suggests its possible use in active agricultural management. This review establishes considerations that are necessary for its possible authorization. The particular characteristics of this substance and its categorization as plant biostimulant are discussed, and also the different legal aspects within the framework of the European Community. The advantages and disadvantages are also described of two of its possible applications, as a plant protector or biostimulant, in accordance with legal provisions.

Keywords: biostimulant; fertilizer; melatonin; phytomelatonin; plant protector; plant stress

---

1. Introduction

Melatonin (*N*-acetyl-5-methoxytryptamine) is a biogenic amine derived from the amino acid tryptophan, which was discovered in 1958 in the cow pineal gland by Lerner and cols. [1]. Two years later, it was detected in humans and its chemical structure was elucidated. This molecule, which was initially only related to changes in the structure of melanocytes in amphibians, fish and reptiles, was soon found to act as a neurohormone in mammals [2,3]. Since its discovery it has become one of the most researched molecules. In animals, it presents a multitude of physiological actions such as a role in the circadian rhythms of several molecules, and its influence on sleep–wake cycles, mood, motor activity and body temperature changes [4–7]. Its influence on food intake and its relationship with metabolic syndrome has also been demonstrated [8–10]. In other more specific situations such as the physiology of the retina, the immune system, sexual behavior and as an anti-cancer effector, melatonin also has a relevant role [11–15]. In addition, interesting and extensive reviews on the role of melatonin in animals and humans can be consulted [16–23].

In 1995, the presence of melatonin in plants was discovered [24–27]. During the following years there was much reluctance on the part of researchers to accept this, since some refused to believe that a neurohormone could be present in plants, and much less that it had any role in their physiology. A key piece was the elucidation of the melatonin biosynthesis route in plants, localized between the mitochondria, chloroplasts and cytoplasm of cells, and which has been studied with great accuracy by K. Back and J. Kong in rice and *Arabidopsis* plants [28–30]. However, it is now fully accepted that melatonin is present in all plant species and that it presents a panoply of interesting actions. Indeed, several studies have demonstrated its role in processes such as seed germination, growth and the development of seedlings, leaves and roots. It takes part in organogenesis processes such as rooting and fruiting, and in processes of leaf and fruit senescence. It acts as a protector of the photosynthetic and stomatic system, and as a regulator of various enzymes of the metabolism of carbohydrates, lipids, amino acids, nitrogen, sulfur and phosphorus. It also has a role in the secondary metabolism, enhancing the synthesis of flavonoids, anthocyanins, and carotenoids, among others. It regulates its own biosynthesis and that of several plant hormones such as auxin, abscisic acid, gibberellins, cytokinins, ethylene, polyamines, jasmonic acid and salicylic acid [31–39].

Of all the aspects investigated, its protective action against stress situations has been the most researched and about which most is known. Melatonin exerts a protective action, mediated by major changes in gene expression, both against abiotic (cold, heat, drought, waterlogging, salinity, alkalinity, acid rain, chemical contamination by heavy metals, UV radiation) and biotic (bacteria, fungi, virus) stressors. As a result, plants are more tolerant and/or resistant to the negative action of such stressors [31,36,40–43] (see below). The term "biostimulant" was first proposed to denote "materials that, in minute quantities, promote plant growth" by Zhang and Schmidt (1997) [44]. Later, the definition was modified by Kaufman et al. (2007) as: "Biostimulants are materials, other than fertilisers, that promote plant growth when applied in low quantities" [45]. According Du Jardin (2015), the following definition is proposed: "A plant biostimulant is any substance or microorganism applied to plants with the aim to enhance nutrition efficiency, abiotic stress tolerance and/or crop quality traits, regardless of its nutrients content", and extended as "plant biostimulants also designate commercial products containing mixtures of such substances and/or microorganisms" [46]. Under the EC (European Community) regulation: "Plant biostimulants will be EC marked as fertilizing products stimulating plant nutrition processes independently of the products' nutrient content with the sole aim of improving one or more of the following characteristics of the plant and the plant rhizosphere or phyllosphere: Nutrient use efficiency, tolerance to abiotic stress, crop quality, availability of confined nutrients in the soil and rhizosphere, humification and degradation of organic compounds in the soil". Extensive revision works on this topic can be consulted [47,48]. The objective of this work is to provide sufficient data to establish the clear protective role of melatonin against adverse environmental situations, and to discuss the possible global use of melatonin as a biostimulant and/or bioprotective agent. Current legislation of the EC, is taken into account and the advantages and disadvantages of its use in plant crops destined for animal and human consumption are analyzed.

2. Melatonin as a Regulator of Plant Stress Physiology

Although there was much evidence in the 1990s that melatonin could exert some role as an antioxidant agent in animal cells and tissues, it was not until 2004 and 2006, in carrot cells and Chinese licorice (*Glycyrrhiza uralensis* Fisch.), that the possible protective role of melatonin in plants was suggested [49–51], although some curious and previous data existed [52]. The initial idea that melatonin in plants, as in animals, could play an important role as an antioxidant was taking shape and results in this regard became ever more plentiful [53–57]. In addition, studies on melatonin as a possible plant regulator were also progressing, especially since the initial studies of Arnao and cols. on the role of melatonin in plant growth and development, and the so-called auxin-like activity [58–64].

It was not until the publication of results on the action of melatonin on changes in gene expression that the extent and potential of melatonin as a regulatory agent of multiple physiological processes in plants became widely known [64–70]. Exceeding previous expectations, melatonin is capable of activating all known molecular stress mechanisms in plants. Thus, gene regulatory factors involved in the response to cold, high temperatures, salinity, drought, chemical toxicity, etc., and also biotic stress, are up-regulated by melatonin [31,38,40,41,43]. Melatonin also regulates the expression of multiple enzymes related to hormonal homeostasis, up-regulating or down-regulating the expression of genes that encode enzymes of the biosynthetic or catabolic pathways of plant hormones including indole-3-acetic acid (auxin), gibberellins such as gibberellin-4 (GA_4), cytokinins, abscisic acid (ABA) and ethylene. It also others regulators such as salicylic acid (SA), jasmonic acid (JA) and polyamines [31–33,35,38,69,71–75]. In general, subjecting plants to a stressful situation—which leads to an increase in endogenous levels of melatonin—or treatment with exogenous melatonin, results in a stress tolerance response mediated by specific stress response factors and changes in the endogenous levels of plant hormones involved in the response [31–36,38,40–43,75–85]. In addition, the recent identification of a melatonin receptor in *Arabidopsis thaliana* has opened new expectations related to its role as a new plant hormone [86]. Figure 1 shows these aspects in a condensed form.

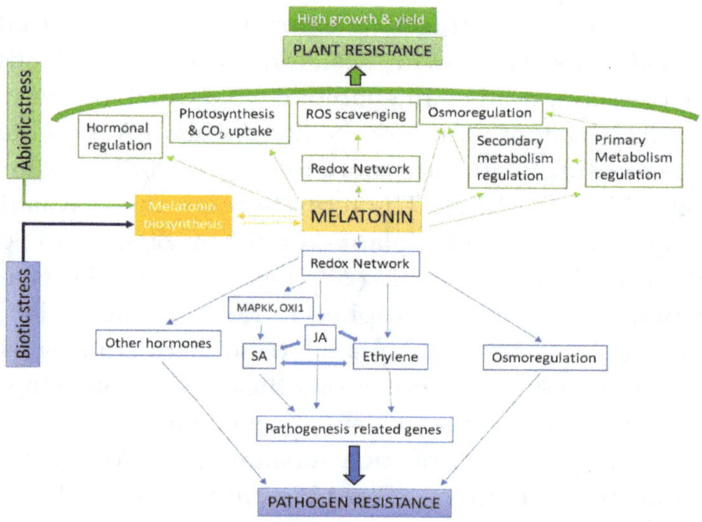

Figure 1. Model of redox network/melatonin action on abiotic and biotic stress responses.

3. Beneficial Responses to Melatonin Treatments in Different Crops in Stress Situations

Studies conducted with melatonin in plants under both abiotic and biotic stress are numerous. Table 1 compiles many of the studies with an agronomic interest since they deal primarily with crop species for human consumption. Table 1 presents studies classified by plant species, where there are many physiological aspects that are investigated in which melatonin exerts some generally beneficial action. These include seed germination, the growth and vegetative development of plants; photosynthesis, its pigments, photorespiration, stomatic conductance and water economy; the yields of seeds and fruits in adverse conditions; osmoregulation, ion exchange and adjustments in osmotic and hydric potentials, and the regulation of the different metabolisms of carbohydrates, lipids, nitrogen compounds, sulfur and phosphorus cycles. In regards to the secondary metabolism, melatonin induces the biosynthesis of flavonoids, anthocyanins and carotenoids, among others; in hormonal homeostasis, it intervenes in the regulation of all plant hormones and its own biosynthesis. It promotes the rooting process of primary, secondary and adventitious roots while during foliar senescence, melatonin regulates the expression of chlorophyll degradation-related and senescence-induced genes. In the postharvest control of fruits, melatonin increases the ethylene and lycopene content, and regulates many enzymes of the cell wall, ethylene biosynthesis, and primary and secondary metabolisms. It also helps preserve cut flowers; in fruiting it induces parthenocarpy. Finally, its role in bacterial, fungal and viral pathogenic infection should be emphasized, slowing damage and stimulating systemic acquired resistance (SAR) to favor crop health.

Obviously, all the above plant physiology aspects are of interest for application in plant production. Indeed, while many of the above studies were at a laboratory level, others have already been put into practice in crops with excellent results.

In general, exogenous melatonin applications are made through the root system, in irrigation water, or by spraying leaves. In the last case, no adjuvant is needed since melatonin is an amphipathic molecule that crosses biological membranes and the waxy cuticles. Melatonin is transported via the xylem from the roots to the rest of the organs of the plant quite effectively [87,88].

Table 1. Studies of different responses to melatonin treatments in different crop species in diverse stress situations.

Plant Species	Stress Type	Melatonin Treatment (μM)	Effects Observed	Reference
Alfalfa	Waterlogging	100	↑ tolerance, growth, photosynthesis, Chls, polyamines, ↓ electrolyte leakage, ROS, ethylene, leaf senescence	[89]
	Metal-Cd	10–200	↑ tolerance, growth, Cd transporters, ↓ Cd in roots, ROS	[90]
	Oxidative	1–100	↑ lateral root formation, cell division	[91]
Apple	Salinity	0.1	↑ shoot height, leaf number, Chls, K^+, ↓ electrolyte leakage, ROS	[92,93]
	Drought	100	↑ tolerance, re-open stomata, water in leaf, photosynthesis, N uptake, N metabolism, growth, ↓ ABA activity, ROS, leaf senescence	[71,94–96]
	Waterlogging	200	↓ chlorosis, wilting of the seedlings, ROS, ↑ tolerance, photosynthesis	[97]
	Alkaline	5	↑ tolerance, root system, redox balance, polyamines	[98]
	Leaf-senescence	10 mM	↓ senescence, ROS burst, ↑ Chls, photosynthesis, sucrose, starch, N	[67,99,100]
	Diplocarpon mali	50–500	↑ resistance to fungal infection, ↓ leaf lesions, cell death, pathogen expansion	[101]
	Apple Replant Disease—ARD	200	↑ growth, photosynthesis, K levels, soil microbial, ↓ ARD effects, ROS	[102]
	Apple Stem Grooving Virus	15	↑ shoot regrowth, 95% shoots virus-free, virus-free area	[103]
Apricot	-	10 ppm	↑ leaf growth, photosynthesis, fruit yield, size and retention, TA, TSS	[104]
Banana	Post-harvest	200–500	↑ shelf life of fruits, ↓ ethylene, ripening, quality sharp changes	[105]
	Anthracnose	10 mM	↑ fruit resistance, banana shelf life, ↓ anthracnose disease	[106]
Barley	Cold, drought	1 mM	↑ photosynthesis efficiency, ABA, water content, ROS	[72]
	Leaf-senescence	0.01–1	↑ Chls, growth, ↓ senescence	[59,107]
Bermudagrass	Cold, salt, drought	20–100	↑ growth, osmoregulation, ↓ ROS burst, cell damage	[108,109]

Table 1. *Cont.*

Plant Species	Stress Type	Melatonin Treatment (μM)	Effects Observed	Reference
Broccoli	-	60 ppm	↑ growth, photosynthetic attributes: LAI, NAR, AGR, CGR, Chls, carotenoids	[110]
Cabbage	Metal-Cu	1–100	↑ germination, growth, ↓ membrane peroxidation	[111]
	-	100–1000	↑ growth, anthocyanins, osmoregulation, redox balance, ↓ ABA, senescence factors, Chls degradation	[112,113]
Cassava	-	100	↓ ROS, postharvest deterioration, starch degradation	[114]
	Xam bacterial bligh		↑ disease resistance, ↓ bacterial propagation in leaves	[115]
Cherry sweet	Rootstocks	0.5–5	↑ number of roots, length, % rooting in 3 cherry rootstocks	[116,117]
			↑ photosynthetic pigments, biomass, total carbohydrates and proline	
	Orchard trees	10	↓ sweet cherries ripening, anthocyanins	[118]
Citrus	Salinity	1	↑ osmoregulation, Chls, ↓ ROS burst, membrane peroxidation	[119]
Coffee	Drought	300	↑ tolerance, root system, photosynthesis, gas exchange, CO_2 fixation, Chls, ASA-GSH cycle, ↓ ROS, MDA	[120]
Cotton	-	20	↑ germination, growth, antioxidant enzymes, GA_3, ↓ ROS, MDA, ABA	[121]
Cucumber	Cold	25–500 mM	↑ germination, ↓ ROS, membrane peroxidation	[122]
	Cold	50–500	↑ GSH pool, ↓ ROS	[123]
	Cold	200	↑ tolerance, photosynthesis, polyamines, ABA, GSH-ASA cycle, ↓ electrolyte leakage, ROS,	[124,125]
	Heat	100	↑ tolerance, N metabol., nitrate, ↓ damage, ammonium	[126]
	N-excess	100	↑ tolerance, growth, NPK balance, Ca, ↓ damage, nitrate, ammonium	[127]
	Salinity	1	↑ germination, GA_4, ↓ ROS, membrane peroxidation, ABA	[69]
		50–150	↑ tolerance, Chls, photosynthesis, GSH-ASA cycle, ↓ ROS	[128]

Table 1. *Cont.*

Plant Species	Stress Type	Melatonin Treatment (µM)	Effects Observed	Reference
		1	↑ germination, protein biosynthesis, lipid and carbohydrate metabol., TCA, ATP	[129]
	Drought	100	↑ germination, root growth	[130]
	Oxidative	50	↑ systemic antioxidant defence, GSH, photosynthesis, ↓ ROS	[131]
	Metal-Cu	0.01	↑ tolerance, growth, Cu-sequestration, TCA, ATP, GSH, ↓ ROS	[132]
	Cinnamic acid	100	↑ tolerance, growth, water and nutrient balance, hormonal balance	[133]
Faba bean	Salinity	100–500	↑ plant height, RWC, photosynthetic pigments, osmolites, phenolic	[134]
Grape	Drought	0.05–0.2	↑ seedling growth, osmoregulation, photosynthesis, ↓ ROS burst	[135]
	Salinity-Rhizobacteria	-	↑ root growth, RWC, melatonin in roots, colonization, ↓ damage, ROS	[136]
	Berry ripening	100	↑ anthocyanins, phenols, flavonoids, proanthocyanidins, resveratrol, ↓ ROS	[137,138]
	Berry/Wine	430	↑ size- and ripening-berries, fruity-, spicy- and sweet-wine	[139]
Kiwifruit	Drought	50–200	↑ tolerance, photosynthesis, CO_2 fixation, growth, biomass, roots, osmoregulation, flavonoids ↓ lipid peroxidation, carotenoid degradation	[140,141]
	Heat	200	↑ tolerance, ASA, proline, antioxidant enzymes, ↓ heat damage, ROS	[142]
Leek	Cold, heat	5	↑ tolerance, germination, growth	[143]
Lychee	Post-harvest	400	↑ redox balance, antioxidant enzymes, ↓ pericarp browning, discoloration, ROS, membrane leakage, loss of phenolics, flavonoids and anthocyanins	[144]
Lupin	Several stress	-	↑ germination, growth, rooting, redox balance, ↓ ROS, foliar senescence	[58,60,62,145]
Maize	Salinity, heat	100	↑ photosynthesis, antioxidant enzymes, ↓ ROS, electrolyte leakage	[146,147]
	-	10–1000	↑ root and stem growth, plant height, leaf surface area, protein, carbohydrates, Chls	[148]

Table 1. *Cont.*

Plant Species	Stress Type	Melatonin Treatment (μM)	Effects Observed	Reference
	Drought	100	↑ tolerance, growth, photosynthesis, stoma conductance, transpiration, RWC, antiox enzymes, ↓ ROS, MDA	[149–151]
	Heat	10–90	↑ tolerance, antioxidant enzymes, osmoregulation, ↓ ROS, MDA, electrolyte leakage	[152]
	Metal-Pb	50–100	↑ tolerance, growth, photosynthesis, Chls, RWC, K, Ca levels, ↓ ROS, MDA	[153]
	-	10	↑ sugar metabolism, photosynthesis, sucrose phloem loading	[154]
	-	50–500	protein synthesis, folding, destination and storage, defence, anti-stresses and detoxifying proteins	[155]
	Aging seeds		↑ viability, growth, antioxidant enzymes, carbohydrate-, secondary-, and amino acid metabol., ↓ ROS, MDA	[156]
Melissa (lemon balm)	Metal-Zn-Cd	1000	↑ tolerance, growth, antioxidant enzymes	[157]
Mung bean	Cold	20	↑ tolerance, growth, plastids, ↓ ROS, lipid peroxidation	[158]
Oat	Salinity, drought	50–100	↑ tolerance, growth, Chls, proline, antioxidant enzymes, ↓ ROS, MDA	[59,159,160]
Onion	Cold, heat	5	↑ tolerance, germination, growth	[143]
Peach (fruit)	Cold	50–200	↑ juice, TSS, polyamines, GABA, proline, ↓ chilling injury, ROS	[161]
	Post-harvest	100	↑ firmness, TSS. ASA, ↓ weight loss, decay incidence, respiration rate,	[162]
Pear (tree)	-	100	↑ photosynthesis, fruit size, TSS, sucrose, sorbitol, starch	[163]
	Parthenocarpy	100	↑ parthenocarpy with expansion, division mesocarp cells, unviable seeds, GAs	[73]
Pear (fruit)	Post-harvest	100	↑ firmness, commercial value, ↓ weight loss, ethylene, softening, core browning	[164,165]
Pepper	Salinity, Fe-low	100	↑ growth, Chls, photosynthesis, fruit yield, Fe, K uptake, antioxidant enzymes	[166]
	Cold	1–5	↑ germination, growth, antioxidant enzymes, ↓ ROS, MDA	[167]

Table 1. *Cont.*

Plant Species	Stress Type	Melatonin Treatment (μM)	Effects Observed	Reference
	Boron-high	1	↑ tolerance, growth, photosynthesis, ↓ B in leaf and fruit, toxicity	[168]
Pea	Oxidative	50–200	↑ photosynthesis efficiency, pigments, water content, ↓ ROS	[169,170]
	Metal-Cu	5	↑ plant survival	[171]
Plum (fruit)	Cold	1–1000	↑ firmness, postharvest life, ASA, phenols, antioxidant activity, ↓ weight loss	[172]
Pomegranate	Cold	100	↑ tolerance, antioxidant enzymes, membrane integrity, phenols, ↓ ROS	[173]
Poplar	Oxidative		↑ redox balance, proline, ↓ ROS, MDA, membrane damage, electrolyte leakage	[174]
Potato	Salinity	0.1–200	↑ tolerance, K^+/Na^+ homeostasis, ATPase, triacylglycerol breakdown, fatty acid β-oxidation, energy turnover	[175]
	Phytophthora infestans (potato late blight)	1–10 mM	↑ plant innate immunity, fungicide resistance and virulence, synergistic anti-fungal effects of melatonin with fungicides	[176]
Radish	Heat	50–300	↑ biomass, quality, antioxidant enzymes, Chls, hormone contents	[177]
Rapeseed	Salinity	0.01–100	↑ tolerance, redox balance, ion homeostasis, ↓ ROS, MDA	[178]
	Drought	500	↑ tolerance, germination, Chls, stoma size, osmoregulation, antioxidant enzymes, ↓ ROS, MDA	[179]
Rice	Cold	20–100	↑ tolerance, growth, photosynthesis, redox balance, ↓ ROS	[180]
	Salinity	10–20	↑ Chls, ↓ senescence, ROS, cell death	[181]
	Metal-Cd		↑ tolerance, growth, photosynthesis, redox balance, panicle number, grain yield	[182,183]
	Bacterial blight	200	↓ bacterial proliferation, motility	[184]
	Salt, cold, Blast fungus	-	↑ tolerance, melatonin induction, hormones, ↓ fungi proliferation	[185]
	-	0.5–1	↑ seminal roots, lateral roots, root growth, root biomass	[186]

Table 1. *Cont.*

Plant Species	Stress Type	Melatonin Treatment (μM)	Effects Observed	Reference
	Soil	10–50	↑ number of lateral roots, root growth, shaping root architecture	[187]
Soybean	Salinity, drought	50–100	↑ tolerance, seedling growth, leaf size, biomass, seed yield	[188]
	Drought	100	↑ RWC, Chls, photosynthetic gas-exchange parameters, osmoregulation, antioxidant enzymes, and seed growth-related indicators	[189]
	Metal-Al	0.1–1	↑ tolerance, root growth, antioxidant enzymes, osmoregulation, ↓ ROS,	[190]
Spinach	Boron	100–300	↑ tolerance, growth, photosynthesis, RWC, CO_2 uptake, sugars, carotenoids, redox balance, ↓ ROS, MDA	[191]
Strawberry	Post-harvest	100	↑ nutritional quality, antioxidant enzymes, anthocyanins, phenols, GABA, ATP ↓ fungal decay	[192]
	Post-harvest	0.1–1	↑ color, firmness, TSS, ASA, flavonoids, ↓ weight loss, senescence, ROS, MDA	[193]
Sunflower	Salinity	15	↑ root, hypocotyl growth, antioxidant potential, antioxidant enzymes, GSH	[194–196]
Tea plant	Salinity, cold, drought	100	↑ photosynthesis, GSH, ASA, antioxidant enzymes, ↓ ROS, MDA	[197,198]
Tobacco	Tobacco mosaic virus	100	↑ tolerance, ↓ virus proliferation, virus-RNA, viral disease	[199]
Tomato	Salinity	50–150	↑ photosynthesis, PSII efficiency, D1 protein turnover, ↓ ROS burst	[200]
	Salinity	20–50	↑ growth, photosynthesis, Rubisco, proline, C-metabol., ASA-GSH cycle, ↓ ROS, MDA	[201]
	Salinity	150	↑ tolerance, photosynthesis, PSII repair, ASA-GSH cycle, ↓ ROS	[202]
	Cold	100	↑ antioxidant enzymes, GSA, ASA, CO_2 uptake, sucrose, proline, Calvin cycle, polyamines, ↓ ROS, MDA, electrolyte leakage	[203]
	Cold	100	↑ tolerance, growth, VAZ cycle, photosynthesis, photosystem efficiency, ↓ ROS, MDA, photoinhibition	[204]
	Cold-fruit	100	↑ tolerance, proline, polyamines, membrane integrity	[205]

Table 1. *Cont.*

Plant Species	Stress Type	Melatonin Treatment (μM)	Effects Observed	Reference
	Heat	10	↑ thermotolerance and cell protection	[206,207]
	Heat-pollen	20	↑ thermotolerance, polen germination, antioxidant enzymes, reproductive development	[208]
	Metal-Cd	25–500	↑ Cd tolerance, phytochelatins, ATPase activity	[209]
	Metal-Cd	-	↑ Cd tolerance, heat-shock factor A1a induction by melatonin	[210]
	Metal-Cd-Se	-	↑ growth, photosynthesis, electrolyte leakage, phytochelatins, GSH, ↓ ROS, Cd leaf,	[211]
	Alkalinity	0.25–1	seedling growth, photosynthesis, ion homeostasis, burst	[212]
	Acid rain	100	↑ tolerance, growth, chloroplast integrity, photosynthesis, antioxidant enzymes, ↓ ROS, MDA	[213]
	Drought	100	↑ tolerance, waxes-cutin leaf, RWC, Chls,	[214]
	Drought	100	↑ tolerance, Chls, antioxidant enzymes, p-coumaric acid, ↓ ROS, MDA	[215]
	S-low	100	↑ S uptake, assimilation, transport and metabolism, peroxiredoxins, redox homeostasis, ↓ ROS, DNA damage	[216]
	Rooting	50	↑ adventitious root formation, auxin, auxin transport and signal transduction	[217]
	On vine-ripening		↑ fruit yield and quality, ASA, citric acid, lycopene, TSS, Ca, P ↓ N, Mg, Cu, Zn, Fe, Mn,	[218]
	Post-harvest	50	↑ fruit ripening, fruit quality, colour, carotenoids, polygalacturonase and related, biosynthesis, perception and signalling of ethylene, anthocyanins, ↓ weight loss	[74,219]
	Mosaic virus	100	↑ tolerance, ↓ virus proliferation, virus-RNA, viral disease	[199]
Valerian	Metal-Zn-Cd	1000	↑ tolerance, growth, antioxidant enzymes	[157]
Watermelon	Cold	150	↑ photosynthesis, ↓ cold-related microRNA	[220]
	Salinity	50–500	↑ tolerance, growth, photosynthesis, antioxidant enzymes, GSH, ASA, ↓ ROS, MDA	[221]

Table 1. *Cont.*

Plant Species	Stress Type	Melatonin Treatment (μM)	Effects Observed	Reference
	Metal-V	0.1	↑ tolerance, growth, photosynthesis, antioxidant enzymes, ↓ V level, V transport, ROS, MDA	[222]
Wheat	Cold	1000	↑ redox balance, Chls, osmoregulation, ↓ ROS	[223]
	Cold	1000	↑ tolerance, growth, Chls, photosynthesis, CO_2 uptake, grain filled	[224]
	Cold	1000	↑ photosynthesis, stomatal conductance, antioxidant enzymes, membrane stability	[225]
	Salinity	1	↑ tolerance, growth, photosynthesis, IAA, polyamines, ↓ ROS	[226]
	Salinity	50–500	↑ growth, yield, antioxidant enzymes, ↓ ROS, MDA	[227]
	Drought	500	↑ tolerance, RWC, photosynthesis, antioxidant enzymes, ASA, GSH, ↓ ROS, membrane damage	[228]
	Metal-Cd	100	↑ tolerance, antioxidant enzymes, ASA, GSH, ↓ ROS	[229]
	Metal-Cd	50–100	↑ tolerance, growth, Chls, photosynthesis, RWC, Ca, K, antioxidant enzymes, ↓ ROS, MDA, Cd	[230]
	Metal-Zn	1000	↑ tolerance, Chls, photosynthesis, Rubisco, ATPase	[231]
	N-low	1	↑ N and nitrate, N absorption, N metabolism, growth, yield, in shoots and roots	[232]

↑, Increased content or increased action. ↓, decreased content or decreased action; ABA, abscisic acid; AGR, absolute growth rate; ASA, ascorbic acid; CGR, crop growth rate; Chls, chlorophylls; CMC, component materials categories of fertilizers; EC, European Community; ECHA, European Chemical Agency; EU, European Union; GA_4, gibberellin-4; GABA, γ-aminobutyric acid; GSH, glutathione; JA, Jasmonic acid; LAI, leaf area index; MDA, malondialdehyde; MAPKK, mitogen-activated protein kinase cascade; NAR, net assimilation rate; OXI1, oxidative signal-inducible1 kinases; PFC, product function categories of fertilizers; ROS, reactive oxygen species; RWC, relative water content; SA, salicylic acid; SAR, systemic acquired resistance; TA, total valuable acidity; TCA, Krebs cycle; TSS, total solid soluble.

4. Melatonin in the Health and Environment of EC

In accordance with the Classification, Labelling and Packaging (CLP, EC-No 1272/2008) regulation, which is based on the United Nations' Globally Harmonized System, which has a purpose to ensure a high level of protection of health and the environment, as well as the free movement of substances, mixtures and articles, the European Chemical Agency (ECHA) classified melatonin (EC No. 200-797-7 (CAS 73-31-4), *N*-(2-(5-methoxyindol-3-yl)-ethyl)-acetamide), as a non-hazardous substance in terms of physical and chemical hazards. With respect to human health, it is classified as a non-hazardous substance in the oral, dermal, inhalation and irritation categories, and in regards to mutagenicity and carcinogenicity. However, melatonin is classified as a health hazard substance (code H-361) in terms of reproductive toxicity because it is suspected of damaging fertility or an unborn child. This classification reflects one of its multiple functions as an animal hormone, in which its participation in the modulation of sexual behavior in mammals has been demonstrated, and also, it is believed, the same of fertility [233,234]. In fact, it is usually applied to sheep as a hormonal regulator of sexual zeal to homogenize the reproductive process in ovine, with demonstrated higher conception and pregnancy rates when applied [235]. Nevertheless, melatonin is classified as non-hazardous in terms of its possible damage to the environment and atmosphere.

5. Melatonin as an Active Substance or as a Plant Biostimulator/Protector in Crops: Concepts and Legal Considerations in EC

After many changes and adaptations, the EC finally seems to have established its policy regarding the authorization, classification, use, distribution, importation, management, etc., of plant protectors and fertilizers, in an attempt to improve agricultural production, while minimizing risks and hazards for humans, animals and the environment. In order to establish the minimum basis for the possible use of melatonin in plant production and post-harvest application, several requirements regarding its human consumption must be taken into account:

(i) Melatonin is a highly studied substance that has given rise to abundant physicochemical and biological data; (ii) there are numerous studies in animals and humans regarding its beneficial effects on health, in aspects as diverse as neurodegenerative, immunological, liver, renal, heart, skin and gastrointestinal diseases, in addition to osteopathy, retinopathy, etc. It also helps in the treatment of various cancers, particularly, chemical and radiological therapies; (iii) in regards to melatonin for human consumption, although it is classified as a drug in the EC, there are some cases in which it does not need a medical prescription, such as those where the amount of melatonin is less than 1 mg. Generally, these are used for jet-lag and sleep disorders. In many other countries (e.g., USA, Canada) melatonin is not treated as a drug, but as a food supplement; (iv) in no case has melatonin been declared as toxic, even at the intake of 1 g/day. Only some slight side effects such as migraine and headache have been described.

The possible use of melatonin in plant production involves particular aspects such as: (i) Melatonin is a molecule that exists in all living things, from bacteria to humans, but also in plants, algae, fungi, etc.; (ii) its action in animals and humans is well known since it has been investigated for many years. In plants, although many physiological effects of melatonin are known, new data are being acquired every day; (iii) in all cases, only positive effects have been described, all beneficial for the development of plants (the same can be said for animals); (iv) little information is available on its effect on bacteria and fungi, especially those that are part of the soil microbiota (rhizosphere); (v) there are also few or no data on its effect on the environment, in particular on agricultural and aquatic fauna; (vi) the levels of melatonin described in plants, and which appear to be effective in pharmacological treatments known to date, are much higher than those described in animals or humans, which may be a cause for caution.

Council Directive 91/414/EEC of 15 July, 1991 concerning the marketing of plant protection products provides rules governing plant protection products and the active substances contained in those products. This old directive has been replaced by two more current ones that are as follows:

- Regulation #1. Regulation EC 1107/2009 of the European Parliament and of the Council of 21 October 2009 concerning the placing of plant protection products on the market and repealing council directives 79/117/EEC and 91/414/EEC, and;
- Regulation #2. Regulation EU 2019/1009 of the European Parliament and of the Council of 5 June 2019 laying down rules on the making available on the market of EU fertilizing products and amending Regulations (EC) No. 1069/2009 and (EC) No. 1107/2009 and repealing Regulation (EC) No 2003/2003.

If we review the actions confirmed so far for melatonin in plants, we find that melatonin exerts a clear action as a plant protector in situations of biotic stress against bacterial, fungal and viral diseases (Regulation #2), but it can also be used as an agent against situations of abiotic stress (Regulation #1). Thus, Regulation #2 says in point 22:

> "Certain substances, mixtures and micro-organisms, referred to as plant biostimulants, are not as such inputs of nutrients, but nevertheless stimulate plants' natural nutrition processes. Where such products aim solely at improving the plants' nutrient use efficiency, tolerance to abiotic stress, quality traits or increasing the availability of confined nutrients in the soil or rhizosphere, they are by nature more similar to fertilising products than to most categories of plant protection products. They act in addition to fertilisers, with the aim of optimising the efficiency of those fertilisers and reducing the nutrient application rates. Such products should therefore be eligible for CE marking under this Regulation and excluded from the scope of Regulation (EC) No 1107/2009".

These two regulations attempt to classify the substances and products applicable to crops into two large groups: Those that are plant protectors (phyto-sanitary) (Regulation #1) and those that can be used as fertilizers (Regulation #2). As we have seen in the previous section, melatonin is classified as a health hazard substance (code H-361) for its reproductive toxicity in ECHA, so its possible authorization as an active substance by regulation EC 1907/2006 of Registration, Evaluation, Authorisation and Restriction of Chemicals (REACH) could be difficult.

Although Regulation #1 on plant protection products extends the concept of an active substance, since it includes microorganisms and preparations (art. 1 point 2): This Regulation shall apply to substances, including micro-organisms having general or specific action against harmful organisms or on plants, parts of plants or plant products, referred to as 'active substances', some interesting restrictions appeared in:

- Art. 23b: "Basic substances shall be approved in accordance with paragraphs 2 to 6. (...) For the purpose of paragraphs 2 to 6, a basic substance is an active substance which (...), (b) does not have an inherent capacity to cause endocrine disrupting, neurotoxic or immunotoxic effects";
- Annex II, Impact on Human Health, 3.6.5: "An active substance, safener or synergist shall only be approved if, on the basis of the assessment of community or internationally agreed test guidelines or other available data and information, including a review of the scientific literature, reviewed by the Authority, it is not considered to have endocrine disrupting properties that may cause adverse effect in humans, unless the exposure of humans to that active substance, safener or synergist in a plant protection product, under realistic proposed conditions of use, is negligible, ... " and in;
- Annex II. Ecotoxicology, 3.8.2. An active substance, safener or synergist shall only be approved if, on the basis of the assessment of community or internationally agreed test guidelines, it is not considered to have endocrine disrupting properties that may cause adverse effects on non-target organisms unless the exposure of non-target organisms to that active substance in a plant protection product under realistic proposed conditions of use is negligible.

Thus, taking into account all this legal information, and ruling out the possibility of using melatonin as an active substance (pure chemical substance) for agronomic application, the possibility of using plant, bacterial, algae, or fungi extracts rich in melatonin would remain. Thus, a good plan might be to use plant (or other) extracts rich in melatonin as a fertilizer, in the category of biostimulants. A biostimulant could also be defined as a formulated product of biological origin that improves plant productivity as a consequence of the emergent properties of its constituents. Thus, biostimulants could be defined by their demonstrated mode of action and origin, or solely by their demonstrated beneficial impact on plant productivity. The challenges in developing a definition are also complicated by the multi-component and largely undefined composition of many biostimulant products and the possibility that the activity of a biostimulant may not be explained by the presence of any individual constituent, but is a result of the interaction of many constituents in the product. Indeed, most biostimulants in use today are complex mixtures of chemicals derived from a biological process or the extraction of biological materials [236].

According Regulation #2 (EU 2019/1009) on fertilizing products, in Annex I, Product Function Categories (PFCs) of EU fertilizing products, in Category 6, two types of plant biostimulant can be developed: Microbial plant biostimulants (subtype A) and non-microbial plant biostimulants (subtype B). In Annex II, it says: "An EU fertilizing product shall consist solely of component materials complying with the requirements for one or more of the CMCs listed in this Annex", where the different component materials categories (CMC) were defined. Of interest are the following:

- CMC2: Plants, plant parts or plant extracts is described as: "An EU fertilizing product may contain plants, plant parts or plant extracts having undergone no other processing than cutting, grinding, milling, sieving, sifting, centrifugation, pressing, drying, frost treatment, freeze-drying or extraction with water or supercritical CO_2 extraction. For the purpose of this point, plants include mushrooms and algae and exclude blue-green algae (cyanobacteria)."
- CMC6: Food industry by-products, point (e): "Plants, plant parts or plant extracts having undergone only heat treatment or heat treatment in addition to processing methods referred to in CMC 2"
- CMC7: Micro-organisms. "An EU fertilising product belonging to PFC 6A may contain micro-organisms, including dead or empty-cell micro-organisms and non-harmful residual elements of the media on which they were produced".

The strategies to obtain melatonin-rich extracts may involve microorganisms (PFC6A) or plants (PFC 6B). At present, there seem to be no data on the production of melatonin by bacteria or fungal cultures. The objective to obtain melatonin-rich plants (CMC2) is ambitious since phytomelatonin levels in plants are usually very low, and less than 5–10 ng per gram of plant. An exhaustive classification of many plants according to their phytomelatonin content can be consulted [37,237,238]. Generally, medicinal plants have high phytomelatonin content, but this tends to vary widely due to the varied origin of plants, technical conditions of growth, variety, post-harvest treatment, etc. Several strategies can be followed: (i) Selecting plant species with high levels of phytomelatonin which can be extracted and concentrated, and (ii) inducing the biosynthesis of phytomelatonin in in vitro cultured pre-selected plant tissues. A discussion on this aspect can be consulted [239]. Our group is developing a formulation where only aromatic/medicinal plants are used to obtain a botanical mixture rich in phytomelatonin through the application of a simple process. A rigorous plant selection protocol and careful management will ensure high phytomelatonin content in the plant extracts generated. The formulation and its protocol are being patented before being made available to interested companies for commercial exploitation. We are currently characterizing it and conducting the appropriate studies and bioassays in plants to confirm its beneficial biological activity related with its high phytomelatonin content.

Figure 2 shows, according to the legislation analyzed, the pros and cons of melatonin (as a chemical substance) and phytomelatonin-rich extracts and its possible regularization as a plant protector or fertilizer (biostimulant).

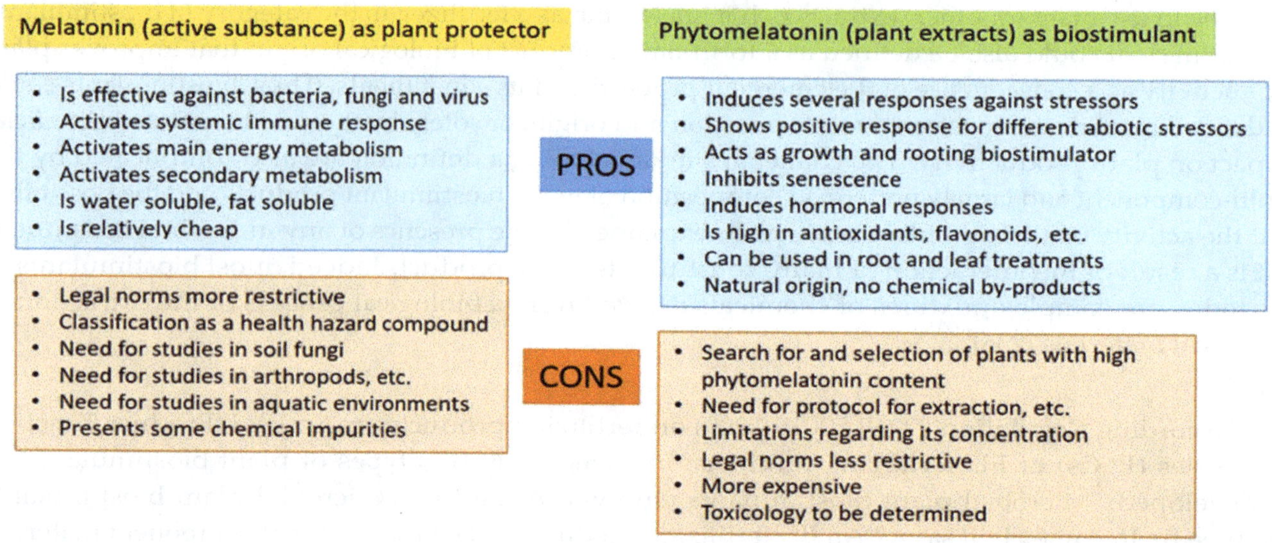

Figure 2. Pros and cons of the possible use of chemical melatonin and rich-phytomelatonin extracts according to EC legislation.

6. Future Prospects

Numerous studies with melatonin have resulted in a set of data that indicate the excellent beneficial effects that this compound has on plants, especially in stress situations. It should not be forgotten that melatonin is a natural compound, endogenous to plants and other organisms including humans. It is this last aspect that makes it more interesting and also more delicate or sensitive, when using it as a plant protective agent or as a biostimulator. However, demonstrating through trials that its use is possible in crops and does not entail risks to human and wildlife health will be the only way forward in this field. The alternative of using phytomelatonin-rich extracts seems more interesting, but also more laborious. The search and selection of plants with high endogenous levels of phytomelatonin is a first requirement for subsequent extraction and preparation. The analysis and study of its potential as a protector against plant stress will throw light on the true effect on crops. However, although many aspects of the mechanism of action of phytomelatonin are already known, there are other relevant aspects to study as: (i) The optimal mode of application, time and rate; (ii) the phenological state; (iii) the effect on rhizosphere; (iv) the persistence in soil or in foliar applications; (v) the synergic or antagonic effects with other plant treatments (pesticides, fertilizers, etc.), among others. Obviously, companies in the phytochemical sector (manufacturers) will need to start field studies and deal with possible legal regularization.

Abbreviations

ABA	abscisic acid
AGR	absolute growth rate
ASA	ascorbic acid
CGR	crop growth rate
Chls	chlorophylls
CMC	component materials categories of fertilizers
EC	European Community
ECHA	European Chemical Agency
EU	European Union
GA_4	gibberellin-4
GABA	γ-aminobutyric acid

GSH	glutathione
JA	jasmonic acid
LAI	leaf area index
MDA	malondialdehyde
MAPKK	mitogen-activated protein kinase cascade
NAR	net assimilation rate
OXI1	oxidative signal-inducible1 kinases
PFC	product function categories of fertilizers
ROS	reactive oxygen species
RWC	relative water content
SA	salicylic acid
SAR	systemic acquired resistance
TA	total valuable acidity
TCA	Krebs cycle
TSS	total solid soluble

Author Contributions: The manuscript was conceived by M.B.A. and written and revised by M.B.A. and J.H.-R.

References

1. Lerner, A.B.; Case, J.D.; Takahashi, Y.; Lee, T.H.; Mori, W. Isolation of melatonin, a pineal factor that lightens melanocytes. *J. Am. Chem. Soc.* **1958**, *80*, 2587. [CrossRef]
2. Lerner, A.; Case, J.; Mori, W.; Wright, M. Melatonin in peripheral nerve. *Nature* **1959**, *183*, 1821. [CrossRef] [PubMed]
3. Lerner, A.; Case, J.; Heinzelmann, R. Structure of melatonin. *J. Am. Chem. Soc.* **1959**, *81*, 6084–6085. [CrossRef]
4. Xie, Z.; Chen, F.; Li, W.A.; Geng, X.; Li, C.; Meng, X.; Feng, Y.; Liu, W.; Yu, F. A review of sleep disorders and melatonin. *Neurol. Res.* **2017**, *39*, 559–565. [CrossRef] [PubMed]
5. Vadnie, C.A.; McClung, C.A. Circadian rhythm disturbances in mood disorders: Insights into the role of the suprachiasmatic nucleus. *Neural Plast.* **2017**, *2017*, 1504507. [CrossRef]
6. Carpentieri, A.R.; Oliva, C.; Díez-Noguera, A.; Cambras, T. Melatonin administration modifies circadian motor activity under constant light depending on the lighting conditions during suckling. *Chronobiol. Int.* **2015**, *32*, 994–1004. [PubMed]
7. Blume, C.; Angerer, M.; Raml, M.; del Giudice, R.; Santhi, N.; Pichler, G.; Kunz, A.B.; Scarpatetti, M.; Trinka, E.; Schabus, M. Healthier rhythm, healthier brain? Integrity of circadian melatonin and temperature rhythms relates to the clinical state of brain-injured patients. *Eur. J. Neurol.* **2019**, *26*, 1051–1059. [CrossRef]
8. Marková, M.; Adámeková, E.; Kubatka, P.; Bojková, B.; Ahlersová, E.; Ahlers, I. Effect of prolonged melatonin administration on metabolic parameters and organ weights in young male and female Sprague-Dawley rats. *Acta Vet. Brno* **2003**, *72*, 163–173. [CrossRef]
9. Bojková, B.; Marková, M.; Ahlersová, E.; Ahlers, I.; Adámeková, E.; Kubatka, P.; Kassayová, M. Metabolic effects of prolonged melatonin administration and short-term fasting in laboratory rats. *Acta Vet. Brno* **2006**, *75*, 21–32. [CrossRef]
10. Cardinali, D.P.; Hardeland, R. Inflammaging, metabolic syndrome and melatonin: A call for treatment studies. *Neuroendocrinology* **2017**, *104*, 382–397. [CrossRef]
11. Kubatka, P.; Bojková, B.; Mociková, K.; Mníchová, M.; Adámeková, E.; Ahlers, I.; Ahlersová, E.; Cermáková, M. Effects of tamoxifen and melatonin on mammary gland cancer induced by N-methyl-N-nitrosourea and by 7,12-dimethylbenz(a)anthracene, respectively, in female Sprague-Dawley rats. *Folia Biologica (Praha)* **2001**, *47*, 5–10.
12. Di Bella, G.; Mascia, F.; Gualano, L.; Di Bella, L. Melatonin anticancer effects: Review. *Int. J. Mol. Sci.* **2013**, *14*, 2410–2430. [CrossRef]
13. Bojková, B.; Kubatka, P.; Qaradakhi, T.; Zulli, A.; Kajo, K. Melatonin may increase anticancer potential of pleiotropic drugs. *Molecules* **2018**, *19*, 3910. [CrossRef]

14. Nabavi, S.M.; Nabavi, S.F.; Sureda, A.; Xiao, J.; Dehpour, A.R.; Shirooie, S.; Silva, A.S.; Baldi, A.; Khan, H.; Daglia, M. Anti-inflammatory effects of melatonin: A mechanistic review. *Crit. Rev. Food Sci. Nutr.* **2019**, *59*, S4–S16. [CrossRef]
15. Carrascal, L.; Nuñez-Abades, P.; Ayala, A.; Cano, M. Role of melatonin in the inflammatory process and its therapeutic potential. *Curr. Pharm. Des.* **2018**, *24*, 1563–1588. [CrossRef]
16. Carrillo-Vico, A.; Guerrero, J.M.; Lardone, P.J.; Reiter, R.J. A review of the multiple actions of melatonin on the immune system. *Endocrine* **2012**, *27*, 189–200. [CrossRef]
17. Reiter, J.R.; Rosales-Corral, A.S.; Tan, D.X.; cuna-Castroviejo, D.; Qin, L.; Yang, S.F.; Xu, K. Melatonin, a full service anti-cancer agent: Inhibition of initiation, progression and metastasis. *Int. J. Mol. Sci.* **2017**, *18*, 843. [CrossRef]
18. Majidinia, M.; Reiter, R.J.; Shakouri, S.K.; Yousefi, B. The role of melatonin, a multitasking molecule, in retarding the processes of ageing. *Ageing Res. Rev.* **2018**, *47*, 198–213. [CrossRef]
19. Boga, J.A.; Caballero, B.; Potes, Y.; Perez-Martinez, Z.; Reiter, R.J.; Vega-Naredo, I.; Coto-Montes, A. Therapeutic potential of melatonin related to its role as an autophagy regulator: A review. *J. Pineal Res.* **2019**, *66*, e12534. [CrossRef]
20. Luo, C.; Yang, Q.; Liu, Y.; Zhou, S.; Jiang, J.; Reiter, R.J.; Bhattacharya, P.; Cui, Y.; Yang, H.; Ma, H.; et al. The multiple protective roles and molecular mechanisms of melatonin and its precursor N-acetylserotonin in targeting brain injury and liver damage and in maintaining bone health. *Free Radic. Biol. Med.* **2019**, *130*, 215–233. [CrossRef]
21. Arnao, M.B.; Hernández-Ruiz, J. Melatonin: Synthesis from tryptophan and its role in higher plants. In *Amino Acids in Higher Plants*; D' Mello, J.P.F., Ed.; CAB Intern.: Boston, MA, USA, 2015; pp. 390–435.
22. Jan, J.E.; Reiter, R.J.; Wasdell, M.B.; Bax, M. The role of the thalamus in sleep, pineal melatonin production, and circadian rhythm sleep disorders. *J. Pineal Res.* **2009**, *46*, 1–7. [CrossRef]
23. McHill, A.W.; Phillips, A.J.; Czeisler, C.A.; Keating, L.; Yee, K.; Barger, L.K.; Garaulet, M.; Scheer, F.A.; Klerman, E.B. Later circadian timing of food intake is associated with increased body fat. *Am. J. Clin. Nutr.* **2017**, *106*, 1213–1219. [CrossRef]
24. Dubbels, R.; Reiter, R.J.; Klenke, E.; Goebel, A.; Schnakenberg, E.; Ehlers, C.; Schiwara, H.W.; Schloot, W. Melatonin in edible plants identified by radioimmunoassay and by HPLC-MS. *J. Pineal Res.* **1995**, *18*, 28–31. [CrossRef]
25. Hattori, A.; Migitaka, H.; Iigo, M.; Yamamoto, K.; Ohtani-Kaneko, R.; Hara, M.; Suzuki, T.; Reiter, R.J. Identification of melatonin in plants and its effects on plasma melatonin levels and binding to melatonin receptors in vertebrates. *Biochem. Mol. Biol. Int.* **1995**, *35*, 627–634.
26. Kolar, J.; Machackova, I.; Illnerova, H.; Prinsen, E.; van Dongen, W.; van Onckelen, H. Melatonin in higher plant determined by radioimmunoassay and liquid chromatography-mass spectrometry. *Biol. Rhythm Res.* **1995**, *26*, 406–409.
27. van Tassel, D.L.; Roberts, N.; O'Neill, S.D. Melatonin from higher plants: Isolation and identification of N-acetyl-5-methoxytryptamine. *Plant Physiol.* **1995**, *108*, 101.
28. Back, K.; Tan, D.X.; Reiter, R.J. Melatonin biosynthesis in plants: Multiple pathways catalyze tryptophan to melatonin in the cytoplasm or chloroplasts. *J. Pineal Res.* **2016**, *61*, 426–437. [CrossRef]
29. Zheng, X.; Tan, D.X.; Allan, A.C.; Zuo, B.; Zhao, Y.; Reiter, R.J.; Wang, L.; Wang, Z.; Guo, Y.; Zhou, J.; et al. Chloroplastic biosynthesis of melatonin and its involvement in protection of plants from salt stress. *Sci. Rep.* **2017**, *7*, 41236. [CrossRef]
30. Wang, L.; Feng, C.; Zheng, X.; Guo, Y.; Zhou, F.; Shan, D.; Liu, X.; Kong, J. Plant mitochondria synthesize melatonin and enhance the tolerance of plants to drought stress. *J. Pineal Res.* **2017**, *63*, e12429. [CrossRef]
31. Arnao, M.B.; Hernández-Ruiz, J. Melatonin: A new plant hormone and/or a plant master regulator? *Trends Plant Sci.* **2019**, *24*, 38–48. [CrossRef]
32. Arnao, M.B.; Hernández-Ruiz, J. Melatonin in its relationship to plant hormones. *Ann. Bot.* **2018**, *121*, 195–207. [CrossRef]
33. Arnao, M.B.; Hernández-Ruiz, J. The multi-regulatory properties of melatonin in plants. In *Neurotransmitters in Plants*; Roshchina, V.V., Ramakrishna, A., Eds.; Taylor & Francis-CRC: New York, NY, USA, 2018; pp. 71–101.
34. Sharif, R.; Xie, C.; Zhang, H.; Arnao, M.; Ali, M.; Ali, Q.; Muhammad, I.; Shalmani, A.; Nawaz, M.; Chen, P.; et al. Melatonin and its effects on plant systems. *Molecules* **2018**, *23*, 2352. [CrossRef]

35. Arnao, M.B.; Hernández-Ruiz, J. Growth activity, rooting capacity, and tropism: Three auxinic precepts fulfilled by melatonin. *Acta Physiol. Plant* **2017**, *39*, 127. [CrossRef]
36. Arnao, M.B.; Hernández-Ruiz, J. Functions of melatonin in plants: A review. *J. Pineal Res.* **2015**, *59*, 133–150. [CrossRef]
37. Arnao, M.B. Phytomelatonin: Discovery, content, and role in plants. *Adv. Bot.* **2014**, *2014*, e815769. [CrossRef]
38. Wang, Y.; Reiter, R.J.; Chan, Z. Phytomelatonin: A universal abiotic stress regulator. *J. Exp. Bot.* **2018**, *69*, 963–974. [CrossRef]
39. Reiter, R.J.; Rosales-Corral, S.; Tan, D.X.; Jou, M.J.; Galano, A.; Xu, B. Melatonin as a mitochondria-targeted antioxidant: One of evolution´s best ideas. *Cell Mol. Life Sci.* **2017**, *74*, 3863–3881. [CrossRef]
40. Debnath, B.; Islam, W.; Li, M.; Sun, Y.; Lu, X.; Mitra, S.; Hussain, M.; Liu, S.; Qiu, D. Melatonin mediates enhancement of stress tolerance in plants. *Int. J. Mol. Sci.* **2019**, *20*, 1040. [CrossRef]
41. Yu, Y.; Lv, Y.; Shi, Y.; Li, T.; Chen, Y.; Zhao, D.; Zhao, Z. The role of phyto-melatonin and related metabolites in response to stress. *Molecules* **2018**, *23*, 1887. [CrossRef]
42. Shibaeva, T.G.; Markovskaya, E.F.; Mamaev, A.V. Phytomelatonin: A review. *Biol. Bull. Rev.* **2018**, *8*, 375–388. [CrossRef]
43. Arnao, M.B.; Hernández-Ruiz, J. Melatonin and reactive oxygen and nitrogen species: A model for the plant redox network. *Melatonin Res.* **2019**, *2*, 152–168. [CrossRef]
44. Zhang, X.; Schmidt, R. The impact of growth regulators on a-tocopherol status of water-stressed *Poa pratensis* L. *Int. Turfgrass Soc. Res. J.* **1997**, *8*, 1364–1373.
45. Kauffman, G.L.; Kneivel, D.P.; Watschke, T.L. Effects of a biostimulant on the heat tolerance associated with photosynthetic capacity, membrane thermostability, and polyphenol production of perennial ryegrass. *Crop Sci.* **2007**, *47*, 261–267. [CrossRef]
46. du Jardin, P. Plant biostimulants: Definition, concept, main categories and regulation. *Sci. Hort.* **2015**, *196*, 3–14. [CrossRef]
47. Rouphael, Y.; Colla, G. Synergistic Biostimulatory Action: Designing the Next Generation of Plant Biostimulants for Sustainable Agriculture. *Front. Plant Sci.* **2018**, *9*, 1655. [CrossRef]
48. Rouphael, Y.; SpÍchal, L.; Panzarová, K.; Casa, R.; Colla, G. High-Throughput plant phenotyping for developing novel biostimulants: From lab to field or from field to lab? *Front. Plant Sci.* **2018**, *9*, 1197. [CrossRef]
49. Afreen, F.; Zobayed, S.M.A.; Kozai, T. Melatonin in *Glycyrrhiza uralensis*: Response of plant roots to spectral quality of light and UV-B radiation. *J. Pineal Res.* **2006**, *41*, 108–115. [CrossRef]
50. Lei, X.Y.; Zhu, R.Y.; Zhang, G.Y.; Dai, Y.R. Attenuation of cold-induced apoptosis by exogenous melatonin in carrot suspension cells: The possible involvement of polyamines. *J. Pineal Res.* **2004**, *36*, 126–131. [CrossRef]
51. Arnao, M.B.; Hernández-Ruiz, J. The physiological function of melatonin in plants. *Plant Sign. Behav.* **2006**, *1*, 89–95. [CrossRef]
52. Fletcher, R.; Sopher, C. Retardation of leaf senescence by melatonin indicates that it acts as an antioxidant in plants. In Proceedings of the 5th Canadian Pineal and Melatonin Symposium, Guelph, ON, Canada, 13–15 July 1997.
53. Hardeland, R.; Pandi-Perumal, S.; Poeggeler, B. Melatonin in plants: Focus on a vertebrate night hormone with cytoprotective properties. *Funct. Plant Sci. Biotechnol.* **2007**, *1*, 32–45.
54. Reiter, R.J.; Korkmaz, A.; Paredes, S.D.; Manchester, L.C.; Tan, D.X. Melatonin reduces oxidative/nitrosative stress due to drugs, toxins, metals, and herbicides. *Neuro Endocrinol. Lett.* **2008**, *29*, 609–613.
55. Paredes, S.D.; Korkmaz, A.; Manchester, L.C.; Tan, D.X.; Reiter, R.J. Phytomelatonin: A review. *J. Exp. Bot.* **2009**, *60*, 57–69. [CrossRef]
56. Posmyk, M.; Janas, K. Melatonin in plants. *Acta Physiol. Plant* **2009**, *31*, 1–11. [CrossRef]
57. Reiter, R.J.; Paredes, S.D.; Manchester, L.C.; Tan, D.X. Reducing oxidative/nitrosative stress: A newly-discovered genre for melatonin. *Crit. Rev. Biochem. Mol. Biol.* **2009**, *44*, 175–200. [CrossRef]
58. Hernández-Ruiz, J.; Cano, A.; Arnao, M.B. Melatonin: Growth-stimulating compound present in lupin tissues. *Planta* **2004**, *220*, 140–144. [CrossRef]
59. Hernández-Ruiz, J.; Cano, A.; Arnao, M.B. Melatonin acts as a growth-stimulating compound in some monocot species. *J. Pineal Res.* **2005**, *39*, 137–142. [CrossRef]
60. Arnao, M.B.; Hernández-Ruiz, J. Melatonin promotes adventitious and lateral root regeneration in etiolated hypocotyls of *Lupinus albus* L. *J. Pineal Res.* **2007**, *42*, 147–152. [CrossRef]

61. Arnao, M.B.; Hernández-Ruiz, J. Melatonin in plants. More studies are necessary. *Plant Sign. Behav.* **2007**, *2*, 381–382. [CrossRef]
62. Hernández-Ruiz, J.; Arnao, M.B. Melatonin stimulates the expansion of etiolated lupin cotyledons. *Plant Growth Regul.* **2008**, *55*, 29–34. [CrossRef]
63. Park, W.J. Melatonin as an endogenous plant regulatory signal: Debates and perspectives. *J. Plant Biol.* **2011**, *54*, 143–149. [CrossRef]
64. Tan, D.X.; Hardeland, R.; Manchester, L.C.; Korkmaz, A.; Ma, S.; Rosales-Corral, S.; Reiter, R.J. Functional roles of melatonin in plants, and perspectives in nutritional and agricultural science. *J. Exp. Bot.* **2012**, *63*, 577–597. [CrossRef]
65. Byeon, Y.; Park, S.; Kim, Y.S.; Back, K. Microarray analysis of genes differentially expressed in melatonin-rich transgenic rice expressing a sheep serotonin N-acetyltransferase. *J. Pineal Res.* **2013**, *55*, 357–363.
66. Arnao, M.B.; Hernández-Ruiz, J. Melatonin: Plant growth regulator and/or biostimulator during stress? *Trends Plant Sci.* **2014**, *19*, 789–797. [CrossRef]
67. Wang, P.; Sun, X.; Xie, Y.; Li, M.; Chen, W.; Zhang, S.; Liang, D.; Ma, F. Melatonin regulates proteomic changes during leaf senescence in *Malus Hupehensis*. *J. Pineal Res.* **2014**, *57*, 291–307. [CrossRef]
68. Weeda, S.; Zhang, N.; Zhao, X.; Ndip, G.; Guo, Y.D.; Buck, G.; Fu, C.; Ren, S. *Arabidopsis* transcriptome analysis reveals key roles of melatonin in plant defense systems. *PLoS ONE* **2014**, *9*, e93462. [CrossRef]
69. Zhang, H.J.; Zhang, N.; Yang, R.C.; Wang, L.; Sun, Q.Q.; Li, D.B.; Cao, Y.Y.; Weeda, S.; Zhao, B.; Ren, S.; et al. Melatonin promotes seed germination under high salinity by regulating antioxidant systems, ABA and GA_4 interaction in cucumber (*Cucumis sativus* L.). *J. Pineal Res.* **2014**, *57*, 269–279. [CrossRef]
70. Zhang, N.; Zhang, H.J.; Zhao, B.; Sun, Q.Q.; Cao, Y.Y.; Li, R.; Wu, X.X.; Weeda, S.; Li, L.; Ren, S.; et al. The RNA-seq approach to discriminate gene expression profiles in response to melatonin on cucumber lateral root formation. *J. Pineal Res.* **2014**, *56*, 39–50. [CrossRef]
71. Li, C.; Liang, D.; Chang, C.; Jia, D.; Ma, F. Melatonin mediates the regulation of ABA metabolism, free-radical scavenging, and stomatal behavior in two *Malus* species under drought stress. *J. Exp. Bot.* **2015**, *66*, 669–680. [CrossRef]
72. Li, X.; Tan, D.X.; Jiang, D.; Liu, F. Melatonin enhances cold tolerance in drought-primed wild-type and abscisic acid-deficient mutant barley. *J. Pineal Res.* **2016**, *61*, 328–339. [CrossRef]
73. Liu, J.; Zhai, R.; Liu, F.; Zhao, Y.; Wang, H.; Liu, L.; Yang, C.; Wang, Z.; Ma, F.; Xu, L. Melatonin induces parthenocarpy by regulating genes in gibberellin pathways in 'Starkrimson' pear (*Pyrus communis* L.). *Front. Plant Sci.* **2018**, *9*, 946. [CrossRef]
74. Sun, Q.Q.; Zhang, N.; Wang, J.; Zhang, H.J.; Li, D.B.; Shi, J.; Li, R.; Weeda, S.; Zhao, B.; Ren, S.; et al. Melatonin promotes ripening and improves quality of tomato fruit during postharvest life. *J. Exp. Bot.* **2015**, *66*, 657–668. [CrossRef]
75. Hernández-Ruiz, J.; Arnao, B.M. Relationship of melatonin and salicylic acid in biotic/abiotic plant stress responses. *Agronomy* **2018**, *8*, 33. [CrossRef]
76. Erland, L.A.E.; Murch, S.J.; Reiter, R.J.; Saxena, P.K. A new balancing act: The many roles of melatonin and serotonin in plant growth and development. *Plant Sign. Behav.* **2015**, *10*, e1096469. [CrossRef]
77. Reiter, R.J.; Tan, D.X.; Zhou, Z.; Coelho-Cruz, M.H.; Fuentes-Broto, L.; Galano, A. Phytomelatonin: Assisting plants to survive and thrive. *Molecules* **2015**, *20*, 7396–7437. [CrossRef]
78. Tan, D.X. Melatonin and plants. *J. Exp. Bot.* **2015**, *66*, 1–2. [CrossRef]
79. Tan, D.X.; Manchester, C.L.; Esteban-Zubero, E.; Zhou, Z.; Reiter, J.R. Melatonin as a potent and inducible endogenous antioxidant: Synthesis and metabolism. *Molecules* **2015**, *20*, 18886–18906. [CrossRef]
80. Zhang, N.; Sun, Q.; Zhang, H.; Cao, Y.; Weeda, S.; Ren, S.; Guo, Y.D. Roles of melatonin in abiotic stress resistance in plants. *J. Exp. Bot.* **2015**, *66*, 647–656. [CrossRef]
81. Hardeland, R. Melatonin in plants. Diversity of levels and multiplicity of functions. *Front. Plant Sci.* **2016**, *7*, 198. [CrossRef]
82. Nawaz, M.A.; Huang, Y.; Bie, Z.; Ahmad, W.; Reiter, R.J.; Niu, M.; Hameed, S. Melatonin: Current status and future perspectives in plant science. *Front. Plant Sci.* **2016**, *6*, 1230. [CrossRef]
83. Erland, L.A.E.; Saxena, P.K. Melatonin in plant morphogenesis. *In Vitro Cell. Dev. Biol. Plant* **2018**, *54*, 3–24. [CrossRef]

84. Fazal, H.; Abbasi, B.H.; Ahmad, N.; Ali, M. Exogenous melatonin trigger biomass accumulation and production of stress enzymes during callogenesis in medicinally important *Prunella vulgaris* L. (Selfheal). *Physiol. Mol. Biol. Plants* **2018**, *24*, 1307–1315. [CrossRef]
85. Kanwar, M.K.; Yu, J.; Zhou, J. Phytomelatonin: Recent advances and future prospects. *J. Pineal Res.* **2018**, *65*, e12526. [CrossRef]
86. Wei, J.; Li, D.; Zhang, J.; Shan, C.; Rengel, Z.; Song, Z.; Chen, Q. Phytomelatonin receptor PMTR1-mediated signaling regulates stomatal closure in *Arabidopsis thaliana*. *J. Pineal Res.* **2018**, *65*, e12500. [CrossRef]
87. Yoon, H.Y.; Kim, M.; Park, J.W. Foliar accumulation of melatonin applied to the roots of maize (*Zea mays*) seedlings. *Biomolecules* **2019**, *9*, 26. [CrossRef]
88. Li, H.; Chang, J.; Zheng, J.; Dong, Y.; Liu, Q.; Yang, X.; Wei, C.; Zhang, Y.; Ma, J.; Zhang, X. Local melatonin application induces cold tolerance in distant organs of *Citrullus lanatus* L. via long distance transport. *Sci. Rep.* **2017**, *7*, 40858. [CrossRef]
89. Zhang, Q.; Liu, X.; Zhang, Z.; Liu, N.; Li, D.; Hu, L. Melatonin improved waterlogging tolerance in alfalfa (*Medicago sativa*) by reprogramming polyamine and ethylene metabolism. *Front. Plant Sci.* **2019**, *10*, 44. [CrossRef]
90. Gu, Q.; Chen, Z.; Yu, X.; Cui, W.; Pan, J.; Zhao, G.; Xu, S.; Wang, R.; Shen, W. Melatonin confers plant tolerance against cadmium stress via the decrease of cadmium accumulation and reestablishment of microRNA-mediated redox homeostasis. *Plant Sci.* **2017**, *261*, 28–37. [CrossRef]
91. Chen, Z.; Gu, Q.; Yu, X.; Huang, L.; Xu, S.; Wang, R.; Shen, W.; Shen, W. Hydrogen peroxide acts downstream of melatonin to induce lateral root formation. *Ann. Bot.* **2018**, *121*, 1127–1136. [CrossRef]
92. Li, C.; Wang, P.; Wei, Z.; Liang, D.; Jia, D.; Fu, M.; Ma, F. The mitigation effects of exogenous melatonin on salinity-induced stress in *Malus hupehensis*. *J. Pineal Res.* **2012**, *53*, 298–306. [CrossRef]
93. Li, C.; Liang, B.; Chang, C.; Wei, Z.; Zhou, S.; Ma, F. Exogenous melatonin improved potassium content in *Malus* under different stress conditions. *J. Pineal Res.* **2016**, *61*, 218–229. [CrossRef]
94. Wang, P.; Sun, X.; Li, C.; Wei, Z.; Liang, D.; Ma, F. Long-term exogenous application of melatonin delays drought-induced leaf senescence in apple. *J. Pineal Res.* **2013**, *54*, 292–302. [CrossRef]
95. Zuo, B.; Zheng, X.; He, P.; Wang, L.; Lei, Q.; Feng, C.; Zhou, J.; Li, Q.; Han, Z.; Kong, J. Overexpression of MzASMT improves melatonin production and enhances drought tolerance in transgenic *Arabidopsis thaliana* plants. *J. Pineal Res.* **2014**, *57*, 408–417. [CrossRef]
96. Liang, B.; Ma, C.; Zhang, Z.; Wei, Z.; Gao, T.; Zhao, Q.; Ma, F.; Li, C. Long-term exogenous application of melatonin improves nutrient uptake fluxes in apple plants under moderate drought stress. *Environ. Exp. Bot.* **2018**, *155*, 650–661. [CrossRef]
97. Zheng, X.; Zhou, J.; Tan, D.X.; Wang, N.; Wang, L.; Shan, D.; Kong, J. Melatonin improves waterlogging tolerance of *Malus baccata* (Linn.) Borkh. seedlings by maintaining aerobic respiration, photosynthesis and ROS migration. *Front. Plant Sci.* **2017**, *8*, 483. [CrossRef]
98. Gong, X.; Shi, S.; Dou, F.; Song, Y.; Ma, F. Exogenous melatonin alleviates alkaline stress in *Malus hupehensis* Rehd. by regulating the biosynthesis of polyamines. *Molecules* **2017**, *22*, 1542. [CrossRef]
99. Wang, P.; Yin, L.; Liang, D.; Li, C.; Ma, F.; Yue, Z. Delayed senescence of apple leaves by exogenous melatonin treatment: Toward regulating the ascorbate-glutathione cycle. *J. Pineal Res.* **2012**, *53*, 11–20. [CrossRef]
100. Wang, P.; Sun, X.; Chang, C.; Feng, F.; Liang, D.; Cheng, L.; Ma, F. Delay in leaf senescence of *Malus hupehensis* by long-term melatonin application is associated with its regulation of metabolic status and protein degradation. *J. Pineal Res.* **2013**, *55*, 424–434.
101. Yin, L.; Wang, P.; Li, M.; Ke, X.; Li, C.; Liang, D.; Wu, S.; Ma, X.; Li, C.; Zou, Y.; et al. Exogenous melatonin improves *Malus* resistance to *Marssonina* apple blotch. *J. Pineal Res.* **2013**, *54*, 426–434. [CrossRef]
102. Li, C.; Zhao, Q.; Gao, T.; Wang, H.; Zhang, Z.; Liang, B.; Wei, Z.; Liu, C.; Ma, F. The mitigation effects of exogenous melatonin on replant disease in apple. *J. Pineal Res.* **2018**, *65*, e12523. [CrossRef]
103. Chen, L.; Wang, M.R.; Li, J.W.; Feng, C.H.; Cui, Z.H.; Zhao, L.; Wang, Q.C. Exogenous application of melatonin improves eradication of apple stem grooving virus from the infected in-vitro shoots by shoot tip culture. *Plant Pathol.* **2019**, *68*, 997–1006. [CrossRef]
104. Mohamed Abd El-Naby, S.; Ahmed Mohamed, A.; Mohamed El-Naggar, Y. Effect of melatonin, GA_3 and NAA on vegetative growth, yield and quality of *Canino* apricot fruits. *Acta Sci. Pol. Hortorum Cultus* **2019**, *18*, 167–174.

105. Hu, W.; Yang, H.; Tie, W.; Yan, Y.; Ding, Z.; Liu, Y.; Wu, C.; Wang, J.; Reiter, R.J.; Tan, D.X.; et al. Natural Variation in Banana Varieties Highlights the Role of Melatonin in Postharvest Ripening and Quality. *J. Agric. Food Chem.* **2017**, *65*, 9987–9994. [CrossRef]
106. Li, T.; Wu, Q.; Zhu, H.; Zhou, Y.; Jiang, Y.; Gao, H.; Yun, Z. Comparative transcriptomic and metabolic analysis reveals the effect of melatonin on delaying anthracnose incidence upon postharvest banana fruit peel. *BMC Plant Biol.* **2019**, *19*, 289. [CrossRef]
107. Arnao, M.B.; Hernández-Ruiz, J. Protective effect of melatonin against chlorophyll degradation during the senescence of barley leaves. *J. Pineal Res.* **2009**, *46*, 58–63. [CrossRef]
108. Shi, H.; Jiang, C.; Ye, T.; Tan, D.; Reiter, R.J.; Zhang, H.; Liu, R.; Chan, Z. Comparative physiological, metabolomic, and transcriptomic analyses reveal mechanisms of improved abiotic stress resistance in bermudagrass [*Cynodon dactylon* (L). Pers.] by exogenous melatonin. *J. Exp. Bot.* **2015**, *66*, 681–694. [CrossRef]
109. Fan, J.; Hu, Z.; Xie, Y.; Chan, Z.; Chen, K.; Amombo, E.; Chen, L.; Fu, J. Alleviation of cold damage to photosystem II and metabolisms by melatonin in *Bermudagrass*. *Front. Plant Sci.* **2015**, *6*, 925. [CrossRef]
110. Gul, M.; Khan, D.; Wani, A.; Bhat, S.; Mir, S.; Malik, A.; Kumar, A.; Narayan, S.; Stephen, K.; Lone, A. Foliar application of melatonin modulates the growth and photosynthetic pigments in broccoli cv. *Palam Samridhi. SKUAST J. Res.* **2018**, *20*, 193–198.
111. Posmyk, M.M.; Kuran, H.; Marciniak, K.; Janas, K.M. Presowing seed treatment with melatonin protects red cabbage seedlings against toxic copper ion concentrations. *J. Pineal Res.* **2008**, *45*, 24–31. [CrossRef]
112. Zhang, N.; Sun, Q.; Li, H.; Cao, Y.; Zhang, H.; Li, S.; Zhang, L.; Qi, Y.; Ren, S.; Zhao, B.; et al. Melatonin improved anthocyanin accumulation by regulating gene expressions and resulted in high reactive oxygen species scavenging capacity in cabbage. *Front. Plant Sci.* **2016**, *7*, 197. [CrossRef]
113. Tan, X.L.; Fan, Z.; Kuang, J.; Lu, W.; Reiter, R.J.; Lakshmanan, P.; Su, X.; Zhou, J.; Chen, J.; Shan, W. Melatonin delays leaf senescence of Chinese flowering cabbage by suppressing ABFs-mediated abscisic acid biosynthesis and chlorophyll degradation. *J. Pineal Res.* **2019**, *67*, e12570. [CrossRef]
114. Hu, W.; Kong, H.; Guo, Y.; Zhang, Y.; Ding, Z.; Tie, W.; Yan, Y.; Huang, Q.; Peng, M.; Shi, H.; et al. Comparative physiological and transcriptomic analyses reveal the actions of melatonin in the delay of postharvest physiological deterioration of cassava. *Front. Plant Sci.* **2016**, *7*, 736. [CrossRef]
115. Wei, Y.; Chang, Y.; Zeng, H.; Liu, G.; He, C.; Shi, H. RAV transcription factors are essential for disease resistance against cassava bacterial blight via activation of melatonin biosynthesis genes. *J. Pineal Res.* **2018**, *64*, e12454. [CrossRef]
116. Sarropoulou, V.N.; Therios, I.N.; Dimassi-Theriou, K.N. Melatonin promotes adventitious root regeneration in in vitro shoot tip explants of the commercial sweet cherry rootstocks CAB-6P (*Prunus cerasus* L.), Gisela 6 (*P. cerasus x P. canescens*), and MxM 60 (*P. avium x P. mahaleb*). *J. Pineal Res.* **2012**, *52*, 38–46. [CrossRef]
117. Sarropoulou, V.N.; Dimassi-Theriou, K.N.; Therios, I.N.; Koukourikou-Petridou, M. Melatonin enhances root regeneration, photosynthetic pigments, biomass, total carbohydrates and proline content in the cherry rootstock PHL-C (*Prunus avium x Prunus cerasus*). *Plant Physiol. Biochem.* **2012**, *61*, 162–168. [CrossRef]
118. Tijero, V.; Muñoz, P.; Munné-Bosch, S. Melatonin as an inhibitor of sweet cherries ripening in orchard trees. *Plant Physiol. Biochem.* **2019**, *140*, 88–95. [CrossRef]
119. Kostopoulou, Z.; Therios, I.; Roumeliotis, E.; Kanellis, A.K.; Molassiotis, A. Melatonin combined with ascorbic acid provides salt adaptation in *Citrus aurantium* L. seedlings. *Plant Physiol. Biochem.* **2015**, *86*, 155–165. [CrossRef]
120. Campos, C.N.; Avila, R.G.; de Souza, K.R.D.; Azevedo, L.M.; Alves, J.D. Melatonin reduces oxidative stress and promotes drought tolerance in young *Coffea arabica* L. plants. *Agric. Water Manag.* **2019**, *211*, 37–47. [CrossRef]
121. Xiao, S.; Liu, L.; Wang, H.; Li, D.; Bai, Z.; Zhang, Y.; Sun, H.; Zhang, K.; Li, C. Exogenous melatonin accelerates seed germination in cotton (*Gossypium hirsutum* L.). *PLoS ONE* **2019**, *14*, e0216575. [CrossRef]
122. Posmyk, M.M.; Balabusta, M.; Wieczorek, M.; Sliwinska, E.; Janas, K.M. Melatonin applied to cucumber (*Cucumis sativus* L.) seeds improves germination during chilling stress. *J. Pineal Res.* **2009**, *46*, 214–223. [CrossRef]
123. Balabusta, M.; Szafranska, K.; Posmyk, M.M. Exogenous melatonin improves antioxidant defense in cucumber seeds germinated under chilling stress. *Front. Plant Sci.* **2016**, *7*, 575.

124. Zhao, H.; Zhang, K.; Zhou, X.; Xi, L.; Wang, Y.; Xu, H.; Pan, T.; Zou, Z. Melatonin alleviates chilling stress in cucumber seedlings by up-regulation of CsZat12 and modulation of polyamine and abscisic acid metabolism. *Sci. Rep.* **2017**, *7*, 4998. [CrossRef]
125. Zhao, H.; Ye, L.; Wang, Y.; Zhou, X.; Yang, J.; Wang, J.; Cao, K.; Zou, Z. Melatonin increases the chilling tolerance of chloroplast in cucumber seedlings by regulating photosynthetic electron flux and the ascorbate-glutathione cycle. *Front. Plant Sci.* **2016**, *7*, 1814. [CrossRef]
126. Zhao, N.; Sun, Y.; Wang, D.Y.; Zheng, J.X. Effects of exogenous melatonin on nitrogen metabolism in cucumber seedlings under high temperature stress. *Zhiwu Shengli Xuebao/Plant Physiol. J.* **2012**, *48*, 557–564.
127. Zhang, R.; Sun, Y.; Liu, Z.; Jin, W.; Sun, Y. Effects of melatonin on seedling growth, mineral nutrition, and nitrogen metabolism in cucumber under nitrate stress. *J. Pineal Res.* **2017**, *62*, e12403. [CrossRef]
128. Wang, L.Y.; Liu, J.L.; Wang, W.X.; Sun, Y. Exogenous melatonin improves growth and photosynthetic capacity of cucumbers under salinity-induced stress. *Photosynthetica* **2015**, *54*, 1–10. [CrossRef]
129. Zhang, N.; Zhang, H.J.; Sun, Q.Q.; Cao, Y.Y.; Li, X.; Zhao, B.; Wu, P.; Guo, Y.D. Proteomic analysis reveals a role of melatonin in promoting cucumber seed germination under high salinity by regulating energy production. *Sci. Rep.* **2017**, *7*, 503. [CrossRef]
130. Zhang, N.; Zhao, B.; Zhang, H.J.; Weeda, S.; Yang, C.; Yang, Z.C.; Ren, S.; Guo, Y.D. Melatonin promotes water-stress tolerance, lateral root formation, and seed germination in cucumber (*Cucumis sativus* L.). *J. Pineal Res.* **2013**, *54*, 15–23. [CrossRef]
131. Li, H.; He, J.; Yang, X.; Li, X.; Luo, D.; Wei, C.; Ma, J.; Zhang, Y.; Yang, J.; Zhang, X. Glutathione-dependent induction of local and systemic defense against oxidative stress by exogenous melatonin in cucumber (*Cucumis sativus* L.). *J. Pineal Res.* **2016**, *60*, 206–216. [CrossRef]
132. Cao, Y.Y.; Qi, C.D.; Li, S.; Wang, Z.; Wang, X.; Wang, J.; Ren, S.; Li, X.; Zhang, N.; Guo, Y.D. Melatonin alleviates copper toxicity via improving copper sequestration and ROS scavenging in *Cucumber*. *Plant Cell Physiol.* **2018**, *60*, 562–574. [CrossRef]
133. Li, J.; Li, Y.; Tian, Y.; Qu, M.; Zhang, W.; Gao, L. Melatonin has the potential to alleviate cinnamic acid stress in cucumber seedlings. *Front. Plant Sci.* **2017**, *8*, 1193. [CrossRef]
134. Dawood, M.G.; El-Awadi, M.E. Alleviation of salinity stress on *Vicia faba* L. plants via seed priming with melatonin. *Acta Biol. Colomb.* **2015**, *20*, 223–235. [CrossRef]
135. Meng, J.F.; Xu, T.F.; Wang, Z.Z.; Fang, Y.L.; Xi, Z.M.; Zhang, Z.W. The ameliorative effects of exogenous melatonin on grape cuttings under water-deficient stress: Antioxidant metabolites, leaf anatomy, and chloroplast morphology. *J. Pineal Res.* **2014**, *57*, 200–212. [CrossRef]
136. Ma, Y.; Jiao, J.; Fan, X.; Sun, H.; Zhang, Y.; Jiang, J.; Liu, C. Endophytic bacterium *Pseudomonas fluorescens RG11* may transform tryptophan to melatonin and promote endogenous melatonin levels in the roots of four grape cultivars. *Front. Plant Sci.* **2017**, *7*, 2068. [CrossRef]
137. Xu, L.; Yue, Q.; Bian, F.E.; Sun, H.; Zhai, H.; Yao, Y. Melatonin enhances phenolics accumulation partially via ethylene signaling and resulted in high antioxidant capacity in grape berries. *Front. Plant Sci.* **2017**, *8*, 1426. [CrossRef]
138. Xu, L.; Yue, Q.; BIAN, F.; Zhai, H.; Yao, Y. Melatonin Treatment Enhances the Polyphenol Content and Antioxidant Capacity of Red Wine. *Hortic. Plant J.* **2018**, *4*, 144–150. [CrossRef]
139. Meng, J.F.; Xu, T.F.; Song, C.Z.; Yu, Y.; Hu, F.; Zhang, L.; Zhang, Z.W.; Xi, Z.M. Melatonin treatment of pre-veraison grape berries to increase size and synchronicity of berries and modify wine aroma components. *Food Chem.* **2015**, *185*, 127–134. [CrossRef]
140. Liang, D.; Ni, Z.; Xia, H.; Xie, Y.; Lv, X.; Wang, J.; Lin, L.; Deng, Q.; Luo, X. Exogenous melatonin promotes biomass accumulation and photosynthesis of kiwifruit seedlings under drought stress. *Sci. Hort.* **2019**, *246*, 34–43. [CrossRef]
141. Liang, D.; Shen, Y.; Ni, Z.; Wang, Q.; Lei, Z.; Xu, N.; Deng, Q.; Lin, L.; Wang, J.; Lv, X.; et al. Exogenous melatonin application delays senescence of kiwifruit leaves by regulating the antioxidant capacity and biosynthesis of flavonoids. *Front. Plant Sci.* **2018**, *9*, 426. [CrossRef]
142. Liang, D.; Gao, F.; Ni, Z.; Lin, L.; Deng, Q.; Tang, Y.; Wang, X.; Luo, X.; Xia, H. Melatonin improves heat tolerance in kiwifruit seedlings through promoting antioxidant enzymatic activity and glutathione S-transferase transcription. *Molecules* **2018**, *23*, 584. [CrossRef]
143. Hanci, F.; Çingi, M.; AKINCI, H. Influence of L-tryptophan and melatonin on germination of onion and leek seeds at different temperatures. *Turk. J. Agric. Res.* **2019**, *6*, 214–221. [CrossRef]

144. Zhang, Y.; Huber, D.J.; Hu, M.; Jiang, G.; Gao, Z.; Xu, X.; Jiang, Y.; Zhang, Z. Delay of postharvest browning in litchi fruit by melatonin via the enhancing of antioxidative processes and oxidation repair. *J. Agric. Food Chem.* **2018**, *66*, 7475–7484. [CrossRef] [PubMed]
145. Arnao, M.B.; Hernández-Ruiz, J. Growth conditions determine different melatonin levels in *Lupinus albus* L. *J. Pineal Res.* **2013**, *55*, 149–155. [CrossRef] [PubMed]
146. Chen, Y.E.; Mao, J.-J.; Sun, L.-Q.; Huang, B.; Ding, C.-B.; Gu, Y.; Liao, J.-Q.; Hu, C.; Zhang, Z.W.; Yuan, S.; et al. Exogenous melatonin enhances salt stress tolerance in maize seedlings by improving antioxidant and photosynthetic capacity. *Physiol. Plant.* **2018**, *164*, 349–363. [CrossRef] [PubMed]
147. Pehlivan, N.; Guler, N.S. Protective effect of a natural ally on simultaneous mild heat and salt episodes in maize seedlings. *Acta Physiol. Plant* **2018**, *40*, 203. [CrossRef]
148. Erdal, S. Melatonin promotes plant growth by maintaining integration and coordination between carbon and nitrogen metabolisms. *Plant Cell Rep.* **2019**, *38*, 1001–1012. [CrossRef]
149. Ye, J.; Wang, S.; Deng, X.; Yin, L.; Xiong, B.; Wang, X. Melatonin increased maize (*Zea mays* L.) seedling drought tolerance by alleviating drought-induced photosynthetic inhibition and oxidative damage. *Acta Physiol. Plant* **2016**, *38*, 48. [CrossRef]
150. Fleta-Soriano, E.; Díaz, L.; Bonet, E.; Munné-Bosch, S. Melatonin may exert a protective role against drought stress in maize. *J. Agron. Crop Sci.* **2017**, *203*, 286–294. [CrossRef]
151. Huang, B.; Chen, Y.E.; Zhao, Y.Q.; Ding, C.B.; Liao, J.Q.; Hu, C.; Zhou, L.J.; Zhang, Z.W.; Yuan, S.; Yuan, M. Exogenous melatonin alleviates oxidative damages and protects Photosystem II in maize seedlings under drought stress. *Front. Plant Sci.* **2019**, *10*, 677. [CrossRef]
152. Li, Z.G.; Xu, Y.; Bai, L.K.; Zhang, S.Y.; Wang, Y. Melatonin enhances thermotolerance of maize seedlings (*Zea mays* L.) by modulating antioxidant defense, methylglyoxal detoxification, and osmoregulation systems. *Protoplasma* **2018**, *256*, 471–490. [CrossRef]
153. Okant, M.; Kaya, C. The role of endogenous nitric oxide in melatonin-improved tolerance to lead toxicity in maize plants. *Environ. Sci. Pollut. Res.* **2019**, *26*, 11864–11874. [CrossRef]
154. Zhao, H.; Su, T.; Huo, L.; Wei, H.; Jiang, Y.; Xu, L.; Ma, F. Unveiling the mechanism of melatonin impacts on maize seedling growth: Sugar metabolism as a case. *J. Pineal Res.* **2015**, *59*, 255–266. [CrossRef] [PubMed]
155. Kolodziejczyk, I.; Dzitko, K.; Szewczyk, R.; Posmyk, M.M. Exogenous melatonin expediently modifies proteome of maize (*Zea mays* L.) embryo during seed germination. *Acta Physiol. Plant* **2016**, *38*, 146. [CrossRef]
156. Su, X.; Xin, L.; Li, Z.; Zheng, H.; Mao, J.; Yang, Q. Physiology and transcriptome analyses reveal a protective effect of the radical scavenger melatonin in aging maize seeds. *Free Radic. Res.* **2018**, *52*, 1094–1109. [CrossRef] [PubMed]
157. Hodzic, E.; Balaban, M.; Suskalo, N.; Galijasevic, S.; Hasanagic, D.; Kukavica, B. Antioxidative response of Melissa officinalis L. and Valeriana officinalis L. leaves exposed to exogenous melatonin and excessive zinc and cadmium level. *J. Serb. Chem. Soc.* **2018**, *84*, 70. [CrossRef]
158. Szafranska, K.; Glinska, S.; Janas, K.M. Ameliorative effect of melatonin on meristematic cells of chilled and re-warmed *Vigna radiata* roots. *Biol. Plant* **2013**, *57*, 91–96. [CrossRef]
159. Gao, W.; Feng, Z.; Bai, Q.; He, J.; Wang, Y. Melatonin-mediated regulation of growth and antioxidant capacity in salt-tolerant naked oat under salt stress. *Molecules* **2019**, *20*, 1176. [CrossRef] [PubMed]
160. Gao, W.; Zhang, Y.; Feng, Z.; Bai, Q.; He, J.; Wang, Y. Effects of melatonin on antioxidant capacity in naked oat seedlings under drought stress. *Molecules* **2018**, *23*, 1580. [CrossRef]
161. Cao, S.; Song, C.; Shao, J.; Bian, K.; Chen, W.; Yang, Z. Exogenous melatonin treatment increases chilling tolerance and induces defense response in harvested peach fruit during cold storage. *J. Agric. Food Chem.* **2016**, *64*, 5215–5222. [CrossRef]
162. Gao, H.; Zhang, Z.K.; Chai, H.K.; Cheng, N.; Yang, Y.; Wang, D.N.; Yang, T.; Cao, W. Melatonin treatment delays postharvest senescence and regulates reactive oxygen species metabolism in peach fruit. *Postharvest Biol. Technol.* **2016**, *118*, 103–110. [CrossRef]
163. Liu, J.; Yue, R.; Si, M.; Wu, M.; Cong, L.; Zhai, R.; Yang, C.; Wang, Z.; Ma, F.; Xu, L. Effects of exogenous application of melatonin on quality and sugar metabolism in *Zaosu* pear fruit. *J. Plant Growth Regul.* **2019**. [CrossRef]
164. Zhai, R.; Liu, J.; Liu, F.; Zhao, Y.; Liu, L.; Fang, C.; Wang, H.; Li, X.; Wang, Z.; Ma, F.; et al. Melatonin limited ethylene production, softening and reduced physiology disorder in pear (*Pyrus communis* L.) fruit during senescence. *Postharvest Biol. Technol.* **2018**, *139*, 38–46. [CrossRef]

165. Liu, J.; Yang, J.; Zhang, H.; Cong, L.; Zhai, R.; Yang, C.; Wang, Z.; Ma, F.; Xu, L. Melatonin inhibits ethylene synthesis via nitric oxide regulation to delay postharvest senescence in pears. *J. Agric. Food Chem.* **2019**, *67*, 2279–2288. [CrossRef] [PubMed]
166. Kaya, C.; Higgs, D.; Ashraf, M.; Alyemeni, M.; Ahmad, P. Integrative roles of nitric oxide and hydrogen sulfide in melatonin-induced tolerance of pepper (*Capsicum annuum* L.) plants to iron deficiency and salt stress alone or in combination. *Physiol. Plant* **2019**. [CrossRef] [PubMed]
167. Korkmaz, A.; Karakas, A.; Kocacinar, F.; Cuci, Y. The effects of seed treatment with melatonin on germination and emergence performance of pepper seeds under chilling stress. *J. Agric. Sci.* **2017**, *23*, 167–176.
168. Sarafi, E.; Tsouvaltzis, P.; Chatzissavvidis, C.; Siomos, A.; Therios, I. Melatonin and resveratrol reverse the toxic effect of high boron (B) and modulate biochemical parameters in pepper plants (*Capsicum annuum* L.). *Plant Physiol. Biochem.* **2017**, *112*, 173–182. [CrossRef] [PubMed]
169. Szafranska, K.; Reiter, R.J.; Posmyk, M.M. Melatonin application to *Pisum sativum* L. seeds positively influences the function of the photosynthetic apparatus in growing seedlings during paraquat-induced oxidative stress. *Front. Plant Sci.* **2016**, *7*, 1663. [CrossRef]
170. Szafranska, K.; Reiter, R.J.; Posmyk, M.M. Melatonin improves the photosynthetic apparatus in pea leaves stressed by paraquat via chlorophyll breakdown regulation and its accelerated *de novo* synthesis. *Front. Plant Sci.* **2017**, *8*, 878. [CrossRef] [PubMed]
171. Tan, D.X.; Manchester, L.C.; Helton, P.; Reiter, R.J. Phytoremediative capacity of plants enriched with melatonin. *Plant Sign. Behav.* **2007**, *2*, 514–516. [CrossRef]
172. Bal, E. Physicochemical changes in *Santa Rosa* plum fruit treated with melatonin during cold storage. *J. Food Meas. Charact.* **2019**, *13*, 1713–1720. [CrossRef]
173. Jannatizadeh, A. Exogenous melatonin applying confers chilling tolerance in pomegranate fruit during cold storage. *Sci. Hort.* **2019**, *246*, 544–549. [CrossRef]
174. Ding, F.; Wang, G.; Zhang, S. Exogenous melatonin mitigates methyl viologen-triggered oxidative stress in poplar leaf. *Molecules* **2018**, *23*, 2852. [CrossRef] [PubMed]
175. Yu, Y.; Wang, A.; Li, X.; Kou, M.; Wang, W.; Chen, X.; Xu, T.; Zhu, M.; Ma, D.; Li, Z.; et al. Melatonin-stimulated triacylglycerol breakdown and energy turnover under salinity stress contributes to the maintenance of plasma membrane H^+-ATPase activity and K^+/Na^+ homeostasis in sweet potato. *Front. Plant Sci.* **2018**, *9*, 256. [CrossRef] [PubMed]
176. Zhang, S.; Zheng, X.; Reiter, R.J.; Feng, S.; Wang, Y.; Liu, S.; Jin, L.; Li, Z.; Datla, R.; Ren, M. Melatonin attenuates potato late blight by disrupting cell growth, stress tolerance, fungicide susceptibility and homeostasis of gene expression in *Phytophthora infestans*. *Front. Plant Sci.* **2017**, *8*, 1993. [CrossRef] [PubMed]
177. Jia, C.; Yu, X.; Zhang, M.; Liu, Z.; Zou, P.; Ma, J.; Xu, Y. Application of melatonin-enhanced tolerance to high-temperature stress in cherry radish (*Raphanus sativus* L. var. radculus pers). *J. Plant Growth Regul.* **2019**. [CrossRef]
178. Zhao, G.; Zhao, Y.; Yu, X.; Felix, K.; Han, H.; Guan, R.; Wang, R.; Shen, W. Nitric oxide is required for melatonin-enhanced tolerance against salinity stress in rapeseed (*Brassica napus* L.) seedlings. *Int. J. Mol. Sci.* **2018**, *19*, 1912. [CrossRef] [PubMed]
179. Khan, M.N.; Zhang, J.; Luo, T.; Liu, J.; Rizwan, M.; Fahad, S.; Xu, Z.; Hu, L. Seed priming with melatonin coping drought stress in rapeseed by regulating reactive oxygen species detoxification: Antioxidant defense system, osmotic adjustment, stomatal traits and chloroplast ultrastructure perseveration. *Ind. Crop Prod.* **2019**, *140*, 111597. [CrossRef]
180. Han, Q.H.; Huang, B.; Ding, C.B.; Zhang, Z.W.; Chen, Y.E.; Hu, C.; Zhou, L.J.; Huang, Y.; Liao, J.Q.; Yuan, S.; et al. Effects of melatonin on anti-oxidative systems and Photosystem II in cold-stressed rice seedlings. *Front. Plant Sci.* **2017**, *8*, 785. [CrossRef] [PubMed]
181. Liang, C.; Zheng, G.; Li, W.; Wang, Y.; Hu, B.; Wang, H.; Wu, H.; Qian, Y.; Zhu, X.G.; Tan, D.X.; et al. Melatonin delays leaf senescence and enhances salt stress tolerance in rice. *J. Pineal Res.* **2015**, *59*, 91–101. [CrossRef]
182. Lee, K.; Back, K. Overexpression of rice serotonin N-acetyltransferase 1 in transgenic rice plants confers resistance to cadmium and senescence and increases grain yield. *J. Pineal Res.* **2017**, *62*, e12392. [CrossRef]
183. Lee, H.Y.; Back, K. Cadmium disrupts subcellular organelles, including chloroplasts, resulting in melatonin induction in plants. *Molecules* **2017**, *22*, 1791. [CrossRef]
184. Chen, X.; Sun, C.; Laborda, P.; Zhao, Y.; Palmer, I.; Fu, Z.Q.; Qiu, J.; Liu, F. Melatonin treatment inhibits the growth of *Xanthomonas oryzae* pv. *oryzae*. *Front. Microbiol.* **2018**, *9*, 2280. [CrossRef] [PubMed]

185. Wei, Y.; Zeng, H.; Hu, W.; Chen, L.; He, C.; Shi, H. Comparative transcriptional profiling of melatonin synthesis and catabolic genes indicates the possible role of melatonin in developmental and stress responses in rice. *Front. Plant Sci.* **2016**, *7*, 676. [CrossRef] [PubMed]
186. Park, S.; Back, K. Melatonin promotes seminal root elongation and root growth in transgenic rice after germination. *J. Pineal Res.* **2012**, *53*, 385–389. [CrossRef] [PubMed]
187. Liang, C.; Li, A.; Yu, H.; Li, W.; Liang, C.; Guo, S.; Zhang, R.; Chu, C. Melatonin regulates root architecture by modulating auxin response in rice. *Front. Plant Sci.* **2017**, *8*, 134. [CrossRef] [PubMed]
188. Wei, W.; Li, Q.; Chu, Y.-N.; Reiter, R.J.; Yu, X.M.; Zhu, D.H.; Zhang, W.K.; Ma, B.; Lin, Q.; Zhang, J.S.; et al. Melatonin enhances plant growth and abiotic stress tolerance in soybean plants. *J. Exp. Bot.* **2015**, *66*, 695–707. [CrossRef] [PubMed]
189. ZOU, J.N.; JIN, X.J.; WANG, M.X.Y. Effects of melatonin on photosynthesis and soybean seed growth during grain filling under drought stress. *Photosynthetica* **2019**, *57*, 512–520. [CrossRef]
190. Zhang, J.; Zeng, B.; Mao, Y.; Kong, X.; Wang, X.; Yang, Y.; Zhang, J.; Xu, J.; Rengel, Z.; Chen, Q. Melatonin alleviates aluminium toxicity through modulating antioxidative enzymes and enhancing organic acid anion exudation in soybean. *Funct. Plant Biol.* **2017**, *44*, 961–968. [CrossRef]
191. Moussa, H.R.; Algamal, S.M.A. Does Exogenous Application of Melatonin Ameliorate Boron Toxicity in Spinach Plants? *Int. J. Veg. Sci.* **2017**, *23*, 233–245. [CrossRef]
192. Aghdam, M.S.; Fard, J.R. Melatonin treatment attenuates postharvest decay and maintains nutritional quality of strawberry fruits (*Fragaria x anannasa* cv. Selva) by enhancing GABA shunt activity. *Food Chem.* **2017**, *221*, 1650–1657. [CrossRef]
193. Liu, C.; Zheng, H.; Sheng, K.; Liu, W.; Zheng, L. Effects of melatonin treatment on the postharvest quality of strawberry fruit. *Postharvest Biol. Technol.* **2018**, *139*, 47–55. [CrossRef]
194. Mukherjee, S.; David, A.; Yadav, S.; Baluska, F.; Bhatla, S.C. Salt stress-induced seedling growth inhibition coincides with differential distribution of serotonin and melatonin in sunflower seedling roots and cotyledons. *Physiol. Plant* **2014**, *152*, 714–728. [CrossRef] [PubMed]
195. Arora, D.; Bhatla, S.C. Melatonin and nitric oxide regulate sunflower seedling growth under salt stress accompanying differential expression of Cu/Zn SOD and Mn SOD. *Free Radic. Biol. Med.* **2017**, *106*, 315–328. [CrossRef] [PubMed]
196. Kaur, H.; Bhatla, S.C. Melatonin and nitric oxide modulate glutathione content and glutathione reductase activity in sunflower seedling cotyledons accompanying salt stress. *Nitric Oxide* **2016**, *59*, 42–53. [CrossRef] [PubMed]
197. Li, J.; Yang, Y.; Sun, K.; Chen, Y.; Chen, X.; Li, X. Exogenous Melatonin Enhances Cold, Salt and Drought Stress Tolerance by Improving Antioxidant Defense in Tea Plant (*Camellia sinensis* (L.) O. Kuntze). *Molecules* **2019**, *24*, 1826. [CrossRef] [PubMed]
198. Li, J.; Arkorful, E.; Cheng, S.; Zhou, Q.; Li, H.; Chen, X.; Sun, K.; Li, X. Alleviation of cold damage by exogenous application of melatonin in vegetatively propagated tea plant (*Camellia sinensis* (L.) O. Kuntze). *Sci. Hort.* **2018**, *238*, 356–362. [CrossRef]
199. Zhao, L.; Chen, L.; Gu, P.; Zhan, X.; Zhang, Y.; Hou, C.; Wu, Z.; Wu, Y.F.; Wang, Q.C. Exogenous application of melatonin improves plant resistance to virus infection. *Plant Pathol.* **2019**. [CrossRef]
200. Zhou, X.; Zhao, H.; Cao, K.; Hu, L.; Du, T.; Baluska, F.; Zou, Z. Beneficial roles of melatonin on redox regulation of photosynthetic electron transport and synthesis of D1 protein in tomato seedlings under salt stress. *Front. Plant Sci.* **2016**, *7*, 1823. [CrossRef]
201. Siddiqui, H.M.; Alamri, S.; Al-Khaishany, Y.M.; Khan, N.M.; Al-Amri, A.; Ali, M.H.; Alaraidh, A.I.; Alsahli, A.A. Exogenous melatonin counteracts NaCl-induced damage by regulating the antioxidant system, proline and carbohydrates metabolism in tomato seedlings. *Int. J. Mol. Sci.* **2019**, *20*, 353. [CrossRef]
202. Yin, Z.; Lu, J.; Meng, S.; Liu, Y.; Mostafa, I.; Qi, M.; Li, T. Exogenous melatonin improves salt tolerance in tomato by regulating photosynthetic electron flux and the ascorbate-glutathione cycle. *J. Plant Interact.* **2019**, *14*, 453–463. [CrossRef]
203. Ding, F.; Liu, B.; Zhang, S. Exogenous melatonin ameliorates cold-induced damage in tomato plants. *Sci. Hort.* **2017**, *219*, 264–271. [CrossRef]
204. Ding, F.; Wang, M.; Liu, B.; Zhang, S. Exogenous melatonin mitigates photoinhibition by accelerating non-photochemical quenching in tomato seedlings exposed to moderate light during chilling. *Front. Plant Sci.* **2017**, *8*, 244. [CrossRef] [PubMed]

205. Aghdam, M.S.; Luo, Z.; Jannatizadeh, A.; Sheikh-Assadi, M.; Sharafi, Y.; Farmani, B.; Fard, J.R.; Razavi, F. Employing exogenous melatonin applying confers chilling tolerance in tomato fruits by upregulating ZAT2/6/12 giving rise to promoting endogenous polyamines, proline, and nitric oxide accumulation by triggering arginine pathway activity. *Food Chem.* **2019**, *275*, 549–556. [CrossRef] [PubMed]
206. Xu, W.; Cai, S.Y.; Zhang, Y.; Wang, Y.; Ahammed, G.J.; Xia, X.J.; Shi, K.; Zhou, Y.H.; Yu, J.Q.; Reiter, R.J.; et al. Melatonin enhances thermotolerance by promoting cellular protein protection in tomato plants. *J. Pineal Res.* **2016**, *61*, 457–469. [CrossRef] [PubMed]
207. Ahammed, G.J.; Xu, W.; Liu, A.; Chen, S. Endogenous melatonin deficiency aggravates high temperature-induced oxidative stress in *Solanum lycopersicum* L. *Environ. Exp. Bot.* **2019**, *161*, 303–311. [CrossRef]
208. Qi, Z.Y.; Wang, K.X.; Yan, M.Y.; Kanwar, K.M.; Li, D.Y.; Wijaya, L.; Alyemeni, N.M.; Ahmad, P.; Zhou, J. Melatonin alleviates high temperature-induced pollen abortion in *Solanum lycopersicum*. *Molecules* **2018**, *23*, 386. [CrossRef] [PubMed]
209. Hasan, M.; Ahammed, G.J.; Yin, L.; Shi, K.; Xia, X.; Zhou, Y.; Yu, J.; Zhou, J. Melatonin mitigates cadmium phytotoxicity through modulation of phytochelatins biosynthesis, vacuolar sequestration and antioxidant potential in *Solanum lycopersicum* L. *Front. Plant Sci.* **2015**, *6*, 601. [CrossRef] [PubMed]
210. Cai, S.Y.; Zhang, Y.; Xu, Y.P.; Qi, Z.Y.; Li, M.Q.; Ahammed, G.J.; Xia, X.J.; Shi, K.; Zhou, Y.H.; Reiter, R.J.; et al. HsfA1a upregulates melatonin biosynthesis to confer cadmium tolerance in tomato plants. *J. Pineal Res.* **2017**, *62*, e12387. [CrossRef]
211. Li, M.Q.; Hasan, M.; Li, C.X.; Ahammed, G.J.; Xia, X.J.; Shi, K.; Zhou, Y.H.; Reiter, R.J.; Yu, J.Q.; Xu, M.X.; et al. Melatonin mediates selenium-induced tolerance to cadmium stress in tomato plants. *J. Pineal Res.* **2016**, *61*, 291–302. [CrossRef]
212. Liu, N.; Jin, Z.; Wang, S.; Gong, B.; Wen, D.; Wang, X.; Wei, M.; Shi, Q. Sodic alkaline stress mitigation with exogenous melatonin involves reactive oxygen metabolism and ion homeostasis in tomato. *Sci. Hort.* **2015**, *181*, 18–25. [CrossRef]
213. Debnath, B.; Hussain, M.; Irshad, M.; Mitra, S.; Li, M.; Liu, S.; Qiu, D. Exogenous melatonin mitigates acid rain stress to tomato plants through modulation of leaf ultrastructure, photosynthesis and antioxidant potential. *Molecules* **2018**, *23*, 388. [CrossRef]
214. Ding, F.; Wang, G.; Wang, M.; Zhang, S. Exogenous melatonin improves tolerance to water deficit by promoting cuticle formation in tomato plants. *Molecules* **2018**, *23*, 1605. [CrossRef] [PubMed]
215. Karaca, P.; Cekic, F.O. Exogenous melatonin-stimulated defense responses in tomato plants treated with polyethylene glycol. *Int. J. Veg. Sci.* **2019**. [CrossRef]
216. Hasn, M.; Liu, C.X.; Pan, Y.T.; Ahammed, G.J.; Qi, Z.Y.; Zhou, J. Melatonin alleviates low-sulfur stress by promoting sulfur homeostasis in tomato plants. *Sci. Rep.* **2018**, *8*, 10182. [CrossRef] [PubMed]
217. Wen, D.; Gong, B.; Sun, S.; Liu, S.; Wang, X.; Wei, M.; Yang, F.; Li, Y.; Shi, Q. Promoting roles of melatonin in adventitious root development of *Solanum lycopersicum* L. by regulating auxin and nitric oxide signaling. *Front. Plant Sci.* **2016**, *7*, 718. [CrossRef] [PubMed]
218. Liu, J.; Zhang, R.; Sun, Y.; Liu, Z.; Jin, W.; Sun, Y. The beneficial effects of exogenous melatonin on tomato fruit properties. *Sci. Hort.* **2016**, *207*, 14–20. [CrossRef]
219. Sun, Q.; Zhang, N.; Wang, J.; Cao, Y.; Li, X.; Zhang, H.; Zhang, L.; Tan, D.X.; Guo, Y.D. A label-free differential proteomics analysis reveals the effect of melatonin on promoting fruit ripening and anthocyanin accumulation upon postharvest in tomato. *J. Pineal Res.* **2016**, *61*, 138–153. [CrossRef] [PubMed]
220. Li, H.; Dong, Y.; Chang, J.; He, J.; Chen, H.; Liu, Q.; Wei, C.; Ma, J.; Zhang, Y.; Yang, J.; et al. High-throughput microRNA and mRNA sequencing reveals that microRNAs may be involved in melatonin-mediated cold tolerance in *Citrullus lanatus* L. *Front. Plant Sci.* **2016**, *7*, 1231. [CrossRef] [PubMed]
221. Li, H.; Chang, J.; Chen, H.; Wang, Z.; Gu, X.; Wei, C.; Zhang, Y.; Ma, J.; Yang, J.; Zhang, X. Exogenous melatonin confers salt stress tolerance to watermelon by improving photosynthesis and redox homeostasis. *Front. Plant Sci.* **2017**, *8*, 295. [CrossRef]
222. Nawaz, M.A.; Jiao, Y.; Chen, C.; Shireen, F.; Zheng, Z.; Imtiaz, M.; Bie, Z.; Huang, Y. Melatonin pretreatment improves vanadium stress tolerance of watermelon seedlings by reducing vanadium concentration in the leaves and regulating melatonin biosynthesis and antioxidant-related gene expression. *J. Plant Physiol.* **2018**, *220*, 115–127. [CrossRef]

223. Turk, H.; Erdal, S.; Genisel, M.; Atici, O.; Demir, Y.; Yanmis, D. The regulatory effect of melatonin on physiological, biochemical and molecular parameters in cold-stressed wheat seedlings. *Plant Growth Regul.* **2014**, *74*, 139–152. [CrossRef]
224. Li, X.; Brestic, M.; Tan, D.X.; Zivcak, M.; Zhu, X.; Liu, S.; Song, F.; Reiter, R.J.; Liu, F. Melatonin alleviates low PS I-limited carbon assimilation under elevated CO_2 and enhances the cold tolerance of offspring in chlorophyll b-deficient mutant wheat. *J. Pineal Res.* **2018**, *64*, e12453. [CrossRef] [PubMed]
225. Sun, L.; Li, X.; Wang, Z.; Sun, Z.; Zhu, X.; Liu, S.; Song, F.; Liu, F.; Wang, Y. Cold priming induced tolerance to subsequent low temperature stress is enhanced by melatonin application during recovery in wheat. *Molecules* **2018**, *23*, 1091. [CrossRef] [PubMed]
226. Ke, Q.; Ye, J.; Wang, B.; Ren, J.; Yin, L.; Deng, X.P.; Shiwen, W. Melatonin mitigates salt stress in wheat seedlings by modulating polyamine metabolism. *Front. Plant Sci.* **2018**, *9*, 914. [CrossRef] [PubMed]
227. Zafar, S.; Hasnain, Z.; Anwar, S.; Perveen, S.; Iqbal, N.; Noman, A.; Ali, M. Influence of melatonin on antioxidant defense system and yield of wheat (*Triticum aestivum* L.) genotypes under saline condition. *Pak. J. Bot.* **2019**, *51*, 1987–1994. [CrossRef]
228. Cui, G.; Zhao, X.; Liu, S.; Sun, F.; Zhang, C.; Xi, Y. Beneficial effects of melatonin in overcoming drought stress in wheat seedlings. *Plant Physiol. Biochem.* **2017**, *118*, 138–149. [CrossRef] [PubMed]
229. Ni, J.; Wang, Q.; Shah, A.F.; Liu, W.; Wang, D.; Huang, S.; Fu, S.; Wu, L. Exogenous melatonin confers cadmium tolerance by counterbalancing the hydrogen peroxide homeostasis in wheat seedlings. *Molecules* **2018**, *23*, 799. [CrossRef] [PubMed]
230. Kaya, C.; Okant, M.; Ugurlar, F.; Alyemeni, M.N.; Ashraf, M.; Ahmad, P. Melatonin-mediated nitric oxide improves tolerance to cadmium toxicity by reducing oxidative stress in wheat plants. *Chemosphere* **2019**, *225*, 627–638. [CrossRef]
231. Zuo, Z.; Sun, L.; Wang, T.; Miao, P.; Zhu, X.; Liu, S.; Song, F.; Mao, H.; Li, X. Melatonin improves the photosynthetic carbon assimilation and antioxidant capacity in wheat exposed to nano-ZnO stress. *Molecules* **2017**, *22*, 1727. [CrossRef]
232. Qiao, Y.; Yin, L.; Wang, B.; Ke, Q.; Deng, X.; Wang, S. Melatonin promotes plant growth by increasing nitrogen uptake and assimilation under nitrogen deficient condition in winter wheat. *Plant Physiol. Biochem.* **2019**, *139*, 342–349. [CrossRef]
233. Hoffman, R.A.; Reiter, R.J. Pineal gland: Influence on gonads of male hamsters. *Science* **1965**, *148*, 1609–1611. [CrossRef]
234. Choi, D. Effects of dietary supplement containing melatonin on reproductive activity in male golden hamsters. *Dev. Reprod.* **2019**, *23*, 101–110. [CrossRef] [PubMed]
235. deNicolo, G.; Morris, S.T.; Kenyon, P.R.; Morel, P.C.H.; Parkinson, T.J. Melatonin-improved reproductive performance in sheep bred out of season. *Anim. Reprod. Sci.* **2008**, *109*, 124–133. [CrossRef] [PubMed]
236. Yakhin, O.I.; Lubyanov, A.A.; Yakhin, I.A.; Brown, P.H. Biostimulants in Plant Science: A Global Perspective. *Front. Plant Sci.* **2017**, *7*, 2049. [CrossRef] [PubMed]
237. Hernández-Ruiz, J.; Arnao, M.B. Phytomelatonin, an interesting tool for agricultural crops. *Focus Sci.* **2016**, *2*, 1–7. [CrossRef]
238. Arnao, M.B.; Hernández-Ruiz, J. The potential of phytomelatonin as a nutraceutical. *Molecules* **2018**, *23*, 238. [CrossRef] [PubMed]
239. Arnao, M.B.; Hernández-Ruiz, J. Phytomelatonin, natural melatonin from plants as a novel dietary supplement: Sources, activities and world market. *J. Funct. Foods* **2018**, *48*, 37–42. [CrossRef]

13

Plant Biostimulants: Importance of the Quality and Yield of Horticultural Crops and the Improvement of Plant Tolerance to Abiotic Stress

Magdalena Drobek, Magdalena Frąc and Justyna Cybulska *

Institute of Agrophysics, Polish Academy of Sciences, Doświadczalna 4, 20-290 Lublin, Poland; m.drobek@ipan.lublin.pl (M.D.); m.frac@ipan.lublin.pl (M.F.)
* Correspondence: j.cybulska@ipan.lublin.pl

Abstract: Biostimulants are among the natural preparations that improve the general health, vitality, and growth of plants and protect them against infections. They can be successfully used in both agri- and horticultural crops. The main active substances used in such preparations are humic and fulvic acids, protein hydrolysates, compounds containing nitrogen, seaweed extracts, beneficial fungi, and bacteria. Biostimulant formulations may be single- or multi-component, but the synergic action of several different components has been observed. Many groups of biostimulants have been distinguished through their method of application (soil, foliar), the material from which they were produced (plant, animal), or the process by which they were created (hydrolysis, fermentation, extraction). Natural soil stimulants can induce the development of beneficial soil organisms that provide substrates for plant growth. The use of natural preparations that are not harmful to the environment is particularly important in connection with the progressive processes of soil degradation and atmospheric pollution. This review gives an overview of the importance and influence of different natural plant biostimulants on both the yield and quality of crops.

Keywords: fruit quality; plants biostimulants; yielding

1. Introduction

The quality and quantity of crops are influenced by both biotic and abiotic factors. Quality may be defined as a set of agronomic (e.g., fruit size, yield, resistance to bacteria and fungi) and organoleptic (e.g., colour, shape, firmness) properties as well as nutrient and vitamin content [1]. The abiotic factors include soil composition, extreme salinity, acidity, high and low temperatures, drought, pollution, humidity, rain, wind, or ultraviolet radiation. Stress caused by unfavourable stimuli can significantly reduce harvest yields because plants respond by using their energy reserves to fight stress instead of concentrating on yielding. Biotic factors include various bacteria, fungi, or viruses that are the cause of numerous plant diseases. Fungal and bacterial infections may not only reduce yield but may also lead to the loss of the entire harvest. To prevent this, various types of plant protection products are used. In accordance with the recommendations of the European Union [2], chemical and mineral plant protection agents are intended to be slowly replaced by natural preparations. The reason for this is the adverse influence of chemical and mineral plant protection agents on the natural environment, as well as on the health benefits of plant crops. Moreover, artificial fertilizers are responsible for the eutrophication of many bodies of water. This results in the formation of dead zones devoid of living organisms. The Baltic Sea alone is distinguished by having oxygen-free zones making up around 60,000 km^2 of area caused by water pollution due to fertilizers. This area constitutes, on average, 3.5% of the catchment area of the Baltic Sea [3]. The effects of fertilizers have an unfavourable effect on algae, plants, animals, and people. Due to the fact that man is a higher-order consumer, people are particularly severely exposed to the harmful effects of fertilizer compounds accumulated at the lower levels of the food chain. Harmful compounds from fertilizers may weaken enzymes or interfere with protein production or vitamin absorption in the human body [3]. Natural preparations called biostimulants increase the efficiency of nutrient utilization and tolerance to abiotic stress and improve the quality of crops [4]. Biostimulants include organic and non-organic substances and/or microorganisms [5]. Farmers who manage organic farms are also eager to use natural stimulants to improve crop quality [6]. Increasing consumer awareness concerning healthy food favours the enhancement of the significance of organic farming [7].

The effects of the stimulators may be multifaceted. The effects of their activities vary depending on the type of biostimulant used and the plant variety. However, it should be noted that most of them have a beneficial effect on crops [8].

2. Definitions and Classification of Biostimulants

Biostimulants can be treated as an additive to fertilizers and support the uptake of nutrients, promote plant growth, and increase tolerance to abiotic stress [9]. The definition of biostimulants is wide and not sufficiently precise. However, there are two main features that distinguish biostimulants from other growth and plant-protection agents. A biostimulant may be any substance or mixture of substances of natural origin or microorganism which improves the condition of crops without causing adverse side effects [10].

Enzymes, proteins, amino acids, micronutrients, and other compounds may be used as biostimulants. Natural stimulants are often included under the term biostimulants, including phenols, salicylic acid, humic and fulvic acids, or protein hydrolases [10,11]. An important group of plant biostimulants are organisms including fungi and bacteria that change the species composition of organisms found in the soil or plants. Their presence may accelerate the rate of degradation processes or limit the number of specific fungal and bacterial groups [12,13]. Popular fungi used as biostimulants include *Glomus intraradices* [14], *Trichoderma atroviride* [14], *Trichoderma reesei*, and *Heteroconium chaetospira* [10,15–19]. Useful bacteria include *Arthrobacter* spp., *Enterobacter* spp., *Acinetobacter* spp., *Pseudomonas* spp., *Ochrobactrum* spp., *Bacillus* spp., and *Rhodococcus* spp. [18,19].

Biostimulants cannot be defined as fertilizers because they do not provide nutrients directly to plants. Biostimulants may facilitate the acquisition of nutrients by supporting metabolic processes in the soil and plants. An example of such an activity is the facilitation of the development of arbuscular mycorrhizal fungi that transport nutrients to the host plant [20].

3. Sources of Biostimulants

Biostimulants are preparations made from natural raw materials. Some of them are plant extracts such as rosemary, which stimulates the growth of tomato plants with a concentration of 1000 ppm. Rosemary oil contributes to improved nutrient uptake and increases the fresh mass of roots [21]. Plant and animal biostimulants are formed, for example, as a result of chemical or enzymatic hydrolysis. The products of hydrolysis are mixtures of peptides and amino acids (protein hydrolysates). Chemical acid or alkaline hydrolysis is used to produce biostimulants of animal origin, from raw materials such as hen feathers, bone meal, casein, collagen from skins, animal tissue or fish waste (Table 1). Biostimulants of plant origin are produced using enzymatic hydrolysis. In the production of plant biostimulants, for example, alfalfa hay, pulses, and vegetable or fruit waste may be used [13,22]. Protein hydrolysates contain amino acids, peptides, and non-protein compounds. Protein hydrolysates stimulate plant growth, reduce the use of inert fertilizers, and are environmentally friendly [22]. The solution, which at the same time allows for a reduction in the amount of organic waste and the creation of biostimulating preparations, is actually a fermentation process. Biostimulants may also be the products of anaerobic digestion (Table 1). Dissolved organic matter is formed in fermentation chambers and has stimulating properties. The source of dissolved organic matter is usually plant, animal, and lignin biomass [23]. Biopreparations from marine algae may contain low-molecular polypeptides and amino acids, vitamins, enzymes, phytohormones, sugars, and antioxidants. These compounds activate the processes of rhizogenesis and lead to positive morphological and anatomical changes in plants (Table 1). Two applications of a biopreparation derived from algae enhanced the development of *Cornus alba* "Aurea" roots by 80% when compared to the control. Such biopreparations can be used for rooting young seedlings or improving the process of adult rhizogenesis [24].

Literature data indicate the positive effect of seaweed extracts as plant biostimulants. The extracts from *Ascophyllum nodosum* are listed as the most frequently used [25,26]. Equally popular are *Solanum lycopersicum* L. [27], *Ecklonia maxima*, *Sargassum* spp. [12], *Laminaria* spp., *Durvillaea potatumum*, *Ulva lactuca*, *Caulerpa sertularioides*, *Padina gymnospora*, *Sargassum liebmannii*, and *Sargassum johnstonii* [25,28].

The group of biostimulants also includes consortia of beneficial fungi or bacteria. Among the fungi used in the cultivation of plants, the following species are noted: *Glomus intraradices*, *Trichoderma atroviride*, *Trichoderma reesei*, and *Heteroconium chaetospira* (Table 1) [10,14–17]. Symbiotic arbuscular mycorrhizal fungi have a positive effect on crop quality. Arbuscular mycorrhizal fungi, along with *Rhizoglomus irregular*, promote the growth of *Stevia rebaudiana* Bertoni [20]. Plant growth bacteria include *Arthrobacter* spp., *Enterobacter* spp., *Acinetobacter* spp., *Pseudomonas* spp., *Ochrobactrum* spp., *Bacillus* spp., and *Rhodococcus* spp. (Table 1) [18,19]. The largest group of beneficial bacteria includes *Rhizobium* spp. and plant growth-promoting rhizobacteria [10]. The plant growth-promoting rhizobacteria group includes *Streptomyces* spp., *Pseudomonas* spp., and *Bacillus* spp. The literature results indicate that *Streptomyces* spp. protect tomato plants against putrefactive bacteria *Pectobacterium carotovorum* subsp. brasiliensis (Pcb). In addition, the volatiles produced by six *Streptomyces* spp. isolates stimulate the growth of tomato roots. The largest fresh root mass obtained due to the volatile substances was 138.2 ± 16.1 mg in comparison to a control mass of 111.5 ± 10.3 mg. The volatile substances also positively affect the dry matter content and root volume as well as the dry and fresh mass of the shoot [29]. Zhao et al. [30] managed to isolate 276 endophytic bacteria from soybean root nodules that protected soybean roots against fungal infections of *Phytophthora sojae*. The antagonistic bacteria included *Enterobacter* spp., *Acinetobacter* spp., *Pseudomonas* spp., *Ochrobactrum* spp., *and Bacillus* spp. Also, lactic acid bacteria, e.g., *Lactobacillus plantarum*, and *Lactobacillus paracasei* are used to promote plant growth and indirectly control diseases [31].

Table 1. Activity of different types of biostimulants.

Source of Biostimulant	Example	Main Activity
Hydrolysis Product	Enzymatic (alfalfa hay, pulses, and vegetable or fruit waste) and chemical (feathers, bone meal, casein, collagen from skins, animal tissue, or fish waste)	Increase in yield [32] Increase in nitrogen and phosphorus content in leaves and macro- and micronutrients [33,34] Increase in protein content in cereal grains [33] Protection against biotic and abiotic stresses [35] Increased soil fertility through the development of soil microorganisms [36]
Anaerobic Digestion Product	Plant, animal, and lignin biomass	Call of the auxin-like effect [23,37] Improving the availability of nutrients [31]
Biopreparations from Marine Algae	*Ascophyllum nodosum*, *Sargassum wightii*, *Ecklonia maxima*, *Enteromorpha intestinalis*, *Gelidium pectinutum*	Antioxidant potential and ability to capture free radicals [12] Chelating effect [12] Increase in plant resistance to fungal and bacterial infections [12] Extension of the shelf-life of fruit for consumption [12] Improve the thermal resistance of plants [12] Protection against drought stress [38]
Consortia of Beneficial Fungi	*Rhizophagus intraradices*, *Rhymbocarpus aggregatus*, *Glomus viscosum*, *Glomus etunicatum*, *Glomus claroideum*, *Trichoderma* sp., *Heteroconium chaetospira*	Increase in the growth and yield of plants alone and in symbiosis with bacteria, e.g., from *Azotobacter* spp. [39] Plant protection against oxidative stress [40]

4. Application Method of Biostimulants

Biostimulants may be used in the form of soil preparations (powders, granules, or solutions added to the soil) or as liquid foliar application products [41]. Biostimulants containing humic substances and nitrogen compounds are often applied directly onto the soil, whereas various types of extracts from plants and seaweed are used in the form of foliar applications. Biostimulants can be introduced into the irrigation system and taken up by plants along with water. One example is the Kelpak SL (*Ecklonia maxima* extract) biostimulant which was sprayed in an aqueous solution of *Phaseolus vulgaris* L. [41]. Biostimulants are used regularly during the whole vegetative period or proactively, i.e., once during the decline of vital forces of the plant. In this case, the biostimulants were administered once during the occurrence of a strong stress factor, e.g., frost [42–44]. The results show that the soil application of the biostimulant was not as effective as foliar application. The foliar application of a biostimulant obtained from sewage sludge increased the level of macro- and micronutrients in the leaves of maize. The nitrogen content in maize leaves increased by 26% (dose 3.6 L/ha) and 46% (dose 7.2 L/ha) [45]. Biostimulants can also be used in the form of biomass or meal from seaweed, however, this method has some limitations. Biomass and meal may be used in areas located close to the source of seaweed acquisition due to transport problems. Biomass or meal is applied directly to the soil long before planting in order to enrich the substrate with nutrients. Agro-technical measures such as ploughing are used to mix biomass or meal with the topsoil [12].

Common forms in which biostimulants occur are ready-to-use extracts or powder to make an aqueous solution. Soil biostimulants often affect the structure of the root, increasing, among other factors, its ability to absorb nutrients. Foliar extracts protect the plant against biotic and abiotic stresses. The circadian rhythm of plants should be taken into consideration. Biostimulants should be applied in the morning when the stomata are open and the assimilation rate is at its peak [12,38]. Biostimulants are also applied directly onto harvested fruits. It was noted that biostimulants containing a combination of extracts from *Sargassum* spp., *Laminaria* spp., and *A. nodosum* (Table 1) significantly extended the shelf-life and the storage life of oranges. After using the biostimulant, the fruits became more resistant to mechanical damage and putrefaction which allowed for an extension of the storage time and suitability for consumption. The extract used produced a better effect than, for example, the calcium chloride normally used in the industry to protect fruit against putrefactive bacteria [12,38].

5. Effect of Biostimulants on Yielding

The popularity of biostimulants in agriculture is associated with the possibility of obtaining higher yields without the need to discontinue the production of ecological crops. According to numerous scientific studies, biostimulants have a positive effect on yielding plants [36]. The yield is usually determined as the amount of fruit obtained from one plant or plot. The yield depends on the type of biostimulant used, the dose, the method of application, and the plant variety. Increased yield is often associated with improving the quality of vegetables or fruit. This is particularly important in organic farming, where artificial fertilizers cannot be used [37,45]. The quality of fruits and vegetables is shaped from the moment of plant growth to the time of harvesting of fruits and vegetables and it consists of the taste and the content of nutrients. The quality is influenced by genetic and agro-environmental factors [46].

The positive influence of biostimulants based on humic, fulvic, and carboxylic acids on the yielding of apricot fruits has been proven [8]. Control trees showed a yield of 12 kg fruit/tree and after the application of humic and fulvic acids together and carboxylic acids in a separate experiment, the yield of the trees increased to 21 kg of fruit/tree and 19 kg of fruit/tree, respectively (Table 2). However, this relationship was observed only in the second year of using the biostimulant. During the first growing season, the yield of the control trees was higher than that of the trees that were treated with biostimulants containing humic and fulvic acids. The biostimulant based on polysaccharides turned out to be ineffective with this variety of apricot trees, which showed a yield comparable to the control in both growing seasons [8]. Preparations containing all amino acids allowed for an increase in mango yield. With a dose of 3 L/ha, the yield increased by 18% compared to the control. At the same time, the biostimulants caused a 15% decrease in fruit weight. The authors explained this phenomenon through the competition of fruit for nutrients [32]. Biostimulants containing phenolic compounds such as sodium para-nitrophenolan, sodium orto-nitrophenolan, and sodium 5-nitroguajakolan proved to be good preparations for raspberry bushes.

As a result of the foliar application of the phenolic compounds, a 20% (Table 2) increase in raspberry yield was obtained [47]. The highest yield was achieved when a biostimulant containing phenol compounds was applied to the "Polka" raspberry variety (yield = 23.03 kg/plot), while the yield in the control was 18.28 kg/plot. Already in the first year after the use of biostimulants, the amount of fruit collected from one bush increased. The best results were produced by 6-benzyladenine in a dose of 100 mg/L, which caused an increase in yield in the first year by about 0.5 kg. In the second year of using the biostimulant 6-benzyladenine at a dose of 100 mg/L and α-naphthaleneacetic acid at a dose of 20 mg/L, the yield increased by more than 1 kg from a single shrub [37].

Strawberry yielding significantly increased after using biostimulants containing herbal and marine plant extracts. In this case, the study proved that soil biostimulants are a source of nitrogen compounds. Moreover, foliar biostimulant application did not produce results as good as biostimulants added to the soil. Biostimulants added to the soil caused a significant increase in the amount of fruit and also improved the condition of the plants. Plants were more resistant to weather conditions and pathogens [42]. Extracts of "Moscatel" vine-shoots improved the yielding of the grapevine variety "Airén". The literature results indicate that two foliar biostimulant variants were prepared, non-toasted and toasted vine-shoots. A significantly higher yield was achieved in the case of two preparations, which were the non-toasted vine-shoots extract (3.09 ± 0.05 kg/plant) and the toasted vine-shoots extract (3.57 ± 0.05 kg/plant) in comparison to the control (2.54 ± 0.03 kg/plant) (Table 2) [48].

A mixture of four biostimulants containing amino acids, polysaccharides, vitamins, humic acids, organic carbon, and enzymatic proteins caused a comparable increase in the yield of two varieties of yellow pepper [49]. The Blondy F1 variety produced a crop at a level of 5.98a 0.23 kg/plot (yield = 5.24 ± 0.30 kg/plot). The Century F1 variety produced a yield of 5.76 ± 0.20 kg/plot (yield in the control = 5.06 ± 0.28 kg/plot) (Table 2). The peptides and amino acids contained in these formulations demonstrated a protective action against excessively high temperatures in the summer season and induced root growth and development, while vitamins and humic acids resulted in fruit growth [49].

Horseradish extract increased pumpkin yield by 12.5% [50] and the beneficial effect of fungal species *Glomus intraradices* and *Trichoderma atroviride* positively influenced the yielding of zucchini, resulting in a yield increase of 0.39 kg per plant on average, probably by increasing the effectiveness of nutrients uptake [14]. A 3% *Moringa oleifera* extract in combination with 0.6% $ZnSO_4$ and 0.25% K_2SO_4 increased the yield of "Kinnow" mandarin plants by 65% (Table 2) compared to the control [51].

The results of the study underline the positive effect of humic acids on the yielding of fruit trees [8]. The use of phenolic compounds [47] resulted in the increased yielding of fruit bushes. A high yield of vegetables may be obtained by using a mixture of amino acids, polysaccharides, vitamins, humic acids, and other compounds. Each of the substrates of the mixture affects another biochemical process occurring in the soil and plant, which allows for the achievement of the desired effect [49].

Table 2. Effect of selected biostimulants on quality of fruit and vegetables.

Biostimulant	Dose	Plant	Type of Claims	Benefitial Effect
Humic and fulvic acids	500 g/100 L	Apricot (*Prunus armeniaca* L.)	Yield	Increase by 75% in the second year compared to the control [8]
Carboxylic acids	200 mL/100 L	Apricot (*Prunus armeniaca* L.)	Yield	Increase 58% in the second year compared to the control [8]
All amino acids	3 L/ha	Mango (*Mangifera indica*)	Yield	Increase by 18% compared to the control [32]
6-benzyladenine	100 mg/L	Blueberry (*Vaccinium corymbosum* L.)	Yield	Increase by 0.5 kg/tree in the first year of using the biostimulant compared to the control [37]
α-naphthaleneacetic acid and 6-benzyladenine	100 mg/L (α-naphthaleneacetic acid) 20 mg/L (6-benzyladenine)	Blueberry (*Vaccinium corymbosum* L.)	Yield	Increase by 1 kg/tree in the second year of using the biostimulant compared to the control [37]
Humic acids	3g/L	Cucumber (*Cucumis sativus* L.)	Diameter	Increase in vegetable diameter by 1.23 cm (first season) and 1.55 cm (second season) compared to the control [52]
Nitrogen, amino acids, auxins	0.45 cm/L	Cucumber (*Cucumis sativus* L.)	Length	Lengthening vegetables by 3.85 cm (first season) and 3.49 cm (second season) compared to the control [52]
Biostimulant	**Dose**	**Plant**	**Type of Claims**	**Benefitial Effect**
Carboxylic acids	200 mL/100 L	Apricot (*Prunus armeniaca* L.)	Diameter	Increase diameter by 2.6 mm (second growing season) compared to the control [8]
Arbuscular mycorrhizal fungi and *Pseudomonas fluorescens* C7	The bacterial inoculum was replicated, watering each plant with 200 mL of bacterial suspension (density about 108 colony forming unit CFU/mL)	Tomato (*Solanum lycopersicum* L.)	Weight	Increase in tomato fruit mass by 6.9 g compared to control [53]
6-benzyladenine	100 mg/L	Blueberry cv. Duke and Bluecrop	Weight	Increase in the weight of blueberry fruit by about 32.4% (first season) and 33.6% (second season) for the blueberry cultivar Duke and 43.5% (first season) and 33.1% (second season) for the blueberry cultivar Bluecrop compared to the control [53]
α-naphthaleneacetic acid	20mg/L	Blueberry cv. Duke and Bluecrop	Weight	Increased the weight of the blueberry cultivar Duke fruit by 41.9% (first season) and 20.0% (second season) and the blueberry cultivar Bluecrop by 55.0% (first season) and 25.4% (second season) compared to the control [37]
Humic Acids	3g/L	Cucumber (*Cucumis sativus* L.)	Length	Increase in vegetable diameter by 1.23 cm (first season) and 1.55 cm (second season) compared to the control [52]
Nitrogen, amino acids, auxins	0.45 cm/L	Cucumber (*Cucumis sativus* L.)	Length	Lengthening vegetables by 3.85 cm in the first and 3.49 cm in the second growing season compared to the control [52]

Table 2. *Cont.*

Carboxylic acids	200 mL/100 L	Apricot (*Prunus armeniaca* L.)	Diameter	Widening of fruit by 2.6 mm on average in the second growing season compared to the control [8]
Arbuscular mycorrhizal fungi and *P. fluorescent* C7	The bacterial inoculum was replicated, watering each plant with 200 mL of bacterial suspension (density about 108 CFU/mL)	Tomato (*Solanum lycopersicum* L.)	Weight	Increase in tomato fruit mass by 6.9 g compared to control [53]
6-Benzyladenine	100 mg/L	Blueberry cv. Duke and Bluecrop	Weight	Increase in the weight of blueberry fruit by about 32.4% (first season) and 33.6% (second season) for the blueberry cultivar Duke and 43.5% (first season) and 33.1% (second season) for the blueberry cultivar Bluecrop compared to the control [53]
α-naphthaleneacetic acid	20 mg/L	Blueberry cv. Duke and Bluecrop	Weight	Increase in weight of the blueberry cultivar Duke fruit by 41.9% (first season) and 20.0% (second season) and the blueberry cultivar Bluecrop by 55.0% (first season) and 25.4% (second season) compared to the control [46]
Biostimulant including, among others, nitrogen, amino acids, auxins	0.45 cm/L	Cucumber (*Cucumis sativus* L.)	Plant height	Increase in plant height of 14.5 cm (in the first growing season) and 19.75 cm (in the second growing season) in comparison with the control plants [52]
Moringa oleifera leaf extract	3% treatments were replicated three times	Pumpkin (*Cucurbita pepo* L.)	Chlorophyll content	34.6% increase in the chlorophyll content compared to the control [50]
Salicylic acid–chitosan nanoparticles	(concentration 0.01%–0.16%)	Maize CV. Surya local	Chlorophyll content	Chlorophyll content in the control was 10.72 mg/g, the chlorophyll content in the maize leaves treated with the biostimulant (concentration 0.01%–0.16%) was in the range of 16.43 to 25.88 mg/g on average [54]
Ascophyllum nodosum Seaweed Extract	1.5 kg/ha	"Sangiovese" grapes	Phenolic content	Phenolic content increased by 1.063 mg/cm^2 compared to the control [55]
A. nodosum Seaweed Extract	3kg/ha	"Sangiovese" grapes	Phenolic content	Phenolic content increased by 0.951 mg/cm^2 compared to the control [55]
Biostimulant (chicken feathers) with Fertilizer (300 kg N/ha + 120 kg K/ha)	3.6 L/ha	Maize (*Zea mays* L. cv PR32W86 Pioneer)	Nitrogen content	Nitrogen content of corn leaves increased by 14.4% (first season) 15% (second season) compared to the control [33]
Biostimulant (chicken feathers) with fertilizer (300 kg N/HA + 120 kg K/HA)	7.2 L/ha	Maize (*Zea mays* L. cv PR32W86 Pioneer)	Nitrogen content	Increase nitrogen content by 39.1% (first season) and 33.3% (second season) compared to the control [33]
Protein hydrolysate	5.0 and 2.5 mL/L	Tomato (*Solanum lycopersicum* L.)	Content of lycopene	Increased by 34.9% and 18.0%, respectively, compared to the control [3]
Protein hydrolysate	2.5 mL/L	Tomato (*Solanum lycopersicum* L.)	Ascorbic acid	Increase ascorbic acid content by 27.3% compared to the control [3]
3% corn seed extract	Soaking and praying 1 mM mg plants	Sunflower seed (*Helianthus annuus* L.)	Enzymatic activity of superoxide dismutase, catalase, and peroxidase	Enzymatic activity of superoxide dismutase, catalase and peroxidase increased by 65.5%, 77.8%, and 84.6%, respectively, as compared to the controls [56]

6. Effect of Biostimulants on the Growth and Size of Plants

The way in which biostimulants work may be defined as multifaceted. The literature describes the positive effect of biostimulants on the growth of fruits and vegetables. At the same time, there are studies in which no effect of biostimulants on fruit size was found. The lack of biostimulant effects is explained by the use of a biostimulant unsuitable for the tested cultivar [37,53].

Fruit producers are interested in biopreparations that allow for the attainment of the largest and healthiest-looking fruit that draw consumers' attention [52]. An increase in the average length and diameter of cucumbers was attained after the use of humic acids and a mixture of nitrogen, amino acids, and auxins. Humic acids in a concentration of 3g/L increased the average length of the fruit in the first and second season by 9.9 cm and 12.2 cm, respectively. The same concentration of humic acids increased the diameter of the cucumbers in the first and second season by an average of 1.23 cm and 1.55 cm, respectively [52]. Three tested biostimulators caused an elongation and increase in the diameter of the vegetables. One biostimulant (containing nitrogen, amino acids, and auxins) led to the elongation of the cucumbers by 3.85 cm in the first and 3.49 cm in the second growing season. The diameter of the fruit increased by 1.12 cm and 1.56 cm, respectively, in the first and second growing season under the influence of this biostimulant. The application of humic acids and biostimulants containing auxins in particular makes it possible to obtain elongated and thickened cucumbers [52]. The use of biostimulants containing humic and fulvic acids as well as carboxylic acids led to a tenfold enlargement of the apricot fruit. The greatest influence on the size of the fruit was carboxylic acids, which contributed to the widening of fruit by 2.6 mm on average in the second growing season [8].

Studies show that the perfect biostimulants that cause the growth of fruits and vegetables are consortia of microorganisms. Examples are arbuscular mycorrhizal fungi and plant growth-promoting bacteria, the use of which resulted in increased tomato weight. A positive effect was obtained through the combination of arbuscular mycorrhizal fungi containing fungi of the following species: *Rhizophagus* spp., *Rhizophagus aggregatus*, *Septoglycus viscosum*, *Claroideoglomus etunicatum*, *Claroideoglomus claroideum* and various types of plant growth-promoting bacteria. All of the biostimulants based on microorganisms (arbuscular mycorrhizal fungi + *Pseudomonas* sp. Strain 19Fv1T, arbuscular mycorrhizal fungi + *Pseudomonas fluorescens* C7, arbuscular mycorrhizal fungi + *Pseudomonas* sp. 19 Fv1T and *Pseudomonas fluorescens* C7) caused an increase in tomato fruit mass, but the most effective result was demonstrated by biostimulants including arbuscular mycorrhizal fungi and *P. fluorescent* C7, which caused an increase in tomato mass to 71.3 ± 0.6 g (weight of control tomatoes = 64.4 ± 0.9 g). The microorganisms used caused a slight elongation of tomatoes as well. The length of the fruit in the controls ranged from 5.49 ± 0.03 cm to 5.81 ± 0.03 cm, while in combination with the biostimulants used, it rose to 5.88 ± 0.03 cm to 6.05 ± 0.02 cm. There was also a slight increase in fruit diameter from 4.24 ± 0.02 cm to 4.62 ± 0.03 cm to 4.64 ± 0.03 cm to 4.78 ± 0.02 cm [53].

The application of 6-benzyladenine as a biostimulant at a dose of 100 mg/L resulted in an increase in the weight of blueberry fruit by about 32.4% (first season) and 33.6% (second season) for the blueberry cultivar Duke and 43.5% (first season) and 33.1% (second season) for the blueberry cultivar Bluecrop compared to the control. The literature results showed that α-naphthaleneacetic acid at a dose of 20mg/L was also an effective biostimulant, which increased the weight of the blueberry cultivar Duke fruit by 41.9% (first season) and 20.0% (second season) and the blueberry cultivar Bluecrop by 55.0% (first season) and 25.4% (second season). As demonstrated in the study, one inefficient biostimulant was gibberellic acid at a dose of 200 mg/L. Gibberellic acid increased the weight of blueberry cultivar Duke fruit by 4.7% (first season) and 14.3% (second season) as well as Bluecrop blueberry cultivar by 0.8% (first season) and 11.5% (second season) compared to the control. For this reason, gibberellic acid is not recommended as a biostimulant for soft fruits [37].

The titanium compounds with which raspberries were treated caused an increase in fruit weight from 4.44 g (control) to an average of 5.4 g, but only at the beginning of the harvest season. At the end of the harvest, a 57% decrease in raspberry weight was observed [47]. The same relationship was observed for phenolic compounds. At the beginning of the harvest, the mass of raspberries treated with phenolic compounds was 5.04 g on average, while at the end of the harvest there was a 44.4% loss in fruit weight. The fact is that the raspberry collection is characterized by the loss of fruit mass at the end of the harvest, but if these three biostimulants were used, the resulting losses were greater than in the control. The control fruits were characterized by a 42.3% weight loss, while the fruits treated with biostimulants showed a mass loss at the end of the harvest in the range from 44.4% to 57.0% [47]. The increase in fruit weight after the use of biostimulants compounds was also observed in the case of cherries. The use of salicylic acid with the addition of calcium resulted in a 15% increase in sour cherries "Sweetheart" (2015) and "Skeena" (2016) [57].

Regarding the multifaceted effect of biostimulants, it should be emphasized that these formulations may affect many of the characteristics of the plant, e.g., fruit size, plant height, and root length [52]. The use of humic acids in a concentration of 3g/L and other biostimulants caused an increase in cucumber plant height, the number of leaves, and the number of stems in both growing seasons [52]. In the first growing season, the height of the control plants was 78.13 cm. Plants treated with 3 g/L of humic acids were 14.25 cm taller on average. In the second growing season plants treated with humic acids were taller by 13.25 cm. Similar dependencies may be observed after using other biostimulators. In the first and second growing season, this biostimulant (including, among others, nitrogen, amino acids, and auxins) was the most effective, causing an increase in plant height of 14.5 cm (in the first growing season) and 19.75 cm (in the second growing season) in comparison with the control plants. The biostimulant including, among others, naphthyl acetic acid was the least effective, and resulted in an increase in the growth of plants by 4.38 cm and 7.92 cm, respectively to the first and second growing seasons. Similar dependencies were noted in the number of leaves and new stems [52].

7. Impact of Biostimulants on Physical Characteristics

Biostimulants also have an influence on mechanical properties, i.e., the firmness of fruits or vegetables. Depending on the type, biostimulants may cause the stiffening of cell walls, thereby reducing their extensibility [8]. Biostimulants that increase the flexibility of cell walls at the same time extend the shelf-life of fruits and vegetables for consumption and facilitate their storage. Biostimulants based on carboxylic, humic, and fulvic acids and also the biopolymers of polysaccharides increased the mechanical strength of apricot fruits during two years of biostimulant use [8]. In turn, biostimulants containing phenolic compounds or chitosan resulted in the loss of fruit firmness of the three raspberry varieties studied. The use of biostimulants based on titanium compounds did not alter the fruit firmness, which was comparable to the firmness in the control test [47]. The use of spic cytozyme containing essential plant nutrients and growth biostimulants in the amount of 4 mL/L significantly reduced the cracking of pomegranate fruit [58]. In addition to improving mechanical properties, biostimulants change the shape and colour of fruits and vegetables. Fruit with larger length and diameter, as well as the right colours, are preferred by consumers [8]. However, consumer preferences are subject to dynamic changes.

An important visual feature that proves the quality of fruit is colour. The colour of the fruit is substantially influenced by the content of anthocyanins. Weber et al. [59] examined the content of anthocyanins in strawberries treated with *Ascophyllum nodosum* extract with silicon. Fruits treated with a biostimulator were characterized by a higher content of anthocyanins in the initial fruiting period, therefore, they were more red than the control fruits [59]. "Sweetheart" cherries treated with glycine and betaine were characterized by a darker skin than the control fruits. Although the mechanism of action of betaine and glycine on the formation of anthocyanins is not fully understood, it is known that the darker colour of the fruit was caused by a higher content of antioxidants [57]. Tarantino et al. (2018) in the second year of using biostimulants obtained apricot fruits with a lighter skin compared to the first year. This could be due to the higher concentration of biostimulants used in the first growing season. There were significant differences in the colour of the fruit. In the second year of the experiment the colour of the apricots was redder than in the first year of fruiting. There were no significant differences in the colour of the fruit produced by the three biostimulants used (1—biopolymers of polysaccharides; 2—humic and fulvic acids; 3—carboxylic acids) [8].

It should be emphasized that the increased mechanical strength or fruit colour change results from the good condition of fruit plants, which in turn is a result of their proper nutrition [50]. Biostimulants are supplied externally, which indirectly, e.g., induced by the photosynthesis process, plays an important role in the nutrition of plants. Indirect induction, for example, consists in increasing the leaf area. Leaves are the main organs in which photosynthesis takes place, therefore increasing the leaf area leads to an increase in photosynthesis. Intensively photosynthetic plants are better nourished. Increasing the leaf area also leads to an increase in the transpiration surface. This phenomenon has the especially important function of protecting the plant from overheating. *Moringa oleifera* leaf extract increased the surface area of the *Cucurbita pepo* L. by 9.7% and simultaneously led to a 34.6% increase in the chlorophyll content of the leaves compared to the control [50]. The positive effect of the *A. nodosum* extract in a period of drought on the growth of spinach has been proven. Seaweed extract increased the relative water content of the leaves from 76% to 82%. The surface area of the leaf was also increased by 16% (foliar spray), 21% (biostimulant in the irrigation system), and 38% (biostimulant in the irrigation system and in a spray). Increasing the area and turgor of the leaf led to an increase in the intensity of photosynthesis and improved the conditions for growing spinach under stress conditions [25].

8. Effect of Biostimulants on Chemical Composition

Biostimulants can affect a number of the chemical properties of fruits and vegetables, including dry mass, acidity or vitamin content. The chemical composition of the fruit directly affects their palatability. It is assumed that fruits with a content of dissolved solids (SSC) above 12°Brix are characterized by an excellent taste [47]. In the first year of using biostimulants containing the biopolymers of polysaccharides, humic and fulvic acids as well as carboxylic acids, the average value of SSC in apricots stood at 10.7°Brix. In the second year of using these biostimulants, fruit taste values improved significantly, as evidenced by the increase in the SSC level to an average of 14.1°Brix [8]. Biostimulants containing phenol compounds or chitosan reduced the dissolved solids content in the fruits of the three raspberry varieties (Pokusa, Polka, and Poranna Rosa). The opposite effect was produced by biostimulants based on titanium compounds, the use of which resulted in an increase in the content of dissolved solids in the raspberry fruit [47]. The quality of the fruit is also demonstrated by the ratio content of dissolved solids to their titratable acidity. Fruit quality is defined as good if the ratio content of dissolved solids to titratable acidity is within the range of 10 to 15. The treatment of fruit trees with biostimulants containing biopolymers of polysaccharides (16.7) and humic and fulvic acids (16.1) leads to an increase in the ratio content of dissolved solids to titratable acidity in relation to the control (14.0) and thus negatively influenced the sensory quality of fruit [8].

It is important to grow fruit that has an appropriate level of acidity but it is difficult to say, however, whether changes in acidity at the level of several percent have a significant impact on the fruit taste, because it is based on the subjective impression of the consumer. Although the literature data present studies on the effect of biostimulants on fruit acidity, there is no explicit interpretation of the results. It is not clear whether the changes in fruit acidity should be understood in terms of the positive or negative effects of the biostimulants used. The use of phenolic compounds and titanium compounds as a biostimulant in the cultivation of raspberries led to an increase in fruit acidity to 2.26% and 2.18%, respectively (control, 2.08%) [47]. A decrease in apricot acidity was noted after the use of biostimulants containing polysaccharides, humic and fulvic acids, and carboxylic acids [8]. In the second year after using these biostimulants, fruit acidity was reduced from an average of 3.45 (control pH) to an average of 3.7–3.8 (pH after using biostimulants) [8].

An important health-related feature of fruits and vegetables is the content of vitamin C and nitrogen compounds. The role of nitrogen in plants results from its influence on growth and development. It is a component of nucleic acids, it participates in the process of photosynthesis, and it builds amino acids that form a part of plant proteins [60]. The content of vitamin C and nitrogen compounds in the fruit depends mainly on the plant variety [47]. The use of selected biostimulants (phenolic compounds, chitosan, and titanium compounds) increased both the levels of ascorbic acid and nitrates. It turned out that the three biostimulators tested positively influenced the level of nitrates in raspberries. It was found that phenolic compounds contained in one of the biostimulants increased the level of vitamin C most effectively [47]. Assuming that phenolic compounds increased the content of vitamin C in fruit, it was necessary to determine how different biostimulators affect the content of phenolic compounds. Zarzecka et al. [61] studied the effect of herbicides and biostimulators on the polyphenol content of potato tubers. The experiment was conducted over a period of two years. Three potato varieties were treated with different substances: Harrier herbicide 295 ZC, Harrier 295 ZC + Kelpak SL growth regulator, and Sencor 70 WG herbicide, 5 Sencor 70 WG + Asahi growth regulator. In the case of Asahi, the active substances were phenolic compounds, while for Kelpak SL, auxins and cytokinins were the active substances. The applied biostimulants and herbicides caused an increase in the polyphenol content in tubers of all potato varieties (on average, 159.8–161.3 mg/kg) compared to the control (average of 156.0 mg/kg). The use of biostimulants and herbicides increased the content of polyphenols in the leaves of the potato to an average of 289.2–291.2 mg/kg compared to the control (287.8 mg/kg). The content of polyphenols in tubers is of particular importance for humans. Polyphenols reduce the risk of numerous diseases, e.g., blocking carcinogenic compounds [61]. It was observed that after using a biostimulant containing seaweed *A. nodosum* and silicon, the content of phenolic compounds in strawberries was slightly lower. Phenols are also defined as compounds produced by plants under stressful conditions, hence the conclusion about the positive effect of the tested biostimulant on strawberries [59]. The foliar spraying of the "Airén" grapevines by non-toasted and toasted biostimulants increased the content of phenolic compounds. In this case, the biostimulants were extracts from the "Moscatel" vine shoots. An important group of phenolic grape buds are hydroxycinnamic acids (trans-caffeic and trans-p-coumaric), which affect the taste of wine. Both non-toasted biostimulants (14.10 ± 0.13) and toasted biostimulants (11.26 ± 0.27) led to a trans-p-coumaric acid increase relative to the control (8.60 ± 0.03). The non-toasted biostimulant (1.14 ± 0.01) and toasted biostimulant (0.95 ± 0.03) also led to a trans-caffeic acid growth compared to the control (0.92 ± 0.03). The effect of the higher content of, among other compounds, hydroxycinnamic acid is a better quality wine [48]. A biostimulant containing *A. nodosum* seaweed extract increased the phenol content of "Sangiovese" grapes. A 1.5 kg/ha dose of biostimulant increased the phenolic content to 1.063 mg/cm^2, while the biostimulant in a dose of 3 kg/ha increased the phenolic content to 0.951 mg/cm^2. The phenol content of the control was 0.753 mg/cm^2. The results were statistically significant [55].

The positive effect of biostimulants is also based on increasing the content of chlorophyll in leaves and thus increasing the efficiency of the process of photosynthesis. Salicylic acid-chitosan nanoparticles used as a biostimulant led to an increase in the content of chlorophyll in leek corn [54]. While the chlorophyll content in the control was 10.72 mg/g, the chlorophyll content in the maize leaves treated with the biostimulant (concentration 0.01%–0.16%) was in the range of 16.43 to 25.88 mg/g on average. In plants treated only with chitosan and salicylic acid, a decrease in the chlorophyll content to an average of 9.24 mg/g and 9.79 mg/g was observed [54]. After the foliar application of the *Moringa oleifera* leaf extract, a 34.6% increase in the chlorophyll content of *Cucurbita pepo* L. leaves was recorded compared to the control (plants sprayed only with water) [50]. The increase in the chlorophyll content combined with the increased intensity of the photosynthesis process was noted during the cultivation of Hibiscus treated with biostimulants formed in the process of the hydrolysis of waste. A 15% increase in the chlorophyll content of leaves resulted in a 24% increase in the photosynthesis rate compared to the control [62].

During the tests to determine the chemical composition of fruits, the content of glucose, fructose, sucrose, ascorbate, proteins, and macro- and micro-elements is often determined. Plants treated with one of the tested biostimulators (arbuscular mycorrhizal fungi + *Pseudomonas* sp. 19 Fv1T and *P. fluorescens* C7) showed an increase in the concentration of glucose in tomatoes at 11.83 g/kg, while in controls the content of glucose was 10.45–11.0 g/kg. After using this biostimulant, the fructose content also increased to about 12.86 g/kg, while in the controls it was 10.77–11.14 g/kg. After using a biostimulator based on *A. nodosum* seaweed extract and a silicon extract, a slight increase in the level of sugars in the strawberry fruit was observed. The most common sugars were glucose and fructose. Sucrose accounted for 11% of total sugars [59]. The use of an extract of *M. oleifera* leaves increased the total soluble sugar content in pumpkin by about 80.6% [50]. An interesting relationship was observed in the case of ascorbate. The use of biostimulant containing arbuscular mycorrhizal fungi, *Pseudomonas* sp. 19 Fv1T and *P. fluorescens* C7 led to an increase in the ascorbate content (10.75 mg/100 g), whereas the use of a biostimulant containing arbuscular mycorrhizal fungi and *P. fluorescens* C7 reduced the ascorbate content in tomatoes (4.30 mg/100 g). In the controls, tomatoes contained about 5.47–7.12 mg/100 g ascorbate. In plants treated with the biostimulant containing arbuscular mycorrhizal fungi and *P. fluorescens* C7, an increase in β-carotene in tomatoes was observed (controls: 2.117–2.224 μg/100 g fresh weight; β-carotene content in the biostimulator study: 2.829 μg/100 g fresh weight). β-carotene may be converted into vitamin A and can protect against the adverse effects of free radicals [53]. The protein content is particularly important in the case of grain plants. The foliarly used biostimulator containing sewage sludge caused an increase in protein content in maize grains by about 30% in both growing seasons [45]. The biostimulator formed as a by-product of the two-stage process of pressing olive oil led to an increase in the protein content of maize grains by 19% [63].

Combining the biostimulant (chicken feathers) with a fertilizer gave a better quality of maize yield than using only the fertilizer. Three combinations of the agent used for spraying the corn were used. In the first variant, only fertilizer was used (300 kg N/ha + 120 kg K/ha). For the second variant, fertilizer was used (300 kg N/ha + 120 kg K/ha) in combination with a biostimulant (3.6 L/ha). In the third variant, fertilizer (300 kg N/ha + 120 kg K/ha) was used in combination with a biostimulant (7.2 L/ha). The treatment was applied during two seasons. The highest level of nitrogen was obtained after the application of a biostimulant containing fertilizer in combination with the highest dose of the biostimulant (7.2 L/ha). The nitrogen content of corn leaves increased by 14.4% (fertilizer with the biostimulant of 3.6 L/ha) and 39.1% (fertilizer with the biostimulant of 7.2 L/ha) in the first vegetative season in comparison to the nitrogen content of maize leaves treated only with fertilizer. In the second growing season, the nitrogen content increased by 15% (fertilizer with the biostimulant of 3.6 L/ha) and 33.3% (fertilizer with the biostimulant of 7.2 L/ha) respectively. The use of a biostimulant in combination with the fertilizer also resulted in an increase in phosphorus content in the leaves of maize. The treatment increased *p* levels by 32.8% (fertilizer with a biostimulant of 3.6 L/ha) and 52.2% (fertilizer with a biostimulant 7.2 L/ha) in the first season and by 43.5% (fertilizer with a biostimulant of 3.6 L/ha) and 51.1% (fertilizer with a biostimulant of 7.2 L/ha) in the second season compared to the control [33]. An effective biostimulant for the "Kinnow" mandarin trees proved to be *Moringa oleifera* extract. The 3% *Moringa oleifera* extract foliar application with 0.6% $ZnSO_4$ and 0.25% K_2SO_4 resulted in a 1.35-fold (first season) and 1.42-fold (second season) increase in nitrogen content compared to the control. Trees sprayed only with 3% *Moringa oleifera* extract showed a 1.09 times (first season) and 1.07 times (second season) higher phosphorus content compared to control trees [51].

The improvement in the chemical properties of fruits may increase not only their pro-health values, but also lead to an improvement in their sensory values. One example may be guaiacol, which was applied in a foliar way to improve the quality of wine. Guaiacol was shown to increase the amount of glycosylated aromatic compounds in "Microvine" grapes. These compounds have influenced the improvement of wine quality in the final step of wine formation. The guaiacol-treated fruits were characterized by a higher aglycone content (534.25 μg/g) compared to the control (157.52 μg/g). Treatment with guaiacol also increased the content of monomethyl alcohols from 2.94 μg/g in control fruits to 170.30 μg/g in guaiacol-treated fruits [64].

Biostimulants are becoming a viable option for solving the problem of the ineffective uptake of nutrients from fertilizers by plants. The fact is that a large proportion of fertilizer nutrients are not taken up by plants. Reducing the amount of mineral fertilizers introduced into the soil limits environmental degradation. It is thought that the development of certain biostimulants has the potential to increase the amount of nutrients taken up by plants [65]. The increase in the amount of nutrients taken up by plants may be achieved through the use of fertilizers and biostimulants in combination [33]. An increased nitrogen content was obtained thanks to the use of a biostimulant formed in the process of the hydrolysis of chicken feathers in combination with nitrogen fertilizer, while the increase in phosphorus content was the result of using a biostimulant formed in the process of the hydrolysis of chicken feathers with phosphate fertilizer [33] and an extract of *Moringa oleifera* as a biostimulant [51].

9. Effect of Biostimulants on Antioxidant Properties

Antioxidant activity is an often-studied property of fruits and vegetables. Antioxidants are listed as compounds that inhibit tumour cell proliferation and protect against oxidative stress caused by excess free radicals. The result of oxidative stress may be, among other factors, damage to DNA, cell membranes, or enzymes [66]. It was shown that the use of biostimulants in plant breeding can change the activity of enzymes and affect the antioxidant properties. Lycopene, ascorbic acid, phenolic compounds and others have antioxidant properties. Reactive oxygen molecules, e.g., OH, O^{2-}, and H_2O_2, are detoxified by antioxidant compounds (e.g., phenols, ascorbic acid) and enzymes (e.g., catalase, peroxidase, superoxide dismutase) [67].

Protein hydrolysate applied as a biostimulant to tomatoes had no effect on the level of phenolic compounds, while its effect on the content of ascorbic acid and lycopene was noted. After using biostimulant doses of 5.0 and 2.5 mL/L, the content of lycopene increased by 34.9% and 18.0%, respectively, compared to the control. The dose of 2.5 mL/L biostimulant increased the content of ascorbic acid by 27.3% [68]. The use of biostimulants on apricot fruit trees increased the antioxidant capacity of fruits. In the first season (average 76.8 mg/100 g), after using the stimulants, the antioxidant capacity of fruit was higher than in the second season (average 66.5 mg/100 g). The observed differences in antioxidative abilities between the two seasons were explained by changes in climatic conditions [8].

Also, a biostimulant based on salicylic acid and chitosan nanoparticles (SA-CS NPs) had an effect on the enzyme and antioxidant activity in maize leaves. The enzyme activity in leaves treated with chitosan, salicylic acid, and a control was comparable. After two days of treating the plants with the biostimulant, the activity of superoxide dismutase increased by two times compared to plants treated with salicylic acid. After three days of treating plants with a biostimulant, superoxide dismutase activity was 3.2 times higher than for plants treated with only salicylic acid. Peroxidase activity in plants treated with a biostimulant was 7.7 (after two days) and 5.2 (after three days) times higher than for plants treated with only salicylic acid. Catalase activity, phenylalanine ammonia lyase, and polyphenol oxidase increased by 2.9, 2.3, and 1.5-fold, respectively, after the second day of treatment with nanoparticles compared to salicylic acid treatment. It should be emphasized that the enzyme activity occurring in the leaves of plants treated with a biostimulant increased during the first three days. After the fourth day of treatment, the enzyme activity decreased in all variants of the experiment. The content of hydrogen peroxide in leaves treated with SA-CS NPs was 1.7 (first day of treatment), 3.6 (second day of treatment), and 1.7 (third day of treatment) times higher than in plants treated with salicylic acid [54].

The use of *M. oleifera* extract as a biostimulant resulted in a decrease in the activity of the antioxidant enzymes (catalase, peroxidase, and superoxide dismutase) in rocket plants (*Eruca vesicaria* subsp. *Sativa*). At the same time, the content of phenol and ascorbic acid was higher with increasing concentrations of the biostimulant [67]. Aqueous garlic extract improved tomato oxidation properties. Superoxide dismutase activity increased in proportion to the aqueous garlic extract concentration. The highest activity of this enzyme was observed with the foliar application of the biostimulant in a volume of 200 μg/L; also, the peroxidase activity was highest after using the biostimulant at this concentration. A lower aqueous garlic extract concentration (50 μg/L) did not affect the activity of these enzymes [69]. Soaking sunflower seed *Helianthus annuus* L. in a 3% corn seed extract and spraying 1 mM Mg plants stimulated the sunflower's antioxidant system. The enzymatic activity of superoxide dismutase, catalase, and peroxidase increased by 65.5%, 77.8%, and 84.6%, respectively, as compared to the controls. The increased level of antioxidant enzymes was related to the foliar application of Mg ions, the use of which also increased the intensity of the photosynthesis process [56].

Biostimulants increased the phenylalanine ammonia lyase enzyme activity. While the phenylalanine ammonia lyase level in the control was 7.9 ± 0.22 IU/mL × min (0.4% *E. maxima extract*), after using biostimulants it increased to 9.0 ± 0.01 and 9.7 ± 0.01 IU/mL × min (10−6 M *E. maxima* extract). Phenylalanine ammonia lyase is an enzyme catalyzing the first step in the synthesis of phenyl compounds. An increased production of phenolic compounds is observed during plant stress. From this, it may be concluded that biostimulants can induce plant stress to increase the production of secondary metabolites [70]. Biostimulatory properties also reveal many components of compost. Depending on the raw materials and methods used for the composting process, the compost may contain, among others, polysaccharides, amino acids, and organic nitrogen. Compost can be used to replace peat in greenhouse cultures. In addition, it may be produced from organic waste, such as wood, plant residue, or other residues. Compost which is considered by the European Union to be ecological must consist solely of natural raw materials, characterized by a limited content of heavy metals and hazardous elements (Se, Mo, S). The product of the composting process must be free of pathogenic agents (*Salmonella* sp. and *Escherichia coli*) [71]. Agroindustrial compost proved to be an alternative to peat in the cultivation of red lettuce. The compost increased the content of antioxidant compounds in lettuce leaves. In the autumn season, lettuce leaves cultivated in the compost showed 1.5 times more antioxidant activity than lettuce grown in compost in the summer season and it was also higher that of lettuce grown in peat in the autumn season [72].

10. Conclusions

Biostimulants are preparations of natural origin that support the pro-ecological cultivation of vegetables and fruits. Although for several years a positive effect of biostimulants has been widely reported, they are rarely introduced into standard cultivation technologies. This is connected with the insufficient knowledge of farmers on functions and usage of biostimulants what results in a fear of an increase in the cost of cultivation and a reduction in the quality and quantity of plants, which would affect the profitability of crops. The problem is also the multitude of preparations and the need to select a proper biostimulant for a specific plant variety in order to obtain the highest and the best quality yields. The market requires the development of preparations with a broad spectrum of functionality, which is easy to apply and has the possibility of combination with other agents.

The use of biostimulants on a commercial scale would limit the amount of mineral fertilizers introduced into the environment, thus reducing the pollution of soils, water, and air. This is especially important in the case of global warming. Global agriculture accounts for an average of 21% [73] of the global greenhouse effect, of which around 13% [74] is concerned with the effect of artificial fertilizers. The newly developed technologies of biopreparations may constitute a significant contribution to environmental protection, but primarily they are closely linked with sustainable agricultural and horticultural production with the aim of obtaining cheap, easily available, and high quality food. The effect of biostimulants depends on many factors, from the raw material and the process as a result of which they arose to the plant varieties, application method, and climate. The positive effect of consortia of microorganisms and plant hydrolysates on growth and yield of crops plants should be particularly emphasized. It is also important to increase the antioxidant potential of plants treated with biostimulants containing algae. A positive impact on crop quality and performance, no negative or harmful impact on people, animals, or the environment, increased biodiversity of beneficial microorganisms, and improvement of soil properties are the main advantages of biostimulants. However, the nature of their positive influence is not fully understood, therefore their mechanisms of action are, in some cases, still a challenge and need to be recognized. For this reason, biostimulants are among the hot topics in agriculture and still require detailed research.

Author Contributions: M.D. collected the data, interpreted the results and wrote the manuscript. M.F. provided ideas and performed reviews and corrections of the manuscript. J.C. designed the study, helped draft the manuscript, interpreted results and corrected the text. All authors read and approved the final manuscript.

References

1. Di Vittori, L.; Mazzoni, L.; Battino, M.; Mezzetti, B. Pre-Harvest Factors Influencing the Quality of Berries. *Sci. Hortic.* **2018**, *233*, 310–322. [CrossRef]
2. European Union. II Non-Legislative Acts. DECISIONS: Commission Implementing Decision (EU) 2017/1442 of 31 July 2017 Establishing Best Available Techniques (BAT) Conclusions, under Directive 2010/75/EU of the European Parliament and of the Council, for Large Combustion Pl. *Off. J. Eur. Union. L* **2017**, *60*, 1–88.
3. Møller, A.P.; Laursen, K. Reversible Effects of Fertilizer Use on Population Trends of Waterbirds in Europe. *Biol. Conserv.* **2015**, *184*, 389–395. [CrossRef]
4. Rouphael, Y.; Spichal, L.; Panzarova, K.; Casa, R.; Colla, G. High-Throughput Plant Phenotyping for Developing Novel Biostimulants: From Lab to Field or from Field to Lab? *Front. Plant Sci.* **2018**, *9*, 1–19. [CrossRef] [PubMed]
5. Rouphael, Y.; Colla, G. Synergistic Biostimulatory Action: Designing the Next Generation of Plant Biostimulants for Sustainable Agriculture. *Front. Plant Sci.* **2018**, *9*, 1–24. [CrossRef]
6. Bradshaw, T.L.; Berkett, L.P.; Griffith, M.C.; Darby, H.M.; Moran, R.E.; Garcia, M.E. Assessment of Kelp Extract Biostimulants on Tree Growth, Yield, and Fruit Quality in a Certified Organic Apple Orchard. In *II International Organic Fruit Symposium 1001*; International Society for Horticultural Science: Brabant, Belgium, 2013; pp. 191–198. [CrossRef]
7. Kyriacou, M.C.; Rouphael, Y. Towards a New Definition of Quality for Fresh Fruits and Vegetables. *Sci. Hortic.* **2018**, *234*, 463–469. [CrossRef]
8. Tarantino, A.; Lops, F.; Disciglio, G.; Lopriore, G. Effects of Plant Biostimulants on Fruit Set, Growth, Yield and Fruit Quality Attributes of 'Orange Rubis®' Apricot (*Prunus armeniaca* L.) Cultivar in Two Consecutive Years. *Sci. Hortic.* **2018**, *239*, 26–34. [CrossRef]
9. The European Parliament and the Council of the European Union. *Regulation (Eu)/Laying down Rules on the Making Available on the Market of EU Fertilising Products and Amending Regulations (EC) No 1069/2009 and (EC) No 1107/2009 and Repealing Regulation (EC) No 2003/2003*; European Commission: Brussels, Belgium, 2019; Volume 0270.
10. du Jardin, P. Plant Biostimulants: Definition, Concept, Main Categories and Regulation. *Sci. Hortic.* **2015**, *196*, 3–14. [CrossRef]
11. Chiaiese, P.; Corrado, G.; Colla, G.; Kyriacou, M.C.; Rouphae, Y. Renewable Sources of Plant Biostimulation: Microalgae as a Sustainable Means to Improve Crop Performance. *Front. Plant Sci.* **2018**, *9*, 1–6. [CrossRef]
12. Battacharyya, D.; Babgohari, M.Z.; Rathor, P.; Prithiviraj, B. Seaweed Extracts as Biostimulants in Horticulture. *Sci. Hortic.* **2015**, *196*, 39–48. [CrossRef]
13. Colla, G.; Nardi, S.; Cardarelli, M.; Ertani, A.; Lucini, L.; Canaguier, R.; Rouphael, Y. Protein Hydrolysates as Biostimulants in Horticulture. *Sci. Hortic.* **2015**, *196*, 28–38. [CrossRef]
14. Colla, G.; Rouphael, Y.; Di Mattia, E.; El-Nakhel, C.; Cardarelli, M. Co-Inoculation of Glomus Intraradices and Trichoderma Atroviride Acts as a Biostimulant to Promote Growth, Yield and Nutrient Uptake of Vegetable Crops. *J. Sci. Food Agric.* **2015**, *95*, 1706–1715. [CrossRef] [PubMed]
15. Mukherjee, P.K.; Horwitz, B.A.; Herrera-Estrella, A.; Schmoll, M.; Kenerley, C.M. *Trichoderma* Research in the Genome Era. *Annu. Rev. Phytopathol.* **2013**, *51*, 105–129. [CrossRef] [PubMed]
16. Nicolás, C.; Hermosa, R.; Rubio, B.; Mukherjee, P.K.; Monte, E. Trichoderma Genes in Plants for Stress Tolerance-Status and Prospects. *Plant Sci.* **2014**, *228*, 71–78. [CrossRef] [PubMed]
17. Usuki, F.; Narisawa, K. A Mutualistic Symbiosis between a Dark Septate Endophytic Fungus, Heteroconium Chaetospira, and a Nonmycorrhizal Plant, Chinese Cabbage. *Mycologia* **2007**, *99*, 175–184. [CrossRef] [PubMed]
18. Gaiero, J.R.; McCall, C.A.; Thompson, K.A.; Day, N.J.; Best, A.S.; Dunfield, K.E. Inside the Root Microbiome: Bacterial Root Endophytes and Plant Growth Promotion. *Am. J. Bot.* **2013**, *100*, 1738–1750. [CrossRef] [PubMed]

19. Zhao, D.; Zhao, H.; Zhao, D.; Zhu, X.; Wang, Y.; Duan, Y.; Xuan, Y.; Chen, L. Isolation and Identification of Bacteria from Rhizosphere Soil and Their Effect on Plant Growth Promotion and Root-Knot Nematode Disease. *Biol. Control* **2018**, *119*, 12–19. [CrossRef]
20. Tavarini, S.; Passera, B.; Martini, A.; Avio, L.; Sbrana, C.; Giovannetti, M.; Angelini, L.G. Plant Growth, Steviol Glycosides and Nutrient Uptake as Affected by Arbuscular Mycorrhizal Fungi and Phosphorous Fertilization in Stevia Rebaudiana Bert. *Ind. Crops Prod.* **2018**, *111*, 899–907. [CrossRef]
21. Souri, M.K.; Bakhtiarizade, M. Scientia Horticulturae Biostimulation Effects of Rosemary Essential Oil on Growth and Nutrient Uptake of Tomato Seedlings. *Sci. Hortic.* **2019**, *243*, 472–476. [CrossRef]
22. Ugolini, L.; Cinti, S.; Righetti, L.; Stefan, A.; Matteo, R.; D'Avino, L.; Lazzeri, L. Production of an Enzymatic Protein Hydrolyzate from Defatted Sunflower Seed Meal for Potential Application as a Plant Biostimulant. *Ind. Crops Prod.* **2015**, *75*, 15–23. [CrossRef]
23. Scaglia, B.; Pognani, M.; Adani, F. The Anaerobic Digestion Process Capability to Produce Biostimulant: The Case Study of the Dissolved Organic Matter (DOM) vs. Auxin-like Property. *Sci. Total Environ.* **2017**, *589*, 36–45. [CrossRef] [PubMed]
24. Pacholczak, A.; Szydo, W.; Jacygrad, E.; Federowicz, M. Effect of auxins and the biostimulator algaminoplant on rhizogenesis in stem cuttings of two dogwood cultivars (*Cornus alba* "Aurea" and 'Elegantissima'). *Hortorum Cultus* **2012**, *11*, 93–103.
25. Xu, C.; Leskovar, D.I. Effects of A. nodosum Seaweed Extracts on Spinach Growth, Physiology and Nutrition Value under Drought Stress. *Sci. Hortic.* **2015**, *183*, 39–47. [CrossRef]
26. Spann, T.M.; Road, E.S.; Alfred, L.; Little, H.A. Applications of a Commercial Extract of the Brown Seaweed Ascophyllum Nodosum Increases Drought Tolerance in Container-Grown 'Hamlin' Sweet Orange Nursery Trees. *HortScience* **2011**, *46*, 577–582. [CrossRef]
27. Hernández-Herrera, R.M.; Santacruz-ruvalcaba, F.; Ruiz-lópez, M.A.; Norrie, J.; Hernández-Carmona, G. Effect of Liquid Seaweed Extracts on Growth of Tomato Seedlings (*Solanum lycopersicum* L.). *J. Appl. Phycol.* **2014**, 619–628. [CrossRef]
28. Kumari, R.; Kaur, I.; Bhatnagar, A.K. Effect of Aqueous Extract of Sargassum Johnstonii Setchell & Gardner on Growth, Yield and Quality of Lycopersicon Esculentum Mill. *J. Appl. Phycol.* **2011**, 623–633. [CrossRef]
29. Dias, M.P.; Bastos, M.S.; Xavier, V.B.; Cassel, E.; Astarita, L.V.; Santarém, E.R. Plant Growth and Resistance Promoted by Streptomyces Spp. in Tomato. *Plant Physiol. Biochem.* **2017**, *118*, 479–493. [CrossRef] [PubMed]
30. Zhao, L.F.; Xu, Y.J.; Lai, X.H. Antagonistic Endophytic Bacteria Associated with Nodules of Soybean (*Glycine max* L.) and Plant Growth-Promoting Properties. *Braz. J. Microbiol.* **2018**, *49*, 269–278. [CrossRef]
31. Lamont, J.R.; Wilkins, O.; Bywater-Ekegärd, M.; Smith, D.L. From Yogurt to Yield: Potential Applications of Lactic Acid Bacteria in Plant Production. *Soil Biol. Biochem.* **2017**, *111*, 1–9. [CrossRef]
32. Morales-Payan, J.P. Influence of Foliar Sprays of an Amino Acid Formulation on Fruit Yield of "Edward" Mango. *Acta Hortic.* **2015**, *1075*, 157–160. [CrossRef]
33. Tejada, M.; Rodríguez-Morgado, B.; Paneque, P.; Parrado, J. Effects of Foliar Fertilization of a Biostimulant Obtained from Chicken Feathers on Maize Yield. *Eur. J. Agron.* **2018**, *96*, 54–59. [CrossRef]
34. Ramírez-Pérez, L.J.; Morales-Díaz, A.B.; Benavides-Mendoza, A.; De-Alba-Romenus, K.; González-Morales, S.; Juárez-Maldonado, A. Dynamic Modeling of Cucumber Crop Growth and Uptake of N, P and K under Greenhouse Conditions. *Sci. Hortic.* **2018**, *234*, 250–260. [CrossRef]
35. Panfili, I.; Bartucca, M.L.; Del Buono, D. The Treatment of Duckweed with a Plant Biostimulant or a Safener Improves the Plant Capacity to Clean Water Polluted by Terbuthylazine. *Sci. Total Environ.* **2018**, *646*, 832–840. [CrossRef] [PubMed]
36. García-Martínez, A.M.; Díaz, A.; Tejada, M.; Bautista, J.; Rodríguez, B.; Santa María, C.; Revilla, E.; Parrado, J. Enzymatic Production of an Organic Soil Biostimulant from Wheat-Condensed Distiller Solubles: Effects on Soil Biochemistry and Biodiversity. *Process Biochem.* **2010**, *45*, 1127–1133. [CrossRef]
37. Milić, B.; Tarlanović, J.; Keserović, Z.; Magazin, N.; Miodragović, M.; Popara, G. Bioregulators Can Improve Fruit Size, Yield and Plant Growth of Northern Highbush Blueberry (*Vaccinium corymbosum* L.). *Sci. Hortic.* **2018**, *235*, 214–220. [CrossRef]
38. Goñi, O.; Quille, P.; O'connell, S. Ascophyllum Nodosum Extract Biostimulants and Their Role in Enhancing Tolerance to Drought Stress in Tomato Plants. *Plant Physiol. Biochem.* **2018**. [CrossRef]

39. Xavier, L.J.C.; Boyetchko, S.M. Arbuscular Mycorrhizal Fungi as Biostimulants and Bioprotectants of Crops. *Appl. Mycol. Biotechnol.* **2002**, *2*, 311–340. [CrossRef]
40. Wu, Q.S.; Zou, Y.N.; Fathi Abd-Allah, E. *Mycorrhizal Association and ROS in Plants*; Elsevier Inc.: Amsterdam, The Netherlands, 2014. [CrossRef]
41. Kocira, A.; Świeca, M.; Kocira, S.; Złotek, U.; Jakubczyk, A. Enhancement of Yield, Nutritional and Nutraceutical Properties of Two Common Bean Cultivars Following the Application of Seaweed Extract (*Ecklonia maxima*). *Saudi J. Biol. Sci.* **2018**, *25*, 563–571. [CrossRef]
42. Filipczak, J.; Żurawicz, E.; Paszt, L.S. Wpływ wybranych biostymulatorów na wzrost i plonowanie roślin truskawki "Elkat" influence of selected biostimulants on the growth and yielding of "Elkat" strawberry plants. *Zesz. Naukowe Inst. Ogrod.* **2016**, *24*, 43–58.
43. Kocira, A.; Kocira, S.; Świeca, M.; Złotek, U.; Jakubczyk, A.; Kapela, K. Effect of Foliar Application of a Nitrophenolate–Based Biostimulant on the Yield and Quality of Two Bean Cultivars. *Sci. Hortic.* **2017**, *214*, 76–82. [CrossRef]
44. Kocira, S.; Sujak, A.; Kocira, A.; Wójtowicz, A.; Oniszczuk, A. Effect of Fylloton Application on Photosynthetic Activity of Moldavian Dragonhead (*Dracocephalum moldavica* L.). *Agric. Agric. Sci. Procedia* **2015**, *7*, 108–112. [CrossRef]
45. Tejada, M.; Rodríguez-Morgado, B.; Gómez, I.; Franco-Andreu, L.; Benítez, C.; Parrado, J. Use of Biofertilizers Obtained from Sewage Sludges on Maize Yield. *Eur. J. Agron.* **2016**, *78*, 13–19. [CrossRef]
46. Rouphael, Y.; Kyriacou, M.C. Quality and Safety of Fresh Fruits and Vegetables at Harvest. *Sci. Hortic.* **2018**, *239*, 78–79. [CrossRef]
47. Grajkowski, J.; Ochmian, I. Influence of three biostymulants on yielding and fruit quality of three primocane raspberry cultivars. *Hortorum Cultus* **2007**, *6*, 29–36.
48. Sánchez-Gómez, R.; Zalacain, A.; Pardo, F.; Alonso, G.L.; Salinas, M.R. Moscatel Vine-Shoot Extracts as a Grapevine Biostimulant to Enhance Wine Quality. *Food Res. Int.* **2017**, *98*, 40–49. [CrossRef] [PubMed]
49. Paradiković, N.; Vinković, T.; Vinković Vrček, I.; Žuntar, I.; Bojić, M.; Medić-Šarić, M. Effect of Natural Biostimulants on Yield and Nutritional Quality: An Example of Sweet Yellow Pepper (*Capsicum annuum* L.) Plants. *J. Sci. Food Agric.* **2011**, *91*, 2146–2152. [CrossRef] [PubMed]
50. Abd El-Mageed, T.A.; Semida, W.M.; Rady, M.M. Moringa Leaf Extract as Biostimulant Improves Water Use Efficiency, Physio-Biochemical Attributes of Squash Plants under Deficit Irrigation. *Agric. Water Manag.* **2017**, *193*, 46–54. [CrossRef]
51. Nasir, M.; Sattar, A.; Basra, S.M.A.; Ullah, A. Foliar Application of Moringa Leaf Extract, Potassium and Zinc Influence Yield and Fruit Quality of 'Kinnow' Mandarin. *Sci. Hortic.* **2016**, *210*, 227–235. [CrossRef]
52. Fawzy, Z.F. Response of Growth and Yield of Cucumber Plants (*Cucumis sativus* L.) to Different Foliar Applications of Humic Acid and Bio-Stimulators. *Int. Res. J. Appl. Basic Sci.* **2012**, *6*, 630–637.
53. Bona, E.; Todeschini, V.; Cantamessa, S.; Cesaro, P.; Copetta, A.; Lingua, G.; Gamalero, E.; Berta, G.; Massa, N. Combined Bacterial and Mycorrhizal Inocula Improve Tomato Quality at Reduced Fertilization. *Sci. Hortic.* **2018**, *234*, 160–165. [CrossRef]
54. Kumaraswamy, R.V.; Kumari, S.; Choudhary, R.C.; Sharma, S.S.; Pal, A.; Raliya, R.; Biswas, P.; Saharan, V. Salicylic Acid Functionalized Chitosan Nanoparticle: A Sustainable Biostimulant for Plant. *Int. J. Biol. Macromol.* **2019**, *123*, 59–69. [CrossRef] [PubMed]
55. Frioni, T.; Sabbatini, P.; Tombesi, S.; Norrie, J.; Poni, S.; Gatti, M.; Palliotti, A. Effects of a Biostimulant Derived from the Brown Seaweed Ascophyllum Nodosum on Ripening Dynamics and Fruit Quality of Grapevines. *Sci. Hortic.* **2018**, *232*, 97–106. [CrossRef]
56. ur Rehman, H.; Alharby, H.F.; Alzahrani, Y.; Rady, M.M. Magnesium and Organic Biostimulant Integrative Application Induces Physiological and Biochemical Changes in Sun Flower Plants and Its Harvested Progeny on Sandy Soil. *Plant Physiol. Biochem.* **2018**, *126*, 97–105. [CrossRef] [PubMed]
57. Queirós, F.; Ribeiro, C.; Vilela, A.; Aires, A.; Barros, A.I.; Schouten, R.; Paula, A.; Gonçalves, B. Scientia Horticulturae Effects of Calcium and Growth Regulators on Sweet Cherry (*Prunus avium* L.) Quality and Sensory Attributes at Harvest. *Sci. Hortic.* **2019**, *248*, 231–240. [CrossRef]
58. Aziz, R.A.; Naira, A.; Moieza, A. Effect of Plant Biostimulants on Fruit Cracking and Quality Attributes of Pomegranate Cv. Kandhari Kabuli. *Sci. Res. Essays* **2013**, *8*, 2171–2175. [CrossRef]

59. Weber, N.; Schmitzer, V.; Jakopic, J.; Stampar, F. First Fruit in Season: Seaweed Extract and Silicon Advance Organic Strawberry (*Fragaria* × *Ananassa* Duch.) Fruit Formation and Yield. *Sci. Hortic.* **2018**, *242*, 103–109. [CrossRef]
60. Jämtgård, S.; Näsholm, T.; Huss-Danell, K. Nitrogen Compounds in Soil Solutions of Agricultural Land. *Soil Biol. Biochem.* **2010**, *42*, 2325–2330. [CrossRef]
61. Zarzecka, K.; Gugała, M.; Sikorska, A.; Mystkowska, I.; Baranowska, A.; Niewęgłowski, M.; Dołęga, H. The Effect of Herbicides and Biostimulants on Polyphenol Content of Potato (*Solanum tuberosum* L.) Tubers and Leaves. *J. Saudi Soc. Agric. Sci.* **2018**. [CrossRef]
62. Massa, D.; Prisa, D.; Montoneri, E.; Battaglini, D.; Ginepro, M.; Negre, M.; Burchi, G. Scientia Horticulturae Application of Municipal Biowaste Derived Products in Hibiscus Cultivation: Effect on Leaf Gaseous Exchange Activity, and Plant Biomass Accumulation and Quality. *Sci. Hortic.* **2016**, *205*, 59–69. [CrossRef]
63. Alberola, C.; Lichtfouse, E.; Navarrete, M.; Debaeke, P.; Souchère, V. Agronomy for Sustainable Development. *Ital. J. Agron.* **2008**, *3*, 77–78. [CrossRef]
64. Sánchez-gómez, R.; Torregrosa, L.; Zalacain, A.; Ojeda, H.; Bouckenooghe, V.; Schneider, R.; Alonso, G.L.; Salinas, M.R. Scientia Horticulturae Behavior of Glycosylated Aroma Precursors in Microvine Fruits after Guaiacol Foliar Application. *Sci. Hortic.* **2019**, *246*, e1–e8. [CrossRef]
65. Halpern, M.; Bar-Tal, A.; Ofek, M.; Minz, D.; Muller, T.; Yermiyahu, U. *The Use of Biostimulants for Enhancing Nutrient Uptake*; Elsevier Inc.: Amsterdam, The Netherlands, 2015; Volume 130. [CrossRef]
66. Han, M.; Li, G.; Liu, X.; Li, A.; Mao, P.; Liu, P.; Li, H.; Journal, H.P.; Li, G.; Liu, X.; et al. Phenolic Profile, Antioxidant Activity and Anti-Proliferative Activity of Crabapple Fruits. *Hortic. Plant J.* **2019**. [CrossRef]
67. Abdalla, M.M. Boosting the Growth of Rocket Plants in Response to the Application of Moringa oleifera Extracts as a Biostimulant. *Life Sci. J.* **2014**, *11*, 1097–8135.
68. Rouphael, Y.; Colla, G.; Giordano, M.; El-Nakhel, C.; Kyriacou, M.C.; De Pascale, S. Foliar Applications of a Legume-Derived Protein Hydrolysate Elicit Dose-Dependent Increases of Growth, Leaf Mineral Composition, Yield and Fruit Quality in Two Greenhouse Tomato Cultivars. *Sci. Hortic.* **2017**, *226*, 353–360. [CrossRef]
69. Hayat, S.; Ahmad, H.; Ali, M.; Ren, K.; Cheng, Z. Aqueous Garlic Extract Stimulates Growth and Antioxidant Enzymes Activity of Tomato (*Solanum lycopersicum*). *Sci. Hortic.* **2018**, *240*, 139–146. [CrossRef]
70. Kulkarni, M.G.; Rengasamy, K.R.R.; Pendota, S.C.; Gruz, J.; Plačková, L.; Novák, O.; Doležal, K.; Van Staden, J. Bioactive Molecules Derived from Smoke and Seaweed Ecklonia Maxima Showing Phytohormone-Like Activity in *Spinacia oleracea* L. *New Biotechnol.* **2019**, *48*, 83–89. [CrossRef] [PubMed]
71. López-López, N.; López-Fabal, A. Scientia Horticulturae Compost Based Ecological Growing Media According EU Eco-Label Requirements. *Sci. Hortic.* **2016**, *212*, 1–10. [CrossRef]
72. Giménez, A.; Fernández, J.A.; Pascual, J.A.; Ros, M. Scientia Horticulturae an Agroindustrial Compost as Alternative to Peat for Production of Baby Leaf Red Lettuce in a Floating System. *Sci. Hortic.* **2019**, *246*, 907–915. [CrossRef]
73. FAO. *The State of Food And Agriculture*; FAO: Rome, Italy, 2016.
74. FAO. *Greenhouse Gas Emissions from Agriculture, Forestry and Other Land Use*; FAO: Rome, Italy, 2014.

PERMISSIONS

All chapters in this book were first published in MDPI; hereby published with permission under the Creative Commons Attribution License or equivalent. Every chapter published in this book has been scrutinized by our experts. Their significance has been extensively debated. The topics covered herein carry significant findings which will fuel the growth of the discipline. They may even be implemented as practical applications or may be referred to as a beginning point for another development.

The contributors of this book come from diverse backgrounds, making this book a truly international effort. This book will bring forth new frontiers with its revolutionizing research information and detailed analysis of the nascent developments around the world.

We would like to thank all the contributing authors for lending their expertise to make the book truly unique. They have played a crucial role in the development of this book. Without their invaluable contributions this book wouldn't have been possible. They have made vital efforts to compile up to date information on the varied aspects of this subject to make this book a valuable addition to the collection of many professionals and students.

This book was conceptualized with the vision of imparting up-to-date information and advanced data in this field. To ensure the same, a matchless editorial board was set up. Every individual on the board went through rigorous rounds of assessment to prove their worth. After which they invested a large part of their time researching and compiling the most relevant data for our readers.

The editorial board has been involved in producing this book since its inception. They have spent rigorous hours researching and exploring the diverse topics which have resulted in the successful publishing of this book. They have passed on their knowledge of decades through this book. To expedite this challenging task, the publisher supported the team at every step. A small team of assistant editors was also appointed to further simplify the editing procedure and attain best results for the readers.

Apart from the editorial board, the designing team has also invested a significant amount of their time in understanding the subject and creating the most relevant covers. They scrutinized every image to scout for the most suitable representation of the subject and create an appropriate cover for the book.

The publishing team has been an ardent support to the editorial, designing and production team. Their endless efforts to recruit the best for this project, has resulted in the accomplishment of this book. They are a veteran in the field of academics and their pool of knowledge is as vast as their experience in printing. Their expertise and guidance has proved useful at every step. Their uncompromising quality standards have made this book an exceptional effort. Their encouragement from time to time has been an inspiration for everyone.

The publisher and the editorial board hope that this book will prove to be a valuable piece of knowledge for researchers, students, practitioners and scholars across the globe.

LIST OF CONTRIBUTORS

Lucia Rocchi, Antonio Boggia and Luisa Paolotti
Department of Agricultural, Food and Environmental Science, University of Perugia, Borgo XX Giugno, 74, 06121 Perugia, Italy

Matteo Caser, Sonia Demasi and Dario Donno
Department of Agricultural, Forest and Food Sciences, University of Torino. Largo Paolo Braccini 2, 10095 Grugliasco (TO), Italy

Valentina Scariot
Department of Agricultural, Forest and Food Sciences, University of Torino. Largo Paolo Braccini 2, 10095 Grugliasco (TO), Italy
Institute for Sustainable Plant Protection, National Research Council, Viale Mattioli 25, 10125 Torino, Italy

Antonella Faccio, Erica Lumini and Valeria Bianciotto
Institute for Sustainable Plant Protection, National Research Council, Viale Mattioli 25, 10125 Torino, Italy

Íris Marisa Maxaieie Victorino
Institute for Sustainable Plant Protection, National Research Council, Viale Mattioli 25, 10125 Torino, Italy
Department of Life Sciences and Systems Biology, University of Turin. Viale Mattioli 25, 10125 Torino, Italy
Biological Science Department, Science Faculty, Eduardo Mondlane University (UEM), Av. Julius Nyerere-Campus Universitário, Maputo 3453, Mozambique

Trifone D'Addabbo
Institute for Sustainable Plant Protection, CNR, Via Giovanni Amendola 122/d, 70126 Bari, Italy

Sebastiano Laquale and Vincenzo Candido
School of Agricultural, Forest, Food and Environmental Sciences–UNIBAS, Viale dell'Ateneo Lucano 10, 85100 Potenza, Italy

Michele Perniola
Department of European and Mediterranean Cultures, UNIBAS, Via Lanera 20, 75100 Matera, Italy

Szilvia Kisvarga
Research Institute for Fruit growing and Ornamentals, National Agricultural Research and Innovation Center (NARIC) Szent-Györgyi Street 4, 2100 Gödöllő, Hungary

Döme Barna, Szilvia Kovács, Gábor Csatári, Ibolya O. Tóth, Miklós Gábor Fári, Péter Makleit, Szilvia Veres and Nóra Bákonyi
Department of Agricultural Botany, Plant Physiology and Biotechnology, Institute of Crop Sciences (MEK), Debrecen University, Böszörményi Street 138, 4032 Debrecen, Hungary

Tarek Alshaal
Department of Agricultural Botany, Plant Physiology and Biotechnology, Institute of Crop Sciences (MEK), Debrecen University, Böszörményi Street 138, 4032 Debrecen, Hungary
Soil and Water Department, Faculty of Agriculture, Kafrelsheikh University, Kafr El-Sheikh 33516, Egypt

Spyridon A. Petropoulos and Sofia Plexida
Department of Agriculture, Crop Production and Rural Environment, University of Thessaly, 38446 N. Ionia, Magnissia, Greece

Ângela Fernandes, João C. M. Barreira, Lillian Barros and Isabel C. F. R. Ferreira
Centro de Investigação de Montanha (CIMO), Instituto Politécnico de Bragança, Campus de Santa Apolónia, 5300-253 Bragança, Portugal

Antonios Chrysargyris and Nikos Tzortzakis
Department of Agricultural Sciences, Biotechnology and Food Science, Cyprus University of Technology, 3036, Lemesos, Cyprus

Alicja Pohl, Aneta Grabowska, Andrzej Kalisz and Agnieszka Sękara
Department of Vegetable and Medicinal Plants, University of Agriculture in Krakow, 29-Listopada 54, 31-425 Kraków, Poland

Murugesan Chandrasekaran
Department of Food Science and Biotechnology, Sejong University, Gwangjin-gu, Seoul 05006, Korea

Se Chul Chun
Department of Environmental Health Science, Konkuk University, Gwangjin-gu, Seoul 05029, Korea

Jae Wook Oh
Department of Stem Cell and Regenerative Biotechnology, Konkuk University, Gwangjin-gu, Seoul 05029, Korea

Manivannan Paramasivan
Department of Microbiology, Bharathidasan University, Tiruchirappalli 620024, India

Ramesh Kumar Saini
Department of Crop Science, Konkuk University, Gwangjin-gu, Seoul 05029, Korea

Jesudoss Joseph Sahayarayan
Department of Bioinformatics, Alagappa University, Karaikudi 630 003, India

Henrietta Allaga, Bettina Bóka, Viktor Dávid Nagy, Attila Szűcs Miklós Takó, László Manczinger, Csaba Vágvölgyi and László Kredics
Department of Microbiology; Faculty of Science and Informatics; University of Szeged; Közép fasor 52, 6726 Szeged, Hungary

Péter Poór
Department of Plant Biology, Faculty of Science and Informatics, University of Szeged, Közép fasor 52, 6726 Szeged, Hungary

Istvánné Stankovics
Felső-Bácskai Agrolabor Ltd., Rákóczi Ferenc u. 19., 6430 Bácsalmás, Hungary

Péter Körmöczi
Department of Field Crops Research, National Agricultural Research and Innovation Centre, Alsó kikötő sor 9, 6726 Szeged, Hungary

Hye-Ji Kim and Seunghyun Choi
Department of Horticulture and Landscape Architecture, Purdue University, West Lafayette, IN 47907, USA

Kang-Mo Ku
Division of Plant and Soil Sciences, West Virginia University, Morgantown, WV 26505, USA

Mariateresa Cardarelli
Consiglio per la ricerca in agricoltura e l'analisi dell'economia agraria, Centro di Ricerca Orticoltura e Florovivaismo, 84098 Pontecagnano, Italy

Alicja Niewiadomska, Zyta Waraczewska and Agnieszka Wolna-Maruwka
Department of General and Environmental Microbiology, Poznań University of Life Sciences, ul. Szydłowska 50, 60-656 Poznań, Poland

Hanna Sulewska and Karolina Ratajczak
Department of Agronomy, Poznan University of Life Sciences, ul. Dojazd 11, 60-632 Poznań, Poland

Anna Budka
Department of Mathematical and Statistical Methods, Poznań University of Life Sciences, ul. Wojska Polskiego 28, 60-637 Poznań, Poland

Mohamed Ashour
Invertebrate Aquaculture Lab., Aquaculture Division, National institute of Oceanography and Fisheries (NIOF), Alexandria Branch, Alexandria 21576, Egypt

Ahmed A. El-Shafei
Department of Agricultural Engineering, College of Food and Agriculture Sciences, King Saud University, Riyadh 11451, Saudi Arabia
Prince Sultan Bin Abdulaziz International Prize for Water Chair, Prince Sultan Institute for Environmental, Water and Desert Research, King Saud University, Riyadh 11451, Saudi Arabia

Abed Alataway
Prince Sultan Bin Abdulaziz International Prize for Water Chair, Prince Sultan Institute for Environmental, Water and Desert Research, King Saud University, Riyadh 11451, Saudi Arabia

Hanan M. Khairy
Hydrobiology Lab., Marine Environmental Division, National Institute of Oceanography and Fisheries, Alexandria 21545, Egypt

Doaa Y. Abd-Elkader and Shimaa M. Hassan
Department of Vegetable, Faculty of Agriculture (El-Shatby), Alexandria University, Alexandria 21545, Egypt

Mohamed A. Mattar
Department of Agricultural Engineering, College of Food and Agriculture Sciences, King Saud University, Riyadh 11451, Saudi Arabia
Agricultural Engineering Research Institute (AEnRI), Agricultural Research Centre, Giza 12618, Egypt

Marino B. Arnao and Josefa Hernández-Ruiz
Department of Plant Physiology, University of Murcia, 30100-Murcia, Spain

Magdalena Drobek, Magdalena Frąc and Justyna Cybulska
Institute of Agrophysics, Polish Academy of Sciences, Doświadczalna 4, 20-290 Lublin, Poland

Index

Printed in the USA
CPSIA information can be obtained
at www.ICGtesting.com
LVHW080052211023
761596LV00010BA/2023

9 781647 404147